Electrically Based Microstructural Characterization II

MATERIALS RESEARCH SOCIETY
SYMPOSIUM PROCEEDINGS VOLUME 500

Electrically Based Microstructural Characterization II

Symposium held December 1–4, 1997, Boston, Massachusetts, U.S.A.

EDITORS:

Rosario A. Gerhardt
Georgia Institute of Technology
Atlanta, Georgia, U.S.A.

Mohammad A. Alim
Alabama A&M University
Normal, Alabama, U.S.A.

S. Ray Taylor
University of Virginia
Charlottesville, Virginia, U.S.A.

Materials Research Society
Warrendale, Pennsylvania

CAMBRIDGE
UNIVERSITY PRESS

University Printing House, Cambridge CB2 8BS, United Kingdom

One Liberty Plaza, 20th Floor, New York, NY 10006, USA

477 Williamstown Road, Port Melbourne, VIC 3207, Australia

314-321, 3rd Floor, Plot 3, Splendor Forum, Jasola District Centre, New Delhi - 110025, India

79 Anson Road, #06-04/06, Singapore 079906

Cambridge University Press is part of the University of Cambridge.

It furthers the University's mission by disseminating knowledge in the pursuit of
education, learning and research at the highest international levels of excellence.

www.cambridge.org
Information on this title: www.cambridge.org/9781558994058

Materials Research Society
506 Keystone Drive, Warrendale, PA 15086
http://www.mrs.org

© Materials Research Society 1998

First published 1998
First paperback edition 2013

Single article reprints from this publication are available through
University Microfilms Inc., 300 North Zeeb Road, Ann Arbor, MI 48106

CODEN: MRSPDH

A catalogue record for this publication is available from the British Library

ISBN 978-1-558-99405-8 Hardback
ISBN 978-1-107-41355-9 Paperback

CONTENTS

*Invited Paper

PART III: MAGNETIC AND POLYMERIC MATERIALS

PART IV: DIELECTRICS AND FERROELECTRICS

PART V: VARISTORS

*Invited Paper

PART VI: IONIC AND MIXED CONDUCTORS

*Invited Paper

PART VII: COMPOSITES AND PERCOLATION SYSTEMS

PREFACE

This volume includes many of the papers that were presented at the symposium entitled 'Electrically Based Microstructural Characterization II,' which was held at the 1997 MRS Fall Meeting in Boston, Massachusetts, December 1–4, 1997. This was the second symposium sponsored by MRS which covered the application of electrical measurements for the detection of microstructural features at all length scales (atomic to macroscopic). In addition to the topics covered in the first symposium (dc and ac resistivity measurements, immittance (impedance/admittance) analysis, multiplane analysis and various other methods such as electron energy loss spectroscopy, ellipsometry, and capacitance voltage measurements), there were several papers which combined electrical measurements with STM, AFM, NSOM and electroluminescence techniques so that more localized information may be obtainable. The papers in this volume cover all classes of materials including semiconductors, electroceramics, biological materials, polymers, metals, geomaterials and a variety of composites. This book has been arranged much like the symposium program where papers were presented according to material types and applications.

The organizers wish to thank all of the symposium participants, in particular our invited speakers, for excellent presentations and after-hours discussions. We would also like to acknowledge the careful work of our many reviewers and the promptness with which authors made requested changes. Dr. Julie R. Kokan of the Georgia Institute of Technology deserves special recognition for her instrumental help in the book preparation. Finally, we would like to thank the MRS staff for their efforts, and all of our financial sponsors for their support.

Financial Contributions made by:

Hewlett-Packard
Keithley Instruments
Solartron Instruments

Rosario A. Gerhardt
Mohammad A. Alim
S. Ray Taylor

September 1998

MATERIALS RESEARCH SOCIETY SYMPOSIUM PROCEEDINGS

MATERIALS RESEARCH SOCIETY SYMPOSIUM PROCEEDINGS

Prior Materials Research Society Symposium Proceedings available by contacting Materials Research Society

Part I

Advances in Localized
Electrical Testing

LOW-FREQUENCY SCANNING CAPACITANCE MICROSCOPY

Š. LÁNYI *, M. HRUŠKOVIC **
*Institute of Physics, Slovakian Academy of Sciences, Dúbravská cesta 9, 842 28 Bratislava, Slovakia, lanyi@savba.sk
**Faculty of Electrical Engineering and Information Technology, Slovak University of Technology, Ilkovičova 3, 812 19 Bratislava, Slovakia,

ABSTRACT

The operation principle and main properties of a Scanning Capacitance Microscope (SCM) are described. It is called low-frequency, because in its design typical low-frequency techniques are utilised. The main attention is focused on its lateral resolution, signal-to-noise ratio and the possibility to detect dielectric losses.

Mapping the electrostatic field of a shielded microscope probe was used to calculate the stray capacitance, flux density, sensitivity and contrast obtained on a flat conducting surface, as well as on a surface covered by a thin dielectric film. The effect of dielectric losses, represented by a parallel conductance, on the detected capacitance and the resulting phase shift has been derived.

Using the results of mapping, the requirements on a SCM input stage and the possible solutions are discussed. From the point of view of frequency range and noise the best is an electrometric input stage, with input impedance represented by its capacitance.

The achieved signal-to-noise ratio of the low frequency Scanning Capacitance Microscope renders the extension of the working frequency range to lower frequencies. The input stage can be optimised for a frequency range from about 1 kHz to a few MHz, with the possibility to extend it to about 10 MHz at the cost of reduced sensitivity at the lowest frequencies.

INTRODUCTION

The Scanning Capacitance Microscope, though one of the first scanning probe microscopes [1] is among the less known and less used arts of scanning probe techniques. In the last few years its importance has been recognised in connection with the needs of the semiconductor industry, which is in search of high-resolution analytical tools for the next generation of integrated circuits [2]. The main advantage of SCM for such purposes is the ability to image not only conducting surfaces but to sense also the properties of dielectric films or semiconductor depletion layers beneath the surface.

In a SCM the capacitance between a sharp conducting tip and the imaged conducting surface, or a base coated by an insulator, is detected and used to reconstruct its topography. Since the electrostatic field between them is poorly localised, the attainable resolution is lower than can be achieved by scanning tunnelling microscopes (STM) or scanning/atomic force microscopes (SFM, AFM).

In the past three basically different approaches were used to capacitance imaging. The microscopes based on the RCA videodisk pickup [1, 3 - 5] use the change of the resonant frequency of a microstrip resonator, caused by changes of the probe/sample capacitance, connected to it. They operate at about 1 GHz. A similar heterodyne solution, based on a lumped element circuit, was followed in reference [6]. It had a working frequency of 90 MHz. The second possibility was the application of a scanning force microscope, with a charged tip attracted to the surface by Coulomb interaction [7]. The representative of the third concept is the low-frequency capacitance microscope [8], using phase-sensitive demodulation of the ac current flowing through the probe/sample capacitor. Until now it is the only concept that makes the imaging of the

3

components of the complex capacitance possible. It operates in the MHz region. By means of active shielding of both probe and input stage - a typical low frequency technique - the parasitic stray capacitance of the probe could be significantly reduced. The measurement principle itself helped to separate it from the capacitance of the input of the electronics, also suppressed by bootstrapping. By further modifications the probe/sample capacitance was reduced to a few hundred aF [9].

Recently SCMs became commercially available [10, 11]. They combine the SCM with AFM and use a metal-coated AFM cantilever as SCM probe. This approach makes the separation of contributions of surface topography, detected by AFM, and of inhomogeneity of dielectric film, possible [12].

Most of the important properties, like the lateral resolution, signal-to-noise ratio, etc., strongly depend on the shape of the probe. Its stray capacitance and its dependence on the distance from the probe axis play a crucial role [13]. In this paper we shall present an analysis of a shielded, conical probe with spherical apex. The assumed imaged surface was a conducting plane or a conducting plane covered by an insulating film. The lateral sensitivity to capacitance and to dielectric losses has been derived. The obtained data have been used for a discussion of optimal input stage of the capacitance microscope, designed to work at still lower frequencies.

MAPPING THE ELECTROSTATIC FIELD

The exact calculation of the capacitance between a surface with arbitrary topography and a needle perpendicular to it, is a rather complex problem. It would require the solution of the Laplace equation $\Delta V = 0$ pertinent to the space between electrodes and the electrostatic field at the electrodes, coupled to the charge density through the Poisson equation, with complicated boundary conditions. The problem is further complicated by the local presence of dielectric, properly described by Maxwell's 1st equation. Therefore numerical computation, using the Finite Element Method (FEM), has been used [9, 14]. The employed program [15] is able to calculate the flux and capacitance between electrodes with high accuracy [16].

Four types of arrangements have been solved: i) the whole probe with the shield over a conducting plane, ii) detail of the tip over a conducting plane, iii) both configurations with the conducting plane covered by dielectric films and iv) the same assuming dielectric films with losses. The dependence of capacitance and dielectric losses on increasing distance from the probe axis was estimated by cutting out reduced details of the tip (zooming) and computing their capacitance.

Though we could confirm the high accuracy of the program, nevertheless, it was not sufficient to calculate reliably very small changes of the capacitance if the distance of the whole probe (with actual size of half the axial section 240 μm x 300 μm) from the planar electrode changed by 5 or 10 nm. The change in capacitance was then comparable with the computational errors or effected by possible changes of the automatically generated triangular grid. Therefore, in such cases a thin slice with separately generated grid, of height equal to the change in separation, has been added. Thus the shape of the grid in the rest of the field was retained. In this way the computational error has become a systematic one and minute changes of the order of 10 ppm could be computed. Utilising the rotational symmetry, axial half-sections have been solved.

Boundary conditions

Previous analysis has shown that proper shielding can significantly reduce the stray capacitance of the probe [9]. The experience with the microscope has lead to some modifications. A flat face of the shield with a radius of 0.3 mm, placed a few μm above the sample, was found to be not convenient [8]. Therefore, the shape of the shield was changed to slightly conical one, with the opening of the orifice just sufficient to let a 0.05 mm diameter wire through. The tip is let to protrude by 5 to 10 µm. This does not increase the stray capacitance significantly, however, the tip may follow also rough surfaces and its adjustment is much easier. The dimensions used in the model correspond to the real probe, except the assumption of spherical apex of the tip. It has frequently a less regular shape and sometimes it is truncated. From the point of view of achieved lateral resolution and contrast, the spherical shape is probably the least advantageous.

The shielding reduces the area, with which the core of the probe interacts, to less than 1/2 of the radius of the hole [9]. Up to about 1/3 of this distance the force lines from a conical tip to a planar electrode were circular arcs, perpendicular to both the tip and the plane. The detailed mapping of the tip took place using the boundary conditions consisting of an equipotential of the tip shape (cone with spherical apex, with radius of curvature 25 nm), a linear Neumann boundary condition (force line) at the axis, an equipotential representing the conducting plane and such an arc-shaped Neumann boundary condition. The circular arc appeared as appropriate up to about 3 µm from the axis.

The real length of the probe is about 10 mm. For mapping only its last 240 µm were considered. The task was defined by the same tip-shaped equipotential, a force line at the probe axis, the conducting plane, circular arc connecting the plane with the outer edge of the shield (Neumann) and the silhouette of the shield as equipotential. In contrast to earlier modelling [9], the shield was connected with the tip rather by means of a linear Neumann boundary condition then directly, to enable independent computation of the flux and capacitance between the tip and the plane, and between the shield and the plane. This modification is without any consequence, since it was already known that there is virtually no electric field in the depth of the probe. Also the presence of the insulation in the shield was neglected [9]. Alternatively, also a larger part of the outer surface was included into the model, which resulted in larger capacitance between the shield and sample but, as expected, without any effect on the tip-to-sample capacitance.

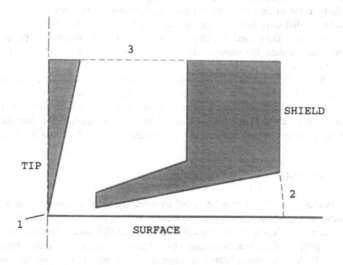

Fig. 1. The boundary conditions used in modelling the shielded probe. 1, 2 and 3 are Neumann boundary conditions.

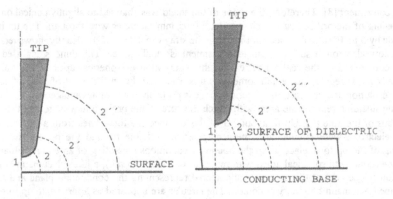

Fig. 2. The boundary conditions used in the modelling of details of the tip. Left: tip over a conducting surface, right: over a dielectric film.

The dielectric film was added as a separate region with relative permittivity $\varepsilon_r = 5$. The assumed thickness was 20, 50, 100 and 200 nm. The correct shape of the Neumann boundary condition at the perimeter is more complex. The force lines are perpendicular to the surface of the tip and to the conducting base, while at the dielectric/air interface only their tangential component is continuous. Therefore the boundary condition consisted of two circular arcs, meeting at the interface under (approximately) proper angles. Sometimes differences in the potential at the dielectric/air interface, have been obtained from sections of different size. They indicated that the boundary conditions were not chosen properly. However, the resulting errors were not significant. The employed boundary conditions are seen in Fig. 1 and 2.

Since a proven way of increasing the lateral resolution is the modulation of the position of the tip perpendicularly to the surface [4, 6, 8], the distance between the tip apex and the surface of the conducting plane or dielectric film was taken 10, 5 and 15 nm. The change of the distance, similarly as the differently thick dielectric films result in changes of spreading of the electric field. To enable comparison of data, the radius r represents the extent of the model at the air/conductor or air/dielectric interface.

The effect of dielectric losses was simulated by assuming the dielectric film to be partly conducting, with conductance equal to its susceptance, resulting in $\tan\delta = 1$. From a given configuration, for which the capacitance was known both without and with dielectric an apparent series capacitance, represented by the film, was calculated. This series capacitor was then replaced with that having dielectric losses and a new capacitance and $\tan\delta$ of the configuration was calculated.

RESULTS OF MAPPING

The accuracy of the method may be tested by comparison of computed results with exact solution of some arrangement, which can be treated analytically. Such tests can be found in reference [16]. Our calculation of the capacitance of a parallel plate capacitor with changing separation of the plates yielded results accurate to at least five digits. The errors appeared to be rounding errors. With more complicated arrangements the errors were up to one order of magnitude larger. Therefore in the cases with dimension ratios 1 to 100, in which relatively uniform grids could be generated, the errors were negligible. Larger sections were divided into subsections, in which finer grids were used. In this way the accuracy could be increased. In the extreme case of the whole probe, with dimensions ranging from 5 nm, the smallest tip-to-surface distance, up to 240 μm, the considered depth of the probe, the systematic errors resulting from coarser grid could be kept constant, while the fine grid near the tip apex rendered sufficiently exact evaluation of small differences possible.

Conducting surface

The most important property of the probe is the increase of the tip/sample capacitance with increasing radius of the considered sample area. The capacitance between two charged bodies is defined by the ratio $C = Q/U$, where Q is the induced charge and U the voltage between them. It cannot be really divided into partial capacitors. Therefore, it is more correct to handle the problem of tip/sample capacitance in terms of flux, which can be defined or measured locally.

The results must be evaluated in two ways. With increasing diameter the distance between the tip and the perimeter of the analysed area increases, which results in decreasing contribution of each surface element to the overall capacitance. However, due to the rotational symmetry the affected area increases with the square of radius. As a result, the capacitance increases monotonically up to the limit imposed by the shield. On the other hand, the contribution of a distinct surface element, e.g. an inhomogeneity that would eventually contribute to the contrast, is decreasing with its distance from the tip apex. Fig. 3. shows the capacitance as a function of radius for different tip/sample distances. The limiting values range from 310 to 313 aF. As it was shown in [9], the total capacitance depends on the angle of the tip apex. Nevertheless, the mean features of the analysis remain valid. The capacitance of the analysed part of the shield with respect to the plane is more than two orders of magnitude larger.

The area contributing to the tip/sample capacitance is comparable to or even larger than a typical scanned area. Therefore the integral capacitance will contribute to an image by a mostly featureless background.

An important feature and deficiency of a SCM is the distortion of surface features. Its cause is the long-range Coulomb interaction. Though the same problem exists also in other microscopes for example in AFM or STM, it is much less expressed and can be explained by a convolution of

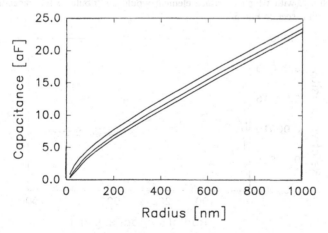

Fig. 3. The increase of the capacitance with the radius of the model. The curves correspond to a tip-to-surface distance of 5 (top), 10 (middle) and 15 nm (bottom).

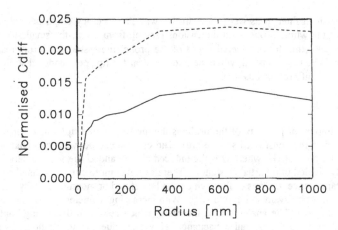

Fig. 4. The differencial capacitance obtained from tip-sample distances 5 nm/10 nm: dashed and 10 nm/15 nm: full

the topography with the tip shape. In an attempt to quantify the imaging ability of SCM we have defined contrast as an objective, measurable property, in analogy with optical imaging. It must not be confused with the contrast of an image, obtained by processing the data. It is understood as the contribution to¹ the signal (capacitance or flux) from areas close to the tip axis, where it is influenced by small-scale inhomogeneities, to the overall capacitance It is then the fraction of the capacitance, seen by a certain part of the probe, i.e. of a circular area with radius r, to the inegral capacitance up to r_{max}. For small radii it would be similar to results in Fig. 3. The contribution to the capacitance is increasing rather slowly.

By changing the probe/sample distance the capacitance $vs.$ radius dependences in Fig. 3 are nearly parallel, except at small radii. Their difference divided by d, the difference in probe/sample separation, related to the difference of integral capacitances is seen in Fig. 4. By dividing the data by r the sensitivity, with which a surface element would contribute to the capacitance, can be derived. The result is seen in Fig. 5.

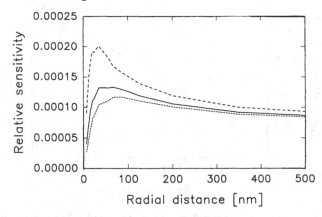

Fig. 5. The lateral sensitivity of the probe

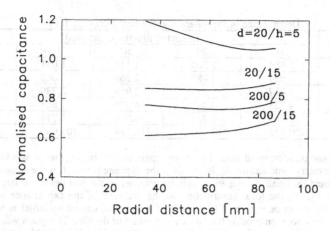

Fig. 6. The local contribution of the dielectric film to the capacitance.

Ideal dielectric film

The evaluation of lateral resolution and contrast in insulating films is more complicated, since due to the three-dimensional geometry of the problem the electric field in them is inhomogeneous. Characterisation of these properties was first attempted in [13]. At present we shall restrict ourselves to the discussion of sensitivity and expected signal-to-noise ratio. Therefore we show only the effect of a dielectric film of certain thickness as a whole. Some data are summarised in Table I. From the results obtained on conducting surface and those on dielectric films an effective series capacitance, represented by the film was calculated. The results are seen in Table II. It is

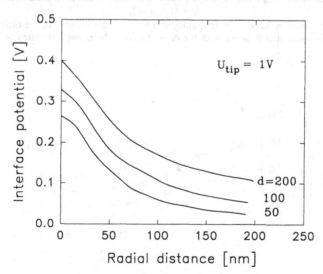

Fig. 7. Potential at the surface of dielectric film.

9

Dependence of capacitance on the film thickness [aF] TABLE I.

Radius [nm]	Dielectric film thickness [nm]				
	0	20	50	100	200
34	1.484	1.269	1.131	1.031	0.911
67	2.924	2.485	2.266	2.062	1.860
87	3.637	3.219	2.959	2.739	2.509
123	4.723	4.367	4.070	3.803	3.506
198	6.745	6.362	6.030	5.711	5.342
12000	320.122	319.493	316.334	307.980	306.830

interesting to note that for small areas the series capacitance is larger, then it would be for one-dimensional geometry, whereas for the larger areas the opposite is true. It is the result of spreading of the electric field for small r, competing with the decrease of the flux density at large radii. As it can be seen in Fig. 6, the local contribution to the decrease of the capacitance may be non-monotonous, which can be explained as a result of a decrease, caused by larger separation from the electrode, partly compensated by the dielectric constant of the film. This point will be discussed in more detail later. The general increase of the contribution of the dielectric film in the vicinity of the probe axis is a result of non-uniform distribution of the applied potential between air and dielectric. The surface potential of the dielectric film is shown in Fig. 7.

Dielectric film with losses

The effect of dielectric losses or non-zero conductivity of the surface film results in a phase shift of the current, sensed by the probe if an ac voltage is connected between the probe and the conducting base. At this point we assume that the phase shift is not influenced in any way by the electronics of the microscope. As already mentioned, we have assumed a $\tan\delta = 1$. This value would be rather high for good dielectrics but still low for many materials, which may become conducting at elevated temperatures. The admittance of the film can be expressed as

$$Y = (1 + i)\,\omega C_s,$$

where $i = \sqrt{-1}$, ω denotes the angular frequency and C_s is the capacitance of the dielectric film. We assume that it is connected in series with the capacitance of the probe, estimated without the film.

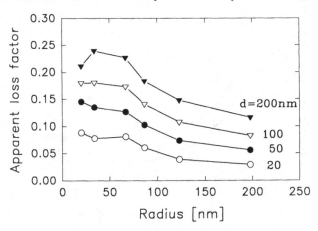

Fig. 8. Radial sensitivity of the probe to dielectric losses.

The resulting phase shift expressed as tanδ is shown in Fig. 8. Note that while the capacitance of the whole tip reaches relatively large values, little effected by the presence of the thin dielectric film, the sensed dielectric losses originate exclusively in the film, with large contribution of the area in the vicinity of the axis.

DISCUSSION

The results of the analysis show that the contrast achieved with the probe in fixed distance from the surface is relatively low. The curves in Fig. 3. would saturate only at large radial distance. The low sensitivity to small features has an important consequence. If the distance over the surface is controlled by keeping the capacitance constant, on small protrusions the tip may be driven into the surface, since the change of the capacitance may be not sufficient to withdraw it properly. The same may happen over more extended convex features.

The curves in Fig. 4 are nearly parallel over most of the radial distance. This suggests that the resolution is much improved by the modulation of the distance, however, at the cost of output signal and the achieved signal to noise ratio. Also the effect of large-scale convex features on the tip control will be suppressed.

The non-monotonous contribution of thin insulating film, seen in Fig. 6. may produce results which must be interpreted with care. The presence of a continuous areas of film would be appear as depressed areas, whereas for very small patches the contrast may be reverted. Therefore an independent information on the topography is needed. An alternative is sequential imaging using gradually decreasing separation of the tip from the surface.

The most local signal is expected from dielectric losses. However, as we shall show later, the expected signal levels are very low.

Our SCM probe has large dimensions. Nevertheless, active shielding reduced its stray capacitance to values probably lower than can be achieved with metal-coated AFM cantilevers. The analysis of typical cantilevers (silicon nitride and silicon) has shown that the silicon nitride pyramids may have lower capacitances to the sample but the resolution is adversely effected by the large tip angle. On the other hand the flat metalised cantilever beam increase the capacitance by orders of magnitude. More slim but longer silicon tips yielded similar capacitance as our probe but again without taking account of the conducting path to it.

The dielectric constant would scale the results. Larger ε_r would have a similar effect as a thinner film.

REQUIREMENTS ON INPUT STAGE

Using the results of modelling we can define the required properties of the input stage of the microscope. The current flowing through the tip/sample capacitor can be measured either using a current-to-voltage converter or measuring the voltage drop on an impedance, connected in series

Apparent capacitance of the dielectric film [aF]				TABLE II.
	Dielectric film thickness [nm]			
Radius [nm]	20	50	100	200
34	8.750	4.746	3.369	2.356
67	16.53	10.07	6.991	5.104
87	28.01	15.86	11.10	8.058
123	57.96	29.45	19.54	13.61
198	111.9	56.85	37.25	25.69
12000	162600	26730	8120	7395

with the probe. Imaging the topography of the surface can be performed virtually at any frequency. The same can be said about the imaging of non-uniformity of an insulating layer on top of a good conductor. However, the spreading resistance of less conducting substrates may cause problems at very high frequencies. The reactance of the film decreases and it may become too small compared with the series resistance. This would be frequently the case of thin insulating films on not heavily doped semiconductors at very high frequencies. Therefore operation in the MHz region might be appropriate.

The possibility to detect dielectric losses with high lateral resolution is an attractive option. However, dielectric losses caused by many kinds of defects usually have a maximum in a limited frequency range. They are usually thermally activated. In principle, they can be tuned to the frequency used for imaging by suitable choice of temperature. This may be frequently inconvenient, first of all because of problems with thermal drifts. Therefore the possibility to select the frequency is advantageous. We shall analyse the available options. We shall use simple examples for illustration.

The obtained capacitance of the probe is around 0.33 fF. Its reactance at 1 MHz is 482.3 MΩ. With 1 V applied voltage the short-circuit current flowing through the tip would be 2.07 nA.

Current-To-Voltage Converter

Using state of the art FET-input operational amplifiers it is easy to build a current-to-voltage converter able to amplify the input current. With a 100 MΩ feedback resistor it would be converted into 207 mV output voltage. However, the parasitic capacitance of such resistors are usually not less than 0.1 pF. The parasitic reactance at 1 MHz is just 1.59 MΩ, i.e. the actual conversion would be only into 3.295 mV. The loss contribution to U_2 is 0.065 μV: The thermal noise of a 100 MΩ resistor at room temperature, limited to tens of kHz by the parasitic capacitance, is approximately 34 μV in a 1-kHz band. The situation can be improved only by reducing the parasitic capacitance of the feedback resistor. In reality the situation is more complex, because the roll-off of open-loop gain of the operational amplifier would introduce an apparent input capacitance. With a parasitic capacitance of 5 fF the output voltage would increase to 62 mV and its loss component to 23 μV [16].

High-Input-Impedance Buffer

A less elegant possibility is to connect a small impedance into the circuit and to measure the voltage drop on it. A passive resistor can be ruled out because of its parasitic capacitance. However, a high-input-impedance buffer may achieve the required properties [18]. Its input impedance may be both resistive and capacitive. Let us analyse the expected properties of both.

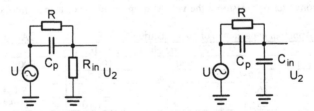

Fig. 9. Equivalent circuits of the input. C_p represents the probe/sample capacitance, R the dielectric losses and R_{in} and C_{in} the load (input resistance or capacitance of the buffer).

We must consider that the stray capacitance of the probe significantly reduces the sensed $\tan\delta$. For illustration we can assume $\tan\delta = 0.006$, the largest value obtained with 50 nm thick film.

A) Resistor as input impedance.

The transfer function for a harmonic voltage U will be

$$U_2/U = R_{in}/R \{(1+ \omega^2 C^2 RR_{in})/(1+ \omega^2 C^2 R_{in}^2) + j [(\omega C(R-R_{in})]/(1+ \omega^2 C^2 R_{in}^2)\}, \quad (1)$$

where U_2 is the output voltage of the voltage divider formed by the probe/sample impedance and the input impedance. $R \gg 1/\omega C$, hence $R \gg R_{in}$. The phase shift between U and the current should be $\pi/2$. In reality it is

$$\varphi = \arctan (1/\omega CR_{in}) \quad (2)$$

The resolution of the in-phase and quadrature components of the current become inconvenient unless R_{in} is not very small compared to $1/\omega C$. Let the permissible phase shift error be $1°$. Then $R_{in} = 8.4$ MΩ, the attenuation $U_2/U = 3.033 \times 10^{-4} + j\ 0.01741$. The assumed $\tan\delta$ of the dielectric film is 0.006, represented by $R = 80$ GΩ would change the attenuation to $U_2/U = 4.083 \times 10^{-4} + j\ 0.01741$ and the resulting phase shift at the input to $\varphi = \pi/2 + 1.343°$. The loss component of the output voltage would change by 105 μV. The signal-to-noise ratio can be improved by increasing R_{in}, however, at the cost of increasing phase error. The inconvenience of the phase error is that it cannot be simply decided whether it results from the change of R_{in} or C.

B) Capacitor as input impedance.

By increasing the real part of the input impedance the input capacitance will dominate the impedance. Then the transfer function will be

$$U_2/U = \omega^2 C(C+C_{in})R^2/(1+ \omega^2(C+ C_{in})^2 R^2) + j\omega CR/(1+ \omega^2(C+ C_{in})^2 R_{in}^2). \quad (3)$$

and $\tan\delta = 1/[\omega(C + C_{in})R]$. Note that with $R \rightarrow \infty$ the phase shift is 0. Assuming an input capacitance $C_{in} = 3.3$ fF and $\tan\delta = 0.006$ the attenuation will be $U_2/U = 0.09091 + j4.98 \times 10^{-5}$ and the resulting $\tan\delta = 5.48 \times 10^{-4}$. The loss contribution to the output voltage is 49.8 μV.

<u>Low Frequencies</u>

The situation would be significantly different at lower frequencies. Since the impedance of the probe is capacitive, the input short-circuit current would be proportional to ω. Assuming the same dielectric losses, the conversion factor of the current-to-voltage converter would be either dominated by the feed-back resistor, or the feed-back resistor would have to be increased, or its effect eliminated in other way. In either way the noise would be increased.

The situation would be much more favourable with a buffer. Using appropriate feed-back it is possible to achieve remarkably high input impedance without the need of very large resistors, hence also relatively low noise. For instance using a 10 MΩ resistor, bootstrapped to 0.999, the input impedance would be increased by a factor of $1/(1-0.999)$, i.e. to 10 GΩ, at the same time retaining the noise of the 10 MΩ resistor (11 μV in a 1-kHz band). Bootstrapping would reduce also its parasitic capacitance.

13

CONCLUSIONS

The analysis has shown that the presence of thin dielectric films gives rise to detectable changes of the capacitance sensed by the probe. The effect of dielectric losses is small but still detectable. They could be detected only with very carefully designed input stage, preferentially a high-input-impedance buffer. Essential is the small input capacitance, preferentially about 1 fF. With such an input stage the imaging of dielectric films can be extended to lower frequencies, approximately to 1 kHz.

ACKNOWLEDGEMENTS

The partial support of the VEGA Grant No. 2/3016/96 is highly appreciated.

REFERENCES

1. J. R. Matey, U.S Patent No. 4 481 616 (6 November 1984)., R.C. Palmer, E.J. Denlinger and H. Kawamoto, RCA Rev. **43**, p. 194 (1982).
2. Two-Dimensional Semiconductor Dopant Profiling Development Project, Funded by Digital Instruments/SEMATECH. Project initiated: 8/1/1995.
3. J. R. Matey and J. Blanc, J. Appl. Phys. **57**, 1437 (1985).
4. H.P. Kleinknecht, J.R. Sandercock and H. Meier, Scanning Microscopy **2**, 1839 (1988).
5. C. C. Williams, W.P. Hough and S.A. Rishton, Appl. Phys. Lett. **55**, 203 (1989).
6. C.D. Bugg and P.J. King, J. Phys. E **21**, 147 (1988).
7. Y. Martin, D.W. Abraham and H.K. Wickramasinghe, Appl. Phys. Lett. **52**, 1103 (1988).
8. Š. Lányi, J. Török and P. Řehůřek, Rev. Sci. Instrum., **65**, 2258 (1995).
9. Š. Lányi and J. Török , J. Elect. Eng. **46**, 126 (1995).
10. Digital Instruments, Santa Barbara, CA, USA.
11. Park Scientific Instruments, Sunnyvale, CA, USA.
12. R.C. Barrett and C.F. Quate, J. Appl. Phys. **70**, 2725 (1991).
13. Š. Lányi, J. Török and P. Řehůřek, J. Vac. Sci. Technol. **B 14**, 892 (1996).
14. O.C. Zienkiewicz, The Finite Element Method, McGraw-Hill, 1977.
15. MEP 5.1, of authors L. Dědek and J. Dědková, Faculty of Electrical Engineering and Computer Science, Technical University of Brno, Brno, Czech Republic.
16. L. Dědek and J. Dědková, Proc. of the 4. ANSYS Users Meeting, Sept. 26-27, 1996, Blansko - Češkovice, Czech Republic.
17. Š. Lányi, will be published.
18. G.B. Picotto, S. Desogus, Š. Lányi, R. Nerino and A. Sosso, J. Vac. Sci. Technol. **B 14**, 897 (1996).

RESISTOMETRIC MAPPING USING A SCANNING TUNNELING MICROSCOPE

C I LANG*, J TAPSON**
*Department of Materials Engineering, University of Cape Town, Private Bag, Rondebosch, 7700 South Africa, clang@engfac.uct.ac.za
**Department of Electrical Engineering, University of Cape Town, Private Bag, Rondebosch, 7700 South Africa, jtapson@eleceng.uct.ac.za

ABSTRACT

We present a method whereby spatial variations in the resistivity of bulk conductive specimens may be detected on the same scale as the microstructural variations from which they arise. This technique, a new development of scanning tunneling potentiometry, offers significant benefits for microstructural characterization and for investigation of microstructure/resistivity relationships in metallic materials.

INTRODUCTION

The electrical resistivity of metals is a consequence of departures from perfect periodicity in the crystal lattice, and consists of temperature-dependent and temperature-independent components which are assumed to be additive. At ordinary temperatures conduction electrons are scattered primarily by the thermal vibration of atoms, which results in the temperature-dependent ideal resistivity. The residual resistivity, normally considered to be temperature-independent, arises from enhanced scattering of conduction electrons by lattice defects such as vacancies, impurities, dislocations and grain boundaries. This additional scattering is not spatially homogeneous, but is localized at or near the site of the defect. In alloys, additional spatial variations in conduction electron scattering may arise from regions of non-random atomic configuration, such as ordered domains. The resistivity of polycrystalline metals and alloys is however generally isotropic (i.e. the resistivity is independent of the direction of current flow), although on a crystallographic scale some materials exhibit anisotropy in their resistivity due to directional variations in their electronic structure. In general then, electrical resistivity may be sensitively dependent on the microstructure of the material under consideration, and resistivity can be dependent on the direction of current flow. These relationships can be utilized to characterize and image microstructure by the spatial mapping of resistance.

The influence of microstructure on the electrical resistivity of metallic materials is conventionally investigated by correlation of microstructural information with resistivity data. Experimentally, microstructure is characterized on a microscopic scale, whereas resistance measurements are typically made on a larger scale. The difficulties associated with the correlation of microstructure with resistivity are predominantly a consequence of this difference in scale. As a further consequence, information regarding the presence of microstructural features is not directly obtainable from macroscopic resistance measurements since the total resistance arises from a number of different structural influences as well as from thermal effects. Clearly clarification of microstructural influences on resistivity, and the utilization of resistivity to probe microstructure, would be aided by a reduction in the scale upon which resistance is measured. Muralt and Pohl [1] pioneered a technique whereby a scanning tunneling microscope (STM) probe was used to detect potentiometric discontinuities in thin films in the presence of an in-plane electrical potential. The method, known as scanning tunneling potentiometry (STP), was shown to detect spatial variations in potential with nanometer resolution. The technique has been applied to thin metal films and to semiconductor structures; local discontinuities in potential have been detected and associated with metal/insulator interfaces, with grain boundaries and with changing dopant concentrations [1, 2, 3, 4]. The output of this STP method is a map of the specimen area scanned, showing the variation in potential as a function of position (x,y). Sequential or simultaneous topographic scanning produces a conventional STM map showing topographic features which may under certain circumstances be correlated with features in the STP map. (This technique should not be confused with the use of STM to measure electrochemical potentials [6], or Kelvin probe mapping, in which the surface potential variations resulting from changes in work function are mapped [7].)

In this paper we present a development of the STP technique which expands the scope of application to include bulk conductive specimens and also materials exhibiting resistive anisotropy. Our method of resistometric mapping (so called to emphasis our interest in resistivity and to distinguish it from other potentiostatic techniques, mentioned above, which are becoming common in probe microscopy) has been successfully applied to a range of materials. Results are presented from representative specimens. The technique lends itself not only to investigations of microstructural influences on resistivity, but also to non-destructive detection of subsurface features. In addition, resistometric scanning offers promise as a technique for identifying low-resistivity paths in a material, for characterization or optimization purposes.

EXPERIMENT

The technique

The technique uses a standard STM with the addition of AC bias and demodulation circuitry, as shown in figure 1. A conventional DC bias voltage is applied between the specimen surface and the STM tip. An AC voltage is imposed in the plane of the specimen; a second in-plane AC voltage, geometrically orthogonal to the first and offset in frequency; may be simultaneously imposed. The tip is scanned across the specimen surface in the normal manner. Topographic information (the conventional STM scanned image) is extracted from the DC current which flows across the tunneling junction. Potentiometric information is extracted simultaneously from the in-plane current (or simultaneously from two orthogonal in-plane currents) by synchronous demodulation of the tunneling current with the AC voltages. This is similar in principle to the methods used previously, but the novel use of AC potentiometric voltages significantly improves the versatility of the technique. It is possible to perform resistometry in two in-plane dimensions; and the potential and topographic information is acquired simultaneously rather than sequentially, leading to lower noise levels [5]. The system was designed to drive large, high-frequency currents through the specimen, so that detectable potential variations would be created despite the low resistance of bulk metallic specimens. It is important that the tunneling tip be positioned with exactly the same tip-to-sample separation at each point on the surface, in order to avoid artifacts arising from changes in the tunneling junction resistance. This is achieved by continuous use of conventional STM-type feedback, using a tunneling signal mutually orthogonal to the two STP signals.

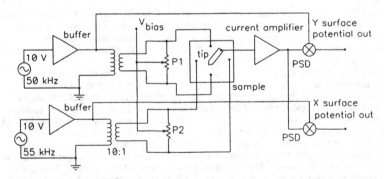

Figure 1: This diagram shows the resistometric scanning system. The tunneling current has three components, derived from the conventional DC bias voltage and the two in-plane AC voltages. The AC voltage at the tip is a sensitive function of local resistivity in the material being scanned. The effect of each (mathematically orthogonal) AC voltage is resolved using synchronous demodulation.

Two conflicting factors must be optimized for in the electronic design – the necessity for high frequency potentials, in order to minimize the current penetration depth, and the high gain necessary in the tunneling current amplifiers. High gain amplifiers generally introduce substantial phase shifts which must be matched in both signal and carrier paths before demodulation. Our approach has been to start at relatively low frequencies (50-55 kHz) to verify the technique, and then to incrementally increase the frequency to reduce the skin depth.

We have used transformer coupling to the specimen to float the AC (resistometric) potentials on a DC (tunneling) bias voltage. This enables us to step up the current in the specimen plane, so that in-plane currents of 1A or more can be obtained. This is necessary to obtain measurable potential variations on the imaged surface, given the extremely low resistivity of bulk metal specimens.

The AC potentials which result from the in-plane currents are synchronously demodulated and low-pass filtered to maintain signal orthogonality. The actual noise bandwith is only 72 Hz; this allows typical potentiometric noise levels of 15 – 75 μV rms (depending on surface-dependent Johnson noise); the noise levels in the ADC system are of the order of 15 μV rms, so the noise limit of the potentiometric system is probably not the limiting factor in imaging [5]. The system is currently being adapted to allow variation of the frequencies between images of the same surface, so that varying penetration depths can be used, in order to explore the influence of penetration depth on the resistometric information acquired. It is conjectured that controlling the penetration depth will provide a new contrast mechanism which should throw some light on subsurface structure and interconnectivity.

In summary, our technique is distinguished by allowing simultaneous measurements in two dimensions; by increasing the in-plane currents to Ampere levels, so as to create measurable potentials in low resistivity specimens; and by the use of synchronous demodulation of AC potentials, which has reduced the noise levels and introduces variations in penetration depth as a contrast mechanism.

Mapping

All specimens are used in bulk form, subject to maximum dimensions of 1 cm x 1 cm x 1 cm imposed by the available volume on the specimen stage. The specimen surface is prepared to a mirror finish where possible. Scanning may be performed over relatively wide areas of 10 μm x 10 μm. A range of specimens was prepared to assess the capabilities of the technique; results are presented from the following materials. 1. Pure palladium: the specimen was prepared as for transmission electron microscopy (TEM) by jet electropolishing; resistometric mapping in air was performed subsequent to imaging in a JEOL 200CX TEM. 2. Polycrystalline graphite: The specimen was polished to a 0.25 μm finish; mapping was performed in air.

RESULTS

Resistometric maps have been successfully obtained from materials with a wide range of resistivities; these maps show resistive variations of a scale comparable to the scale of the microstructure of the material investigated. Two-dimensional resistometric mapping simultaneously produces pairs of maps from current flow in two orthogonal directions. Results are presented for a pure metal and for a material having anisotropic resistivity.

Figure 2 (a) is a TEM image of pure palladium, showing a specimen region adjacent to the foil perforation (P). Figure 2(b) is an STM image of the same region, showing the featureless surface topography to be expected from an electropolished surface. Figure 2(c) is a resistometric map acquired simultaneously with the STM image in 2(b). The resistometric map shows features which indicate localized variations in resistivity within the specimen; these variations must accordingly arise from spatial variations in microstructure. Also notable are the apparent variations in resistivity *outside* the specimen edge. These are tip-related artifacts which occur as the STM tip moves over the edge of the specimen: the tunneling location moves to the side of the tip, continuing to produce a signal although the tip is no longer directly above the specimen surface.

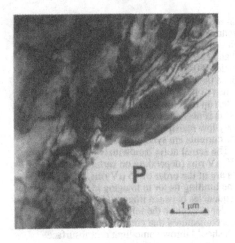

(a)

(b)

Figure 2: Images of pure palladium
electropolished foil. (a) TEM image;
(b) STM topographic image;
(c) resistometric map showing spatial
variations in resistivity.

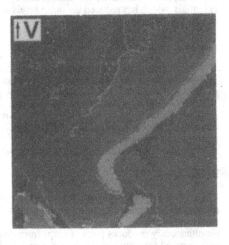

(c)

Figure 3(a) is a topographic image of polished polycrystalline graphite, grain size ±5 μm, showing nearly featureless surface topography. Figures 3(b) and 3(c) are resistometric maps of the same area; the current sampled in figure 3(b) is orthogonal to the current sampled in figure 3(c). Each resistometric map reveals features, not evident in the topographic map, on a scale comparable with the grain structure. Although some individual features are observed in both resistometric maps, their intensity differs from one map to the other. This is expected, since the resistance of graphite has a sensitive dependence on crystal orientation relative to the direction of the imposed current. For a given current direction, this can result in large resistive differences between neighboring grains; for a given grain, each current direction can produce a different resistance.

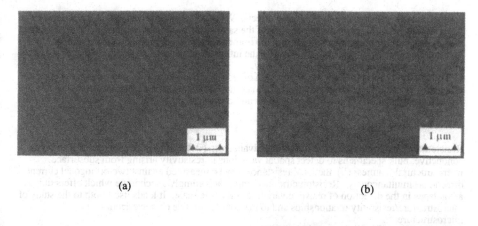

(a)   (b)

Figure 3: STM/resistometric image triplet
from graphite specimen. (a) STM image
showing surface topography;
(b) resistometric map of same region;
(c) resistometric map acquired using current
orthogonal to that used for (b).

(c)

DISCUSSION

Resistometric mapping provides a fine-scale map of local variations in the resistance of bulk conductive specimens. It has previously been shown that in STP, tip artifacts can compromise potentiometric information obtained from rough surfaces [8]; such artifacts are observed at the specimen edge in fig. 2(c). It is accordingly necessary to exercise care in the interpretation of resistive variations which are associated with topographic features. In the present work, local variations in resistance are observed in specimens without surface topographic irregularities: the observed variations must therefore arise from localized resistivity variations due to subsurface microstructural features. Resistometric mapping is thus a technique which may be used to probe subsurface microstructure.

Clearly, it is necessary to consider the depth beneath the surface from which information may be extracted. The use of alternating current at an appropriate frequency results in confinement of most of the applied current to a region beneath the surface of the specimen (the "skin effect"). The effective penetration depth then represents an upper limit to the depth from which potentiometric information may be acquired. In the materials investigated in this study, however, the penetration depth is considerably greater than the grain size. If the observed resistive variations arose from all microstructure-related scattering within the volume defined by the (AC) current path, many subsurface microstructural features might be expected to contribute to the potential at each point (x,y). This summing of these influences at each point (x,y) might be expected to lead to a spatially homogeneous resistance due to averaging effects. What is in fact observed is that the resistometric scans differentiate sharply between regions of the same scale as the microstructure. It appears then that resistive information is extracted from the immediate subsurface region of the specimen. Investigations to clarify this and to determine the influence of penetration depth on the resistometric information acquired, are presently under way.

The capability for simultaneous acquisition of resistometric maps from two orthogonal currents offers the opportunity to further characterize materials exhibiting anisotropic resistivity, by providing information on low-resistivity paths and connectivity.

CONCLUDING REMARKS

Resistometric mapping offers the following advantages: (i) the technique may be used on highly conductive, bulk specimens to detect spatial variations in resistivity arising from subsurface microstructural features; (ii) the local resistance may be measured against two orthogonal current directions simultaneously. Resistometric mapping is accordingly a technique which offers unique advantages in the detection of resistive variations on a fine scale. It lends itself both to the study of microstructure/resistivity relationships and to the non-destructive characterization of microstructure.

ACKNOWLEDGMENTS

The financial support of Mintek and the assistance of the Electron Microscope Unit at the University of Cape Town is gratefully acknowledged.

REFERENCES

1. P. Muralt and D.W. Pohl, Appl. Phys. Lett. **48**, 514 (1986).
2. P. Muralt, H. Meier, D.W. Pohl and H.W.M. Salemink, Appl. Phys. Lett. **50**, 1352 (1987).
3. J.R. Kirtley, S. Washburn and M.J. Brady, Phys. Rev. Lett. **60**, 1546 (1988).
4. M.A. Schneider, M. Wenderoth, A.J. Heinrich, M.A. Rosentreter and R.G. Ulbrich, Appl. Phys. Lett. **69**, 1327 (1996).
5. J. Tapson, PhD thesis, UCT (1994).
6. M.M. Dovek, M.J. Heben, C.A. Lang, N.S. Lewis and C.F. Quate, Rev. Sci. Instrum. **59** 2333 (1988).
7. H.O. Jacobs, H.F. Knapp, S. Muller and A. Stemmer, Ultramicroscopy 69, 39 (1997).
8. J.P. Pelz and R.H. Koch, Phys. Rev. B **41**, 1212 (1990).

MICROMACHINED SFM PROBES FOR HIGH-FREQUENCY ELECTRIC AND MAGNETIC FIELDS

D. W. VAN DER WEIDE, V. AGRAWAL, P. NEUZIL*, T. BORK
Department of ECE, University of Delaware, Newark, DE 19716-3130, dan@ee.udel.edu
*Ginzton Laboratory, Stanford University, Stanford, CA 94305

ABSTRACT

We discuss micromachined localized high-frequency electric (coaxial) and magnetic (loop) field probes integrated with scanning force microscopes. Our approach enables simultaneous acquisition of both field and topography in the radio frequency (RF) through millimeter-wave regime, enabling more complete characterization of materials, devices and circuits.

INTRODUCTION

Current directions in characterizing the microstructure of materials in the RF through mm-wave regime are evolving from using large waveguide structures that measure bulk properties[1-5] toward smaller geometries for *localized* spectroscopy[6-14]. Here the probe/waveguides concentrate the fields of interest to dimensions much smaller than the sample size with the possibility of molecular or even atomic-scale field resolution on the horizon.

We have focused on building multifunctional tips for scanning force microscopy (SFM) to achieve broadband performance and high spatial resolution of both field and topography at the same time. Here we review our progress, present new calculations on coaxial tips and outline some future directions for this work.

HIGH FREQUENCY MULTIFUNCTIONAL PROBES

Electric Field/SFM Probes

Electric field probing in the near field region (at distances and probe apertures much smaller than the wavelength of probing energy) is conveniently done with open-ended coaxial waveguides[4,5]. We have combined this standard approach with a shielded SFM tip/cantilever to apply this technique to submicrometer dimensions[13,14]. At these dimensions, tip-sample distance control becomes increasingly important, and is facilitated by using a vibrating cantilever ("non-contact") approach in our experiments. Millimeter-scale efforts in microwave microscopy have begun to yield important results on material properties in reflection[9-11] as well as in transmission[8]. While we have confined our efforts so far to probing samples such as circuits and waveguides to better understand the field patterns we measure, results from other groups will help establish work on materials at even smaller scales offered by our probes.

These electric field probes are built in two ways: batch fabricated using a modified single-crystal SFM probe process[13] and modified from commercially available SFM probes[14]. As shown in Fig. 1, the batch fabrication process yields a coaxial tip that appears ideal for simultaneous acquisition of high frequency electric field and topography; integrating this coaxial tip onto a cantilever with a broadband transmission line running along the cantilever, as depicted in Fig. 2 is still ongoing. Current results such as picosecond waveforms spatially resolved along ultrafast GaAs integrated circuits (Fig. 3) have been achieved with the modified commercial cantilevers[14].

Figure 1. Micromachined coaxial SFM tip with ~10 nm inner conductor tip radius and ~100 nm shield diameter[13].

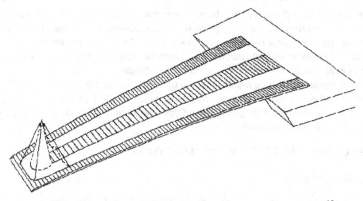

Figure 2. Integration of coaxial tip and broadband transmission line along SFM cantilever[13].

For these measurements we chose an ultrafast distributed circuit, a nonlinear transmission line (NLTL) which not only exhibits clearly distinct waveforms along its structure but also approximates the switching waveforms of advanced digital circuits[15].

We prepared the tip/cantilever assembly with a 90 °C bake to drive off water, followed by a drop of HMDS applied with a brush fiber to the cantilever and body to promote adhesion of photoresist. Using a small drop of AZ-5214 photoresist to cover the cantilever and body provides a thick, low-stress dielectric suitable for insulating the conductive cantilever from a metal shield. We evaporated a 300Å Ti/2000Å Au shield onto the cantilever and tip to create a controlled impedance structure which can be contacted with 0.2 mm diameter flexible coaxial cable[16]. Bringing this contact out to a 50 GHz sampling oscilloscope completes the local probe. We bring the tip into light contact with a Au or alumina surface to remove the shell of metal and photoresist at the tip, exposing the SFM tip as a center conductor. With this assembly we are able to probe local picosecond electric fields along the NLTL using a sampling oscilloscope without intervening amplification. There are limitations on bandwidth imposed by the connectors and cabling, indicated in Fig. 4, which can be overcome with better assembly approaches and localized instrumentation, such as diode samplers and detectors.

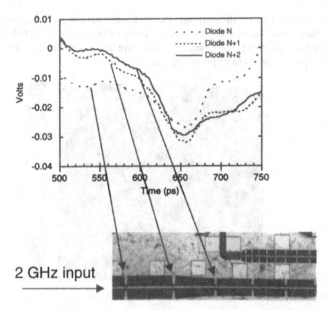

Figure 3. Spatially resolved waveforms along intermediate section of nonlinear transmission line measured with coaxial SFM probe tip.

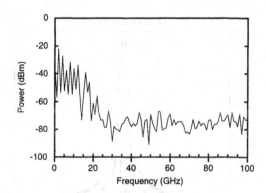

Figure 4. Spectrum of output signal from nonlinear transmission line as measured by coaxial SFM probe.

We expect these frequency limitations to be resolved with fully integrated tips, cantilevers and probe bodies, as discussed in the next section for magnetic field probing.

While we are resolving experimental limitations, it is also important to understand the field strengths and distributions at the probe tip. To do this we have performed full-wave three dimensional field calculations[17] on the tip geometry and material of Fig. 1, as shown in Figs. 5 and 6. In Fig. 5 we plot the total electric field strength integrated over three dimensions as the tip is positioned over a sample with 100 nm conductors and spaces at a 10 nm height measuring a frequency of 20 GHz. In this plot, the tip is located > 5 µm from the energized conductor, so the field strength is low, but it is the relative strength of the field that is most important here. The expected "lightning rod" field-concentrating effect is clear, yet as shown in Fig. 6, a cut through the plane normal to the tip axis shows transverse field component contributions along with the expected and ideal contribution along the tip axis. These extraneous field contributions could smear field images and make precise measurements more difficult if they are not deconvolved from the data, a process we have begun to explore.

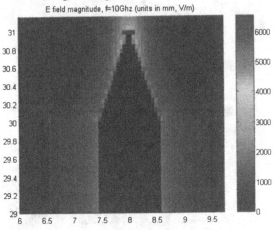

Figure 5. Full-wave calculation of 10 GHz electric field on coaxial tip structure of Fig. 1. Tip and shield metal are dark geometrical shapes.

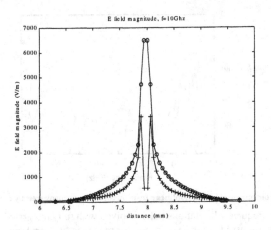

Figure 6. Axial and transverse electric field components at tip of Fig. 5.

Magnetic Field/SFM Probes

To complement the electric field probes we have developed combined μm-level topography and high-frequency magnetic field probes, 2–6 μm diameter loops defined on scanning-force microscope cantilevers[18]. Since these also function as near-zone antennas, they can be used with

24

any suitable detector, and can probe the normal magnetic field component on both active and passive planar samples. We have demonstrated these tips by scanning a coplanar waveguide sample at 10 GHz using a microwave network analyzer. With this instrument, the device noise at this frequency is ~ 2 $\mu\Phi_o/Hz^{1/2}$ at T = 300 K, comparable to that found for a SQUID at DC[19,20]. Unlike the SQUID, however, the thermally limited sensitivity of this system is inversely proportional to frequency, so its sensitivity is only competitive in the GHz regime.

High-frequency magnetic field probes have important applications, from examining superconducting thin films to locating short circuits on IC's, and probing at μm length scales is greatly enhanced by an ability to acquire sample topography along with high-frequency field data, much like the magnetic-force microscope (MFM) can do for static fields[21].

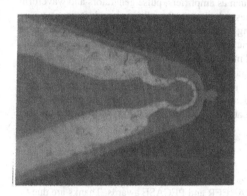

Figure 7. Left: Close-up of loop tip on silicon nitride cantilever. Right: Oblique view of two tips, coplanar waveguide interface and etched V grooves to accommodate microcoaxial cable.

To address this need for a combined instrument, we have designed, fabricated and tested a combined scanning-force microscope (SFM) tip/cantilever and magnetic field loop antenna for simultaneous acquisition of both sample topography and high-frequency magnetic field. Although it collects the field component primarily normal to the sample plane, it can still be a useful tool for characterizing both materials and circuits. We have demonstrated its utility by probing a coplanar waveguide (CPW) transmission line at 10 GHz, a simple structure enabling us to achieve a quantitative field measurement by correlation to a full-wave electromagnetic simulation.

We designed the SFM/loop tip to be compatible with a commercial SFM platform. This required minimizing the additional height of the coaxial cable interface to the loop by etching V-grooves into the silicon body of the probe (Fig. 7, right); this also helps to align the cable for ease of assembly. Instead of forming pyramidal SFM probe tips, we chose to extend a small finger out from the end of the cantilever (Fig. 7, left), simplifying fabrication but limiting our topographical resolution to the radius of this finger, about 0.5 μm.

Previous work probing high-frequency magnetic fields on high-frequency circuits has been done at millimeter dimensions using both loop probes[22] and open-ended rectangular waveguides[2]. In these efforts, having both larger probe dimensions and samples essentially planar on the scale of the probe tips minimized the need for probe-sample distance control, but with μm-scale probes this control is important for field calibration. Building on the precise probe-sample distance control inherent in an SFM platform and the precision available in a microwave vector network analyzer, we can achieve calibrated field measurements at these shorter length scales.

CONCLUSIONS

We have seen how near field probes enable microscopy in the sub-visible spectrum, where phenomena range from IC circuit operation in the MHz-GHz regime to the far-infrared (FIR) "fingerprint" region of molecular spectra. To fully realize the potential of this new capability will take advances in micromachining, instrumentation, and computer-based image handling and presentation. We may, for example, fabricate several antennas onto the same probe to achieve a kind of interferometric depth resolution using a discrete approximation of a lens. We may also begin to integrate high frequency instruments such as amplifiers, pulse generators and waveform detectors directly onto the SPM cantilever bodies, which are, after all, semiconductors. Furthermore, with these probes we are collecting what should be considered a generalized image database, each point having not only topographic data but perhaps a time domain record as well. New ways of visualizing this sub-visible world might include a "fly-through," observing the terrain of the sample as it's painted with field magnitude or phase. Finally, our group is developing a Web-based technique for remotely accessing these specialized instruments so that they can be used by researchers regardless of location. While we are only beginning to build the microscopes for this regime, the range of applications may someday surpass those of the familiar far-field microscope.

ACKNOWLEDGMENTS

This work is supported by the Office of Naval Research, DARPA, W.L. Gore & Associates, the State of Delaware, Topometrix, and NSF CAREER and PECASE awards. Thanks are due to Scott Kee and Jacob Adopley for help with the electromagnetic simulations.

REFERENCES

1   D. Misra, M. Chabbra, B. R. Epstein, M. Microtznik, and K. R. Foster, IEEE Trans. Microwave Theory Tech. **38**, 8-14 (1990).

2   S. S. Osofsky and S. E. Schwarz, IEEE Trans. Microwave Theory Tech. **40**, 1701-8 (1992).

3   Y. Xu, F. M. Ghannouchi, and R. G. Bosisio, IEEE Trans. Microwave Theory Tech. **40**, 143-50 (1992).

4   G. Q. Jiang, W. H. Wong, E. Y. Raskovich, W. G. Clark, W. A. Hines, and J. Sanny, Rev. Sci. Instrum. **64**, 1622-6 (1993).

5   G. Q. Jiang, W. H. Wong, E. Y. Raskovich, W. G. Clark, W. A. Hines, and J. Sanny, Rev. Sci. Instrum. **64**, 1614-21 (1993).

6   E. A. Ash and G. Nichols, Nature **237**, 510-12 (1972).

7   M. Fee, S. Chu, and T. W. Hansch, Optics Comm. **69**, 219-24 (1989).

8   F. Keilmann, D. W. van der Weide, T. Eickelkamp, R. Merz, and D. Stockle, Optics Comm. **129**, 15-18 (1996).

9   T. Wei, X. D. Xiang, W. G. Wallace-Freedman, and P. G. Schultz, Appl. Phys. Lett. **68**, 3506-8 (1996).

10 C. P. Vlahacos, R. C. Black, S. M. Anlage, A. Amar, and F. C. Wellstood, Appl. Phys. Lett. **69**, 3272-4 (1996).

11 M. Golosovsky, A. Galkin, and D. Davidov, IEEE Trans. Microwave Theory Tech. **44,** 1390-2 (1996).

12 M. Golosovsky and D. Davidov, Appl. Phys. Lett. **68,** 1579-81 (1996).

13 D. W. van der Weide and P. Neuzil, JVST-B **14,** 4144-7 (1996).

14 D. W. van der Weide, Appl. Phys. Lett. **70,** 677-79 (1997).

15 D. W. van der Weide, Appl. Phys. Lett. **65,** 881-883 (1994).

16 W.L. Gore & Associates, private communication.

17 MicroStripes 2.4 (Nottingham, KCC Ltd.).

18 V. Agrawal, P. Neuzil, and D. W. van der Weide, Appl. Phys. Lett. **71,** 2343-45 (1997).

19 J. R. Kirtley, M. B. Ketchen, K. G. Stawiasz, J. Z. Sun, W. J. Gallagher, S. H. Blanton, and S. J. Wind, Appl. Phys. Lett. **66,** 1138-40 (1995).

20 R. C. Black, F. C. Wellstood, E. Dantsker, A. H. Miklich, D. Koelle, F. Ludwig, and J. Clarke, Appl. Phys. Lett. **66,** 1267-9 (1995).

21 P. Gruetter, H. J. Mamin, and D. Rugar, in *Scanning tunneling microscopy II. Further applications and related scanning techniques. Second edition*, edited by R. Wiesendanger and H. J. Guntherodt (Springer-Verlag, Berlin, Germany, 1995), p. 151-207.

22 G. Yingjie and I. Wolff, IEEE Trans. Microwave Theory Tech. **44,** 911-18 (1996).

MEASUREMENT OF STRATIFIED DISTRIBUTIONS OF DIELECTRIC PROPERTIES AND DEPENDENT PHYSICAL PARAMETERS

A.V. MAMISHEV, Y. DU, B.C. LESIEUTRE, M. ZAHN
Laboratory for Electromagnetic and Electronic Systems, Department of Electrical Engineering and Computer Science, Massachusetts Institute of Technology, Cambridge, MA, 02139, mamishev@mit.edu

ABSTRACT

Recent advances in ω-k (frequency-wavenumber) interdigital dielectrometry are described. Using this technology, information about the microstructure of dielectric materials is obtained by applying to the sensor-dielectric interface a spatially periodic electric potential swept in frequency from 0.005 Hz to 10,000 Hz. The penetration depth of the electric field is proportional to the spatial wavelength of the electric potential. Application of multi-wavelength electrode arrangements allows measurement of stratified distributions of complex dielectric permittivity. Calibration techniques serve to relate the distributed dielectric properties of materials to other physical variables, such as density, porosity, cracking, lamination, and diffusion of contaminants into the material. The specific problem treated in this paper is in the measurement of moisture concentration distribution in transformer pressboard during the diffusion of water molecules from ambient transformer oil. The output of interdigital sensors is strongly influenced by the microgranularity of the material's surface. Although this dependence complicates interpretation of the measurements in some applications, the variation of the output may also be used to characterize the shape of the surface on the microscale.

INTRODUCTION

Incremental advances in the theoretical background and measurement techniques of interdigital dielectrometry have been made for more than a decade. Previous work resulted in several algorithmic approaches to the inverse problem of material characterization [1, 2]. The multi-wavelength approach has a potential of measuring property profiles of dielectric materials, their dimensions, or both, depending on the complexity of the algorithms. Depending on the application, the size of electrodes may vary from meters to microns. For example, measurement of properties of thin films may be accomplished using VLSI fabrication techniques [3]. This paper describes theoretical and experimental techniques developed for the measurement of one-dimensional diffusion processes.

Continuous monitoring of the moisture diffusion process has an immediate application in the electric power industry. Flow static electrification of transformer pressboard sometimes leads to catastrophic explosions of large power transformers. The charge buildup processes in the course of electrification depend on the surface and volume conductivity of pressboard. The conductivity of the cellulose based pressboard is a function of two major factors: temperature and moisture content. Thus, monitoring of the pressboard state helps to understand, predict, and prevent explosions due to significant charge accumulation with negligible leakage when a highly insulating surface dry zone forms as increases in temperature at the pressboard drives moisture out of pressboard into the oil. Each failed transformer entails costs of about 50 million dollars, including the costs of repair, spillover cleanup, and purchase of the replacement power.

INSTRUMENTATION

Three-wavelength sensor

Figure 1 shows the three-wavelength sensor, which consists of three structurally similar sets of interdigitated electrodes (one driven and one sensing) deposited on a flexible Teflon substrate. The electrodes are driven with sinusoidal frequency. The frequency range used in most of our experiments is from 0.005 Hz to 10 kHz, which implies electroquasistatic fields. The material samples are placed in contact with the sensor, and the fringing electric field penetrates into the sample. The depth of penetration is equal to approximately 1/5 of the spatial wavelength defined as the distance between two neighboring fingers of the same electrode. The transadmittance between the driven and the sensing electrode varies with changes of the complex dielectric permittivity of the adjacent measured sample.

EXPERIMENTAL ARRANGEMENT

Diffusion process

The process of diffusion of water molecules into transformer pressboard was simulated and staged experimentally in a controlled environment chamber. The controlled parameters include temperature, pressure, relative humidity, internal flow rates and amounts of injected substances. The electrical properties of the transformer oil are monitored with a parallel-plate capacitor cell. For the purposes of the spatial distribution description, the 1 mm thick pressboard is modeled as a three-layer medium with the thickness of the layers of 200 μm, 300 μm, and 500 μm as shown in Figure 2. The sensor is pressed against the pressboard surface with enough force to ensure a quality interfacial contact. These distances from the sensor-pressboard interface to the layer boundaries are chosen to be 1/5 of the spatial wavelengths of the sensor. The area of the pressboard sample is approximately equal to the area of the sensor head.

Initially, the transformer oil and the oil-impregnated pressboard are dried under vacuum to be water-free. Then, moist transformer oil is introduced from a separate vessel. The moisture molecules in the oil diffuse from the top in Figure 2 towards the sensor shown at the bottom of the same diagram. Since the experimental chamber is hermetically sealed, the total amount of moisture in the chamber remains constant throughout the process. The relative humidities of oil and pressboard determine the absorption or desorption exchange of moisture between the two media at a given temperature.

Frequency-domain spectroscopy

Oil-impregnated transformer pressboard is frequency dispersive, that is, its dielectric permittivity and conductivity vary with the frequency of the driving signal.

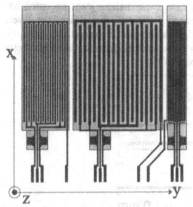

Figure 1. The three-wavelength interdigital sensor with wavelengths of 2.5 mm, 5 mm, and 1 mm built on a common flexible substrate allows measurement of material property profiles at different depths from the surface.

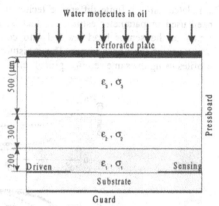

Figure 2. The moisture or chemical contaminant molecules diffuse from the top. The pressboard above the sensor can be discretized into three layers, with the distances from the sensor to the layer upper boundaries equal to 1/5 of the respective wavelengths.

The optimal range of frequencies at which measurement should be taken depends on several factors. Normally, the most information about the system can be extracted at frequencies which lead to a comparable magnitude of the real and imaginary components of the signal gain, which is defined here as the ratio of the voltage on the sensing and the driven electrodes.

Figure 3 shows a typical frequency sweep of the driving voltage signal from 0.005 Hz to 10 kHz. In the high frequency range, the conduction currents are negligibly small, and only capacitance between the electrodes can be measured reliably. At the low frequency limit, the displacement currents are negligibly small and only the conductance between the electrodes can be measured reliably. So, the range from 10 Hz to 0.1 Hz is usually chosen for continuous measurements in order to view the transition from high to low frequency behavior.

NUMERICAL SIMULATION

Numerical simulation of the moisture diffusion process was done using classical Fick's diffusion law with a constant diffusion coefficient, and assuming that the total amount of moisture in the system remains constant. When moist oil is introduced to dry pressboard, moisture in the oil diffuses into the pressboard. As the oil moisture concentration decreases, the pressboard moisture at the upper pressboard boundary in Figure 2 must also decrease to maintain moisture equilibrium at the interface. Then, the curves of the concentration of moisture were transformed into spatial profiles of the complex dielectric permittivity using a semi-empirical functional dependence determined earlier [4].

A numerical algorithm was developed to relate the response of the three wavelength sensor to the spatial distribution of dielectric permittivity and conductivity. This algorithm provides estimates of the complex dielectric permittivity in the stratified media adjacent to the sensor. In order to verify the validity of the algorithm, the process of diffusion of water molecules was simulated using a finite difference technique. In this case, the marching steps approach has been used in parameter estimation, that is, the estimated dielectric properties of the layers closer to the sensor head are fed into consecutive calculations to estimate the dielectric properties of the upper layers. The moisture profiles are calculated from the dielectrometry profiles using the same semi-empirical relationships from [4].

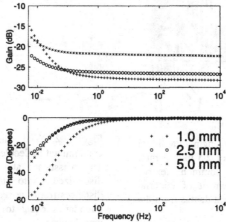

Figure 3. Frequency spectrometry results for the oil-impregnated Hi-Val 1.0 mm thick pressboard at room temperature.

Figure 4 shows the intermediate step in the calculation of moisture profiles. The reduction of the moisture concentration at the upper boundary is not reflected in this plot due to the limited resolution of the three-wavelength sensor. The approximate stair-step profiles shown in Figure 4 do not reflect well the actual distribution of the moisture. In reality, the moisture concentration profile is a continuous function. A piecewise linear approximation would be better than a stair-step approximation. The transformation from the discontinuous to the continuous function should be made taking the mass balance of the moisture into account. The outside moisture boundary concentration is known from separate moisture measurements in the oil using an oil moisture sensor and from standard equilibrium curves between the oil and pressboard [5]. The moisture profile at each moment of time must satisfy the following minimum requirements: external boundary value is known, the profile is continuous, and it is differentiable everywhere except across boundaries.

In order to satisfy both mass conservation and continuity, the currently employed profile shaping algorithm moves the collocation point between the two adjacent regions towards the exterior interface . The new coordinate of the collocation point is found as follows:

$$\text{If } y_{e,n} > 2y_{e,o}, \quad \text{then} \quad x_{i,n} = x_{e,o} + \frac{2y_{e,o}}{y_{e,n}}\left(x_{i,o} - x_{e,o}\right), \quad \text{and} \quad y_{i,n} = 0, \quad (1)$$

where the x coordinate refers to the distance axis, the y coordinate refers to the moisture concentration axis, subscript "i" stands for interior boundary, subscript "e" stands for exterior boundary, subscript "o" refers to the old collocation point, and subscript "n" refers to the new collocation point. For example, for the moisture profile at the leftmost region in Figure 4 and Figure 5, $x_{e,o}=0$, $x_{i,o}=500$ μm, $x_{e,n}=0$, $x_{i,n}=395\mu$m, $y_{e,o}=3.2$ %, $y_{i,o}=y_{e,o}=3.2$ %, $y_{e,n}=8$ %, and $y_{i,n}=0$.

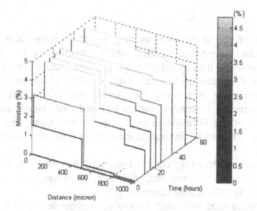

Figure 4. Moisture diffusion process monitored by the three-wavelength sensor. The moisture diffuses from left to right. The averaged value of moisture concentration in each region is found for nine distinct moments of time. The shade in the gray-scale bar indicates the moisture concentration. Penetration depths of 200 μm, 500 μm, and 1000 μm correspond to 1/5 of the wavelengths of 1.0 mm, 2.5 mm, and 5.0 mm, respectively. The driving frequency is 1 Hz.

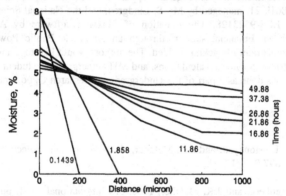

Figure 5. The continuity conditions at the interlayer boundaries and the mass balance requirement help to transform the stair-step output response of the three-wavelength sensor into a physically meaningful description of moisture diffusion. The number next to the curve indicates time in hours At each region, the area under the curves in Figure 4 and Figure 5 are equal.

When the average value of the moisture concentration is more than a half of the pre-computed exterior boundary value, the collocation point does not shift and the rectangular profile is replaced with an equal area trapezoid that is:

$$\text{If } y_{e,n} < 2y_{e,o}, \quad \text{then} \quad x_{i,n} = x_{i,o}, \quad \text{and} \quad y_{i,n} = 2y_{e,o} - y_{e,n}. \qquad (2)$$

Figure 5 shows the same process as in Figure 4 after the continuity and mass conservation conditions were applied. The change of the moisture concentration at the outer boundary (x=0) is now clearly visible.

CONCLUSIONS

The measurement of spatial distributions of physical properties is possible with multi-wavelength interdigital dielectrometry. First, the distribution of the complex dielectric permittivity is measured, and then by calibration it can be related to other physical properties, such as concentration of moisture and contaminants, porosity, density, structural integrity, etc.

This paper presents the theoretical background and algorithmic approach to the measurement of moisture concentration changes in transformer pressboard due to diffusion. Numerical simulation of the diffusion process is performed using empirically determined process parameters. Solution of the inverse problem using synthetic data confirms viability of the chosen approach.

Once developed, this methodology can be directly applied for measurement of the spatial concentration of chemical contaminants, diffusion processes, and for non-destructive characterization of functionally graded materials.

The immediate future work will involve application of the described algorithmic approach to a variety of experimental data of the diffusion processes taken in a controlled environment setup.

ACKNOWLEDGMENTS

The authors would like to acknowledge the support of the Electric Power Research Institute, under grant WO 8619-21, managed by Mr. S. Lindgren, and the National Science Foundation under grant No. ECS-9523128. The donation of Maxwell software by Ansoft Corp. is gratefully appreciated. Financial support through an American Public Power Association DEED scholarship is gratefully acknowledged. The authors would like to thank MIT graduate student Julio Castrillon for initial calculations, and MIT undergraduate student John Miller for computer simulations done as a part of the Undergraduate Research Opportunities Program at MIT.

REFERENCES

1. M.C. Zaretsky, L. Mouayad, and J.R. Melcher, IEEE Trans. on Dielectrics and Electrical Insulation, 23 (6), 897-917 (1988).

2. P.A. von Guggenberg, and J.R. Melcher, Proc. 3rd International Conf. on Properties and Applications of Dielectric Materials, 1262-5 (1991).

3. T.M. Davidson and S.D. Senturia, International Reliability Physics Symposium, 20th Annual Proceedings, p. vii+324, 249-52 (1982).

4. Y.K. Sheiretov and M. Zahn, IEEE Trans. on Dielectrics and Electrical Insulation, 2, 329-351 (1995).

5. T.V. Oommen, Proc. of the 16th Electrical/Electronics Insulation Conference, 162-6 (1983).

UNDERSTANDING COATING AND SUBSTRATE HETEROGENEITIES USING ELECTROCHEMICAL IMPEDANCE METHODS

A. M. MIERISCH, S.R. TAYLOR
Center for Electrochemical Science and Engineering, Dept. of Materials Science,
University of Virginia, Charlottesville, VA 22903, srt6p@s1.mail.virginia.edu

ABSTRACT

This study examines natural breakdown events on organic coated AA 2024-T3 coated using Local Electrochemical Impedance (LEI) mapping (M) and spectroscopy (S). LEIM was able to identify not only different types of defects on this system, but also provided information about the kinetics and stages of development of these defects. Supportive evidence regarding the impedance characteristics of these defects was provided by Capillary Electrophoresis (CE). Data from early stages of defect development indicate that an increased impedance develops at the site. This is related to either the development of aggregated water or electrolyte, as well as corrosion product development. Direct evidence of defect healing is also provided.

INTRODUCTION

The corrosion protection provided by organic coatings on metals has traditionally been investigated with electrical or electrochemical methods (e.g. open circuit potential, electrochemical impedance spectroscopy, electrochemical noise) which provide surface averaged data [1-6]. Frequently, the time course of these measurement clearly indicate breakdown events which are often metastable [7], and believed to occur at discrete sites on the surface [8,9]. While it is logical to think of breakdown events on organic coated alloys as occurring at local sites, particularly on multi-phase engineering alloys, clear demonstration of these events on contiguous coatings has been forthcoming. A method which could locate and evaluate these breakdown sites *in situ* could also be invaluable in understanding the mechanistic issues which control the life/death cycle of these defects, as well as the source of these defects. Questions remain as to whether the defects originate in the coating, the substrate, or whether they exist in both coating and substrate and must be juxtapositioned. Knowledge of the events at these sites will be important to understanding the parameters which control the performance of coated metal substrate.

A method for mapping the local impedance of coated metal substrates was introduced by Lillard, Moran, and Isaacs [10] and later expanded upon in the investigation of a variety of intentional chemical and physical defects [11,12]. Although numerous electrochemical methods are available for the mapping of local electrochemical phenomenon [13-18], Local Electrochemical Impedance Spectroscopy (LEIS) is particularly well suited for the investigation of organic coated substrates due to the AC nature of the excitation which lowers the impedance of the dielectric interface.

As mentioned above, this technique has been used thus far to investigate simulated chemical and physical defects on organic coated steels. The objective of the present research is to locate and characterize local electrochemical breakdown events that occur <u>naturally</u> on a coated <u>aluminum</u> alloy when exposed to an aqueous chloride environment. An understanding of the nature of these defects will eventually lead to an understanding of the factors which limit the service life of coated engineered products, and thus to ways for improvement.

EXPERIMENTAL METHODS

Local electrochemical impedance measurements were made with a five electrode arrangement shown in Figure 1 and discussed elsewhere [11,12]. The interface is excited potentiostatically using a conventional three electrode technique, however the local current is assessed by two vertically displaced micro-reference electrodes. The potential difference between these two electrodes is amplified, then converted into a local current density knowing the solution conductivity and using Ohms law (see Figure 1), and compared to the excitation voltage via a Frequency Response Analyzer (FRA) to generate the local impedance. A 15 mV sine wave excitation at a fixed frequency (500 to 700 Hz) was used in the case of LEIM, and a swept frequency (from 1 Hz to 10 kHz) in the case of LEIS. The DC potential and AC excitation of the coated samples were established with a Solartron 1286 Electrochemical Interface and Solartron 1255 HF FRA all under computer control.

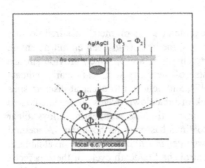

Figure 1. Schematic of five electrode configuration used to acquire LEIS data. Note current lines (dashed) and local field lines (solid) associated with local defect.

The local impedance system can be used in two operational modes. In one mode, the probe and sample substrate can be fixed at a single position while the excitation frequency is swept through a range. These authors prefer to call this mode Local Electrochemical Impedance Spectroscopy (LEIS). In the second mode, the excitation frequency can be fixed as the local admittance magnitude is mapped over the surface. This mode is called Local Electrochemical Impedance Mapping (LEIM). The local admittance is mapped for convenience, so that local breakdown events appear as peaks rather than valleys. In the present system, the mapping is accomplished by holding the electrodes fixed and moving the substrate with a positioning table. The positioning table used in this study consisted of a modified Teledyne TAC model PR-52 automatic wafer probe with adjustable X and Y step sizes ranging from 25 to 25,000 μm.

The substrate material used in all experiments was AA 2024-T3 in the form of 1 mm thick sheet. Samples were cleaned in high purity water (18 $M\Omega\ cm^2$), then ultrasonically cleaned in hexane for 5-10 minutes and left to air dry at room temperature.

Model coated substrate systems were made with two coating chemistries, a vinyl (Vinyl VYHH) and polyurethane. The vinyl was a copolymer of polyvinyl chloride and polyvinyl acetate. The polyurethane coatings were either a 100% polyether polyurethane, or a polyurethane composed of 50% polyether and 50% polyester. Both the 100% and 50/50 polyurethanes were mixed with isocyanate, a dibutyl tin dilaurate catalyst and methyl ethyl ketone solvent. All coatings were spin cast onto the cleaned aluminum samples with a resulting dry coating thickness of ca. 10 μm. The application of neat resin chemistries directly to an untreated substrate (i.e. no conversion coating or primer) is a model system and is not meant to represent a commercial system. The long-term goal of this project will be to eventually examine the effects of the various components within a commercial coating, as well as the effects of various surface treatments (e.g. chromates, alternate conversion coatings, primers, etc.).

Electrochemical test cells were established on the coated samples by attaching glass cylinders to the coating. The test electrolyte consisted of room temperature 0.6M NaCl with ambient aeration.

RESULTS AND DISCUSSION

Initial characterizations of "inherent" defects that develop naturally on coated AA 2024-T3 substrates focused on barely visible spots. After making LEI maps as a function of time on several of these spots, it became apparent that two types of defects were evolving. The distinction between these two types of defects can be seen electrochemically in the LEI maps shown in Figures 2 and 3. Differences also became visually apparent with time as one type was more active, growing to a bright red colored blister site 2 to 5 mm in diameter within several days. While the other type of defects were more passive, either black or white, and were smaller in size (1 to 2 mm diameter). For convenience, the active blisters will be referred to as "red" blisters, and the other type

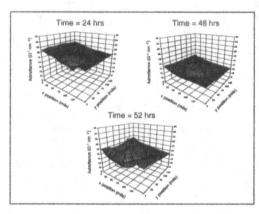

Figure 2. LEI maps of red colored blister at various times. Sample consisted of 10 μm thick Vinyl VYHH on AA 2024-T3. The electrolyte was 0.6 M NaCl.

as "black" blisters. The distinctive active and passive character of the red and black defects respectively is shown in the LEIMs and local impedance spectra in Figure 4. Although the data shown in Figures 2 , 3, and 4 are from samples with different coating chemistries, the statements made concerning their characteristic development, was consistent regardless of the coating material.

Further testing of these defects was performed through the analysis of the solution chemistry beneath the coating using Capillary Electrophoresis (CE). CE is a technique which can identify and determine the concentration of the anions and cations within very small quantities of solution (10 nl) at very low levels of ion concentration (e.g. ppb to ppm). The electropherograms for the black and red blister solutions using two different carrier solutions (UVCat I and UVCat II) are shown in Figure 5. The two background carriers were needed to highlight specific ions.

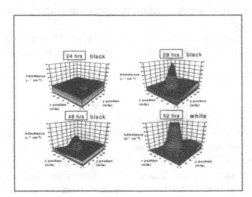

Figure 3. LEIM of black colored blister at various times. Sample consisted of 10 μm thick polyurethane (50/50) file on AA 2024-T3. The electrolyte was 0.6 M NaCl.

Figure 4. Comparison of LEIM and LEIS (below) of red and black defects beneath a vinyl coating at 35 hours. Notes higher admittance in map and lower polarization resistance in spectra for the red site, and converse for black.

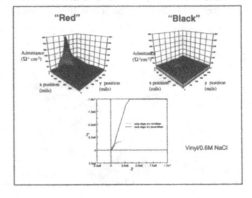

38

It is clear from this data, that the red blisters contain significant concentrations of Cu^{2+}, Al^{3+}, and Mg^{2+} ions, whereas the black blister solution contained essentially no Cu or Al ions, but did contain Mg^{2+} and Zn^{2+}. Although future experiments will utilize other surface analysis techniques (e.g. Scanning Auger Microscopy) to investigate the origins of these defects, it is speculated at this point that the behavior of these defects is related to specific precipitates within in the substrate such as Al_2CuMg (S phase)[19] in the case of red spots, and perhaps a magnesium bearing precipitate for the black/white spot. It should be noted that the red spots formed in the presence of various electrolytes including NaCl, KCl, and Na_2SO_4.

The "black" spot had the character of changing from black to white. The time sequence of LEIMs shown in Figure 3 indicate that the color changes along with the individual maps. This color change may be related to the conversion of a magnesium oxide as the chemistry within the blister changes. It is also interesting to note that the

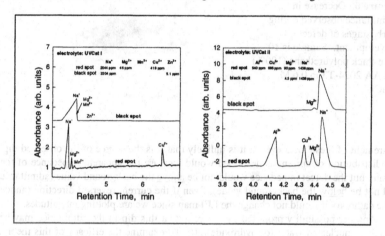

Figure 5. Electropherograms of underfilm solution from red and black defects of Figures 4.

open circuit potential of the sample became significantly more negative at the time of the color change. Further studies will hopefully explore the origins of these changes.

Also of note, is the time sequence of the admittance peaks in the LEIMs for the black defect. It can be seen that at early times the admittance grows, then gets smaller, and then grows catastrophically. This is strongly suggestive of a metastable type defect that "heals" itself with time. It has been postulated by others that "healing" of films can occur through the sealing effects of corrosion products precipitated at the defect site [8]. In this case the defect eventually re-establishes itself.

Another interesting and unexpected result from LEIM studies of early stages of defect development was that the earliest observations of initial defects began with a decrease in the admittance magnitude relative to the background. An example of this is seen in Figure 6. This behavior is counterintuitive. Intuitively, one would expect that the admittance would increase (impedance would decrease) as electrolyte penetrates the film and initiates corrosion.

In Figure 6, the difference between the lowest point in the trough and median point in the background plane is ca. 10 ohm^{-1}. This difference converts to an increase of 20 ohms. It has been suggested by some that this decrease may be artefactual in nature due to nonuniform current distribution. Since the microreference electrodes only sense the vertical component of the local current, a shift of cathodic current from the counter electrode to local surrounding cathodes would alter the measurement. While it is extremely important to consider current distribution issues in the

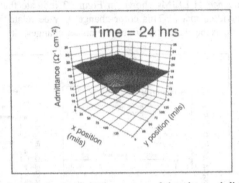

Figure 6. Decrease in admittance observed during early stages of defect development. Sample is 10 mm thick polyurethane (100%) on AA 2024-T3 in 0.6 M NaCl.

measurement of these local events, it is unlikely that it is the source of the observed dip. The redistribution of current as described would increase the apparent impedance of the local site, but the defect impedance would not be <u>above</u> the background value (admittance would not be <u>below</u> the background value). Even if the current reversed direction (anode became cathode), it would not change the LEI map since we are plotting magnitudes.

Another possibility proposed by others is that this dip in the admittance may be the result of nucleated water (or hydroxide)[20]. To examine the efficacy of this theory, calculations based on two models were considered. In the first model, it was assumed that a "nucleus" of pure water in the form of a disk was placed at the metal/polymer interface and only the dielectric properties of the materials were considered, i.e. no ions or charge transfer were considered. In the second model, it was assumed that ions were present in solution and interfacial polarization could occur. The physical representation of the models and values assumed for the calculation are shown in Figure 6 and Table 1 respectively. In Model 2 the assumed double layer capacitance and charge transfer resistance were 25 μF/cm^2 and 1000 Ω cm^2 respectively. Based on these calculations, it is conceivable that either the collection of associated pure water of sufficient thickness or the inclusion of electrochemical processes could cause the observed dip in the admittance.

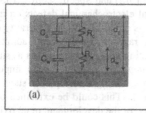 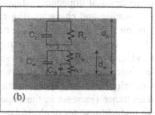

(a)   (b)

Figure 6. Circuit models used to calculate the local impedance associated with the collection of a small mass of pure water (Model I) or the development of an electrochemical interface in ionic solution (Model II).

	d_w	d_c	ρ_w (Ω cm^2)	ρ_c (Ω cm^2)	κ_w	κ_c	Z @ 500 Hz
Model 1	0.1 μm	10 μm	1.8×10^7	10^{14}	80	5	2 Ω cm^2
	1.0 μm	same	same	same	same	same	20 Ω cm^2
Model 2	1.0 nm	same	same	same	same	same	20 Ω cm^2

Table 1. Values of constants and dimensions used in the calculation of local impedance.

Another consideration for the decreased admittance in the initial stages of defect development is that a resistive oxide or corrosion product forms. As stated previously, the eventual goal of this project will be to try to examine the aluminum surfaces of autopsied specimens to characterize the surface chemistry at these early stage defects.

The effect of corrosion product on the LEI map is possibly seen in Figure 7. In this map, the admittance peak is surrounded by an admittance trough. This peak to trough behavior might be explained by the recent proposed mechanism for filiform corrosion on aluminum [21]. In this mechanism, the underfilm corrosion of aluminum is established by an active anode at the periphery of the site. This region is of low pH. The cathode is established more to the center of the site, and becomes alkaline. The low and high pH regions can be established by the development of a precipitated corrosion product which acts as a separator of these two regions.

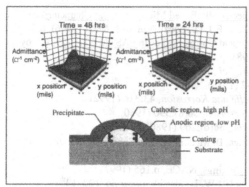

Figure 7. LEI map of 50/50 polyurethane on AA 2024-T3 in 0.6 M NaCl at 24 and 48 hours. Note the trough at the base of the peak. This might be explainable by the development of corrosion product which separates the anodic and cathodic regions of the corrosion process.

CONCLUSIONS

Local Electrochemical Impedance Mapping (LEIM) and Spectroscopy (LEIS) has been successful at locating and characterizing natural electrochemical defects on model coated substrates using vinyl and polyurethane coatings on AA2024-T3. Several types of natural defects have been observed, differing in rate of development, admittance magnitude (severity), and underfilm chemistry suggesting different precipitates as sites of origin. In the development of defects in this coating/alloy system, several important stages of defect development have been documented. Early development starts as a decrease in the admittance (increase in the impedance). This could be explained by the development of an underfilm water or electrolyte layer, or the development of an oxide or corrosion product. It has also been observed that defects can heal and re-grow, possibly from the development of an expansive corrosion product.

ACKNOWLEDGEMENTS

The authors gratefully acknowledge the Air Force Office of Scientific Research for financial support of this research, and J. Yuan and Prof. R.G. Kelly for their contribution of Capillary Electrophoresis data and analysis.

REFERENCES

1. H. Leidheiser, Jr., Corrosion, 39(5):189 (1983).
2. F. Mansfield, M.W. Kendig, S. Tasi, *Corros. Sci.*, **23**(4):317 (1983).
3. S.R. Taylor, *IEEE Trans. Elec. Insul.*, 24(5):787 (1989).
4. S. Haruyama, M. Asari, Tsuru, "Corrosion Protection by Organic Coatings", The Electrochemical Society, Proc. 87-2 (1987), p.197.
5. S. Hirayama, S. Haruyama, *Corrosion*, **47**(12):953 (1991).
6. C.T. Chen and B.S. Skerry, *Corrosion*, **47**(8):598 (1991).
7. J.A. Grandle, S.R. Taylor, *Corrosion*, **50**(10):792 (1994).
8. J.E.O. Mayne, D.J. Mills, *J. Oil Col Chem Assoc.*, **58**:155 (1975).
9. R.C. Bacon, J.J. Smith, F.M. Rigg, *Ind. Engr. Chem.*, **40**(1):161 (1948).
10. R.S. Lillard, P.J. Moran, H.S. Isaacs, *J. Electrochem. Soc.*, **139**(4):1007 (1992).
11. M.W. Wittmann, S.R. Taylor, in Advances in Corrosin Protection by Organic Coatings II, Ed. by J.D. Scantlebury and M.W. Kendig, The Electrochem Soc., **PV 95-13**:158-168 (1995).
12. S.R. Taylor, M.W. Wittmann, in Electrically Based Microstructural Characterization, Ed. by R.A. Gerhardt, S.R. Taylor, and E.J. Garboczi, Vol. 411, MRS (1996), p.31.
13. H.S. Isaacs, B. Vyas, in Electrochemical Corrosion Testing ASTM STP 727, Ed. by F. Mansfeld, U. Betocci, ASTM 1981, p. 3.
14. I.L. Rosenfeld, I.S. Danilov, *Corros. Sci.*, **7**:129 (1967).
15. H.S. Isaacs, *Corros. Sci.*, **28**(6):547 (1988).
16. M. Strattman, H. Streckel, *Werkstoffe und Korrosion*, **43**:316 (1992).
17. J.V. Standish, H. Leidheiser, Jr., *Corroison*, **36**(8):390 (1980).
18. H.S. Isaacs, M.W. Kendig, *Corrosion*, **36**(6):269 (1980).
19. R.G. Buchheit, R.P. Grant, P.F. Hlava, B. Mckenzie, G.L. Zender, *J. Electrochem. Soc.*, **144**(8):2621 (1997).
20. J.E. Castle, in Corrosion Resistant Coatings, NACE, p.165 (1997).
21. J.H.W. deWit, D.H. van der Weijde, H.J.W. Lenderink, paper no. 102, 13th International Corrosion Congress, 1997.

ELECTROMECHANICAL STUDY OF CARBON FIBER COMPOSITES

XIAOJUN WANG, XULI FU AND D.D.L. CHUNG
Composite Materials Research Laboratory, State University of New York at Buffalo
Buffalo, NY 14260-4400

ABSTRACT

Electromechanical testing involving simultaneous electrical and mechanical measurements under load was used to study the fiber-matrix interface, fiber residual stress and marcelling (fiber waviness) in carbon fiber composites. The interface study involved single fiber pull-out while the fiber-matrix contact resistivity was measured. The residual stress study involved measuring the resistance of a single fiber embedded in the matrix while the fiber was tensioned at its exposed ends. The marcelling study involved measuring the resistance of a composite in the through-thickness direction while tension was applied in the fiber direction.

INTRODUCTION

Carbon fibers are used as a reinforcement in composite materials. Much research has been conducted on the structure and properties of these materials. Experimental techniques include mechanical testing, electrical measurements, fiber-matrix bond testing, microscopy, etc. These techniques have their limitations. For example, microscopy and surface analysis give information on the structure, but do not provide bond strength data; fiber pull-out testing provides bond strength data, but does not give information on the structure of the fiber-matrix interface. Electrical measurements provide information on the structure, since carbon fibers are conducting. However, interpretation of the results in terms of the structure is complicated by the multiplicity of factors that affect the electrical behavior. For example, both fiber waviness (marcelling) and delamination affect the through-thickness resistivity of a continuous carbon fiber polymer-matrix composite laminate. Fiber waviness increases the chance that adjacent fiber layers touch, thereby decreasing the through-thickness resistivity. However, delamination increases this resistivity. The decoupling of these factors is difficult. Mechanical testing provides data on the mechanical behavior, but interpretation of the results in terms of the structure is complicated by the multiplicity of factors that affect the mechanical behavior. For example, both marcelling and fiber breakage decrease the tensile strength of a continuous fiber laminate along the fiber direction. To help alleviate this problem, this paper presents electromechanical testing, which involves simultaneous electrical and mechanical measurements on the same sample under load, in contrast to separate measurements in previous work [1,2]. This paper illustrates the use of electromechanical testing in studying the fiber-matrix interface, the fiber residual stress and marcelling.

FIBER-MATRIX INTERFACE

The fiber-matrix interface is critical to a composite material. The bond strength and interfacial structure are related. The desirable interfacial phase depends on the origin of bonding, but interfacial voids are undesirable. The bond strength is commonly determined under shear by single fiber pull-out testing, in which a single fiber is embedded at one end in the matrix and then pulled out. This technique does not provide information on the interfacial structure. It also suffers from the large data scatter, which stems from variation in interface cleanliness. This problem can be alleviated by measuring both bond strength and contact resistivity for each interface sample and seeing how these two quantities correlate among samples that are identically prepared. The sense of the correlation provides information on the origin of bonding. Decrease of the contact resistivity with increasing bond strength means that interfacial voids govern the bond strength; a higher void content causes the contact resistivity to increase and causes the bond strength to decrease. Increase of the contact resistivity with increasing bond strength means that an interfacial phase of high volume resistivity helps the bonding. A change in interfacial structure, as caused by surface treatment of the fiber or by change in composition of the matrix, causes the curve of contact resistivity vs. bond strength to shift. By observing the shift, even a small change in bond strength resulting from the change in interfacial structure can be discerned. This technique [3] is illustrated below for the interface between carbon fiber and cement matrix.

The fibers were isotropic pitch based, unsized and of diameter 15±3 µm, from Ashland Petroleum Co. (Ashland, KY). Fiber surface treatment involved exposure to O_3 gas (0.6 vol.%, in O_2) at 160°C for 5 min. Portland cement (Type 1) from Lafarge Corp. (Southfield, MI) was used to make (i) plain cement paste (water/cement ratio is 0.45), (ii) paste with methylcellulose in the amount of 0.4% by weight of cement (with water reducing agent in the amount of 1% by weight of cement, and water-cement ratio = 0.32), and (iii) paste with latex in the amount of 20% by weight of cement (water/cement ratio = 0.23, without water reducing agent).

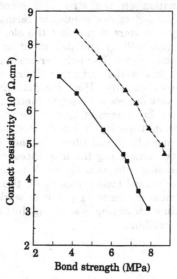

Fig. 1   Variation of contact electrical resistivity with bond strength for plain cement in contact with as-received (■) and treated (▲) carbon fibers.

The contact resistivity was measured at 28 days of curing using the four-probe method and silver paint as electrical contacts. A current contact and a voltage contact were on the fiber, while the other voltage and current contacts were on the cement paste embedding the fiber to a distance ranging from 0.51 to 1.20 mm. Pull-out testing was conducted on the same samples as the contact resistivity measurement. The contact resistivity was the value prior to pull-out testing. The shear bond strength was the maximum load divided by the initial interface area.

Figure 1 shows the correlation of contact resistivity and bond strength for plain cement paste in contact with as-received and treated fibers. Among the samples in each case, a high bond strength is associated with a low contact resistivity, because a high bond strength is associated with a low content of interfacial voids. The treatment increased both bond strength and contact resistivity, due to an interfacial layer of high volume resistivity. Although the bond strength increase due to the treatment was slight (Fig. 1), it caused substantial increases in tensile strength, modulus and ductility of the composite [4]. Both advancing and receding contact angles with water were decreased to zero by the treatment. The improved wetting is due to the increase in surface oxygen concentration and change in surface oxygen from C-O to C=O, as shown by the O_{1s} and C_{1s} peaks in ESCA. Improved wetting by water means improved wetting by cement paste.

FIBER RESIDUAL STRESS

Due to the shrinkage of the matrix during composite fabrication and/or the thermal contraction mismatch between fiber and matrix during cooling; the fibers in a composite can have a residual compressive stress [5-8]. This stress may affect the structure of the fiber so that the fiber properties are affected, often adversely. The measurement of the fiber residual strain by x-ray diffraction, Raman scattering and other optical techniques is difficult due to the anisotropy of the fiber strain and the necessity of embedding the fiber in the matrix. Electromechanical testing provides a simple and effective method for measuring the fiber residual stress along the fiber direction, as illustrated below for the case of carbon fiber in epoxy.

The residual stress σ_f along the fiber direction (one dimension) is

$$\sigma_f = \frac{E_f E_m V_m (\alpha_m - \alpha_f) \Delta T}{(V_m E_m + V_f E_f)} \tag{1}$$

Since $E_f = 221$ GPa, $E_m = 3.7$ GPa, $\alpha_m = 42 \times 10^{-6}$ K^{-1}, $\alpha_f = 0.09 \times 10^{-6}$ K^{-1} and $\Delta T = 155$ K, σ_f is 1438 MPa. The fiber was 10E-Torayca T-300 (unsized, PAN-based), of diameter 7 μm and resistivity $(2.2 \pm 0.5) \times 10^{-3}$ Ω.cm. The epoxy was EPON(R) resin 9405 with curing agent 9470, both from Shell Chemical Co., in weight ratio 70:30.

The resistance of a carbon fiber embedded in epoxy before and after the curing of the epoxy (at 180°C, without pressure, for 2 h), as well as during subsequent tensile loading, was measured. A single fiber was embedded in epoxy for a length of 60 mm and an epoxy coating thickness of 5 mm, such that both ends of the fiber protruded. Four contacts were made using silver paint on the protruded fiber. The resistivity of the fiber increased by ~ 10% after curing and subsequent cooling. The

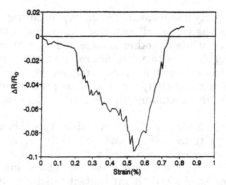

Fig. 2   The fractional resistance change of carbon fiber in epoxy under tension.

fractional resistance increase was also ~ 10%. The observed resistance increase after curing and cooling is attributed to the residual thermal stress, which is compressive in the fiber and probably causes an unidentified microstructural change in the fiber.

Electromechanical testing of a single fiber in cured epoxy was similarly conducted during tension under load control. The crosshead speed was 0.1 mm/min. The strain was obtained from the crosshead displacement. Fig. 2 shows the fractional change in resistance ($\Delta R/R_0$) of fiber up to fracture. Due to the small strains involved, $\Delta R/R_0$ was essentially equal to the fractional change in resistivity. The $\Delta R/R_0$ decreased by up to ~10% upon tension to a strain of ~ 0.5% (a stress of 1320 MPa) and then increased upon further tension. The magnitude of resistance decrease of fiber in initial tension is close to the value of the prior resistance increase during curing and cooling of epoxy.   The stress at which the resistance decrease was complete (1320 MPa) is close to the value of 1438 MPa obtained from Eq. (1). Therefore, the initial decrease in $\Delta R/R_0$ is attributed to the reduction of the residual compressive stress in the fiber. The later increase in $\Delta R/R_0$ is attributed to damage in the fiber. Previous work on the electromechanical behavior of a bare carbon fiber has shown that damage causes the resistivity of the fiber to increase [9]. Although not shown in Fig. 2, $\Delta R/R_0$ decreased upon loading and returned to the initial value upon unloading from a strain of ~ 0.3%, indicating the reversibility of the electromechanical effect.   The $\Delta R/R_0$ per unit strain is -17.   In contrast, that associated with a bare carbon fiber and due to dimensional changes is +2.

MARCELLING

Marcelling causes localized redistribution of stresses [2,10]. It is inherent in composites, as it is formed during manufacture. Detection of marcelling after the manufacture of a composite is necessary. The use of microscopy to detect marcelling is difficult, especially if the marcelling does not occur at the edge of a composite. Moreover, it is tedious and insensitive to small degrees of waviness. Marcelling affects both mechanical and electrical properties, but these effects cannot be used to indicate the degree of marcelling, because other structural aspects (such as fiber breakage and delamination) also affect these properties.

Electromechanical testing for detecting the degree of marcelling involves applying tension (within the elastic regime) to the composite in the direction of the fibers whose waviness is to be determined and simultaneously measuring the resistance in the through-thickness direction. This procedure is performed during tensile loading and unloading. Due to inherent marcelling, the fibers along the stress axis become less wavy as tension is applied, so the chance that adjacent fiber layers touch is decreased and the resistivity (DC) perpendicular to the fiber layers increases. The more severe is the marcelling, the more is the fractional increase in through-thickness resistance per unit strain in the stress direction. Upon subsequent unloading, the resistance returns to its original value and the strain returns to zero. Delamination as well as reduced marcelling cause increase in the through-thickness resistance. Therefore, electrical measurement is not effective for marcel detection. Delamination cannot cause an electromechanical effect, whereas marcelling can.

The electromechanical method of marcel detection is illustrated below for a unidirectional continuous carbon fiber epoxy-matrix composite. Samples were constructed from a unidirectional carbon fiber prepreg tape manufactured by ICI Fiberite (Tempe, AZ), i.e., Hy-E 1076E, which consisted of a 976 epoxy matrix and 10E carbon fibers. A composite laminate (14 laminae) was laid up in a compression mold with configuration $[0]_{14}$. Curing occurred at $179 \pm 6°C$ and 0.61 MPa for 120 min. The density and nominal thickness of the laminate were 1.52 ± 0.01 g/cm^3 and 1.4 mm respectively after curing. The fiber volume fraction was 58%.

The volume resistance R was measured using the four-probe method while cyclic tension was applied in the longitudinal direction (parallel to fibers). The resistance R refers to the sample resistance between the inner probes. The longitudinal and through-thickness R were measured in different samples. For the longitudinal R measurement, the four electrical contacts (silver paint) were around the whole perimeter of the sample in parallel planes perpendicular to the stress axis. For the through-thickness R measurement, the current contacts were centered on the largest opposite faces and in the form of open rectangles in the longitudinal direction, while each of the two voltage contacts was in the form of a solid rectangle surrounded by a current contact (open rectangle). A strain gage was also attached.

Fig. 3 shows the longitudinal stress and strain and the through-thickness $\Delta R/R_0$ obtained simultaneously during cyclic tension to a stress amplitude equal to 35% of the breaking stress. The through-thickness $\Delta R/R_0$ increased upon loading and decreased upon unloading in every cycle, such that R irreversibly decreased slightly after the first cycle. Upon increasing the stress amplitude to 45% of the breaking stress, the effect was similar, except that the reversible part of $\Delta R/R_0$ was larger. The modulus was determined from the slope of the stress-strain curve in the elastic regime. The absolute value of the reversible part of $\Delta R/R_0$ (both longitudinal and through-thickness) decreases with increasing modulus. A sample with a larger absolute value of the reversible part of $\Delta R/R_0$ tended to have a smaller modulus.

A dimensional change without resistivity change would have caused R to increase during tensile loading. In contrast, longitudinal R decreased upon loading. Furthermore, the observed magnitude of $\Delta R/R_0$ was from 9 to 14 times that of $\Delta R/R_0$ calculated by assuming that $\Delta R/R_0$ was only due to dimensional change. The

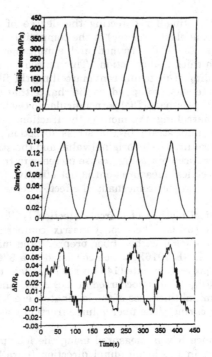

Fig. 3　　Longitudinal stress and strain and the through-thickness $\Delta R/R_0$ obtained simultaneously during cyclic tension to a stress amplitude equal to 35% of the breaking stress for carbon fiber epoxy-matrix composite.

observed decrease in longitudinal $\Delta R/R_0$ and increase in through-thickness $\Delta R/R_0$ upon longitudinal tension are attributed to the decrease in the degree of marcelling. This decrease causes the longitudinal resistivity to decrease and causes the adjacent fiber layers to have less chance of touching, so the through-thickness resistivity increases.

REFERENCES

1.　S.A. Jawad, M. Ahmad, Y. Ramadin, A. Zihlif, A. Paesano, E. Martuscelli and G. Ragosta, Polymer Int. **32**(1), 23 (1993).
2.　H.M. Hsiao and I.M. Daniel, Composites A **27**(10), 931 (1996).
3.　X. Fu and D.D.L. Chung, Composite Interfaces **4**(4), 197 (1997).
4.　X. Fu and D.D.L. Chung, Cem. Concr. Res. **26**(10), 1485 (1996).
5.　A.S. Crasto and R.Y. Kim, Proc. Am. Soc. Composites, 8th Tech. Conf., Technomic Pub. Co., 1994, p. 162-173.
6.　C.F. Fan and S.L. Hsu, J. Polymer Science: Part B, **27**(2), 337 (1989).
7.　D.T. Grubb and Z. Li, J. Mater. Sci. **29**(1), 203 (1994).
8.　K.S. Kim and H.T. Hahn, Composites Sci. Tech. **36**(2), 121 (1989).
9.　X. Wang and D.D.L. Chung, Carbon **35**(5), 706 (1997).
10.　R.S. Feltman and M.H. Santare, Composites Manufacturing **5**(4), 203 (1994).

FORMATION OF DISLOCATIONS IN NiAl SINGLE CRYSTALS STUDIED BY *IN SITU* ELECTRICAL RESISTIVITY MEASUREMENT

Y.Q. SUN[†], P.M. HAZZLEDINE[‡] and D.M. DIMIDUK[‡]
[†]Department of Materials Science and Engineering, University of Illinois at Urbana-Champaign, IL 61801, USA.
[‡]Materials and Manufacturing Directorate, Air Force Research Laboratory, Wright-Patterson AFB, OH 45433-7817, USA

ABSTRACT

This paper reports experiments in which *in situ* electrical resistivity measurements were used to monitor the formation of dislocations in initially dislocation-free NiAl single crystals. The electrical resistivity is found to exhibit an abrupt jump at the onset of plastic yielding. This is interpreted to result from an abrupt nucleation of a massive density of dislocations at the yield point.

INTRODUCTION

In this research we have used *in situ* electrical resistivity to monitor dynamically the formation of dislocations in crystals that are initially dislocation-free, subjected to high stresses and undergoing plastic deformation.

How dislocations are generated in the presence of high stresses is important to a number of materials properties. A well-known case concerns the high tensile and shear stresses concentrated in the vicinity of a loaded crack in semi-brittle materials. Here the generation of dislocations will restore local plasticity and lead to a transition to ductility [1,2]. Another case involves the stress build-up due to lattice and thermal mismatches at interfaces in multilayer materials and thin films. Here the formation of interfacial dislocations under mismatch stresses directly affects the interfacial structure and integrity, and ultimately the functional performance of the materials [3].

The approach of this research is to probe the formation of dislocations in single crystals that are initially free of mobile dislocations and yield at high shear stresses. The probing method used is electrical resistivity which reflects the dislocation content of the crystals because of the scattering of the conduction electrons by the distortion fields of dislocations.

EXPERIMENTAL PROCEDURES

Specimen Preparation

The material selected for this study is single crystal NiAl in the [001] orientation. Besides being a good conductor of electricity and thus suitable for studying the dislocation content with the resistivity method, a NiAl single crystal in the [001] orientation has the following properties that are particularly suitable for studying dislocation processes under high stresses: (1) *[001] is the dislocation-free orientation for NiAl*. In NiAl the grown-in dislocations and slip dislocations under normal deformation conditions have <100> as the Burgers vector [4]. The crystal is effectively dislocation-free along the [001] orientation because the Schmid factor is zero on all the dislocations that may be present after the annealing treatment. (2) *The high yield stress of [001] NiAl*. For NiAl [001] is the so-called 'hard orientation' along which the yield strength is very high [4,5]. At room temperature the yield strength is around 1500 MPa and increases to

2000 MPa at 77 K. The high yield strength makes [001] NiAl a suitable material for investigating the effect of high stress on dislocation nucleation.

The NiAl single crystal was annealed at 1200 C for 7 days in vacua, and was slowly cooled to the room temperature inside the vacuum furnace. The crystal orientation was determined with Laue X-ray back-diffraction. Thin tensile specimens were prepared with mechanical grinding and electro polishing. Figure 1 shows a photograph of a prepared I-shaped tensile specimen. The two broad ends were for load transmission and for attaching the four-probe terminals for the resistance measurement. The gauge length, between the shoulders, is 24 mm. The cross-section over the gauge length is 0.24mm by 0.4 mm. The axial orientation of the specimen is [001].

[001]

Figure 1. A photograph of a prepared tensile specimen.

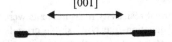

Testing Methods

Testing involves loading the wire specimen in tension at a constant strain-rate and measuring simultaneously the changes in the resistance. For electrical insulation from the mechanical testing system, the transmission of load to the specimen was via two circularly cylindrical alumina adapters each with two perpendicular slits, one across the diameter for supporting the specimens and the other for specimen loading and retrieval. A schematic drawing of the adapter is shown in Figure 2. The adapters are fitted onto the metal fixtures of the tensile testing machine.

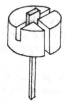

For electrical resistance measurement, pure Ni wires were used as leads and were spot-welded onto the broad faces of the tensile specimens. Spot welding was used because of the expected need to test the specimens at high temperatures. With careful cleaning, NiAl and Ni exhibited excellent weldability. The weld points lay outside the slits in the load adapters and were not strained during the test.

Fig.2. The method used for loading specimens.

The weld spots caused some contact resistance which affected the reading of the absolute resistance of the specimen, but did not affect the measurement of the change of the resistance with load and displacement.

For the resistance measurement, a four-terminal AC (alternating current) micro-Ohmmeter with $\mu\Omega$ resolution was used. The micro-Ohmmeter has a built-in standard that is maintained at 35 °C thus the measurement is independent of the ambient temperature. The AC current is used to eliminate thermal EMF or thermo-couple effect due to the inevitable temperature gradient in the leads. The resistance data was collected at 30 second intervals during the test.

All the tests were conducted in air. The strain rate was 10^{-6} per second. A 25 pound load cell was used to measure the loading force.

RESULTS

The Brittle Regime

In the temperature region up to 242 °C, [001] NiAl is fully brittle with no visible plastic deformation. Figure 3a below shows the result of a test at 229.9 °C. The solid curve is the stress-strain curve and shows brittle fracture at 1320 MPa. The resistance, shown by the symboled curve, increases approximately linearly with strain, up to the point of brittle failure. For a wire specimen undergoing uniform elongation, with initial length L_0 and volume V, the resistance change can be shown to be

$$\Delta R = \rho_E \frac{L_0}{V}(2\varepsilon + \varepsilon^2) = \rho_E \frac{L_0}{V}\lambda \qquad (1)$$

where ρ_E is the resistivity, ε is the axial strain and λ is a parameter given by $\lambda = 2\varepsilon + \varepsilon^2$. So, for a constant resistivity, the resistance variation ΔR should increase linearly with λ. Figure 3b shows the ΔR - λ plot from the same test as Fig.3a. The approximately linear ΔR - λ curve shows that the resistivity remains approximately unchanged. The small but visible upward bend shows that, with increasing load, there is a slight increase in the resistivity.

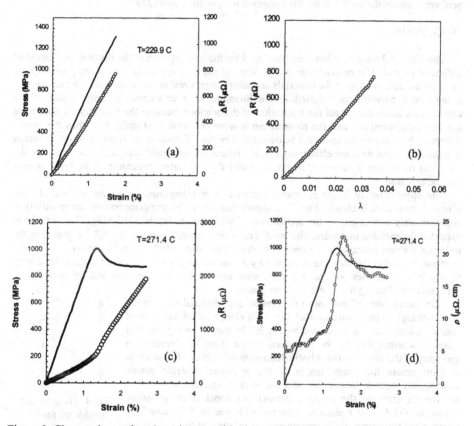

Figure 3. Changes in specimen's resistance with elongation (symboled curves) plotted together with the stress-strain curves (solid curves). (a,b) T=229.9 °C representative of the brittle regime. (c,d) T=271.4 °C representative of the ductile regime. The resistivity of the plastically deforming sections at 271.4 °C is shown in (d).

<u>The Ductile Regime</u>

Above about 242 °C, the deformation of [001] NiAl is ductile. Here a completely different resistance change is observed with deformation. Figure 3c shows the result of a test at 271.4 °C. The stress-strain curve shows a well-defined yield point, at 1007 MPa, followed by a large yield drop. During the stage of elastic deformation, the resistance increased approximately linearly with strain, similar to that observed in the brittle regime. A dominant feature here is that, at the yield point, the rate of resistance increase with strain is abruptly increased, as is reflected by the sudden change in the slope of the resistance curve in Fig.3c. A close examination of the resistance curve of Fig.3c shows that, after the abrupt jump at the yield point, the slope decreases gradually with the falling flow stress during the yield drop. In the fully plastic region, the slope of the resistance curve remains approximately constant. This feature is observed in all the tests performed above the brittle-to-ductile transition temperature, up to 319 °C.

DISCUSSION

The electrical resistivity has long been used for the investigation of the dislocation content of deformed crystals (for reviews, see [6,7]). Most of the measurements were made *post mortem*, i.e. after the deformation of the materials and after the removal of the load. With the exception of a few, most experiments reported in the literature were conducted on polycrystals. These experiments could not reveal the nucleation of dislocations, because the polycrystals used were not free of dislocations and the measurements were not performed under load thus the stress-driven nucleation process could not be detected. The main feature of the research reported here is that the specimens used are effectively free of dislocations and yield at high stresses, and that the electrical resistance is measured *in situ* thus that the dislocation processes under high stresses may be captured.

The approximately linear resistance increase with elongation during the regime of pure elastic deformation, shown by Figs.3a,c, shows that, during elastic deformation, the resistivity is approximately unchanged by the application of high stresses (up to 1320 MPa). This feature is expected for metallic materials. The small but visible upward curve in the ΔR - λ plot (Fig.3b) indicates a slight increase in the resistivity during the stage of elastic deformation. A plausible explanation is that the increase is caused by an increased vacancy concentration; under a high tensile load, vacancies will have a somewhat smaller formation energy and hence a higher concentration, leading to a slightly higher resistivity.

The main feature of interest in the present experimental result is the abrupt change in the resistance variation with elongation at the onset of plastic deformation, as is shown by Fig.3c. In the following we first examine whether this can be explained by the differing deformation geometry of the elastic versus plastic deformation: elastic deformation is uniform across the gauge length of the specimen whereas plastic deformation occurs locally by slip on crystallographic planes. To analyze the effect of slip on the resistance, we consider the geometry shown in Fig.4. The resistance increase with length L is found by viewing the resistance of the specimen as consisting of the sum of the resistances from the sections, connected in a series, shown divided by the dotted lines in Fig. 4. The resistance increase-rate with elongation can be shown to be

Fig.4. The geometry considered for the resistance change with slip.

$$\left.\frac{dR}{dL}\right|_{slip} = \frac{2\rho}{A} \cdot \frac{1}{(1 - \frac{L-L_0}{\sqrt{A}} \operatorname{ctan}\theta)}$$

where L_0 is the length of the specimen just before the onset of slip. At small plastic strains, such as near the yield point, $L-L_0$ is small and hence the resistance increase rate is approximately

$$\left.\frac{dR}{dL}\right|_{slip} \cong \frac{2\rho}{A} \tag{2}$$

The resistance increase rate associated with pure elastic deformation can be obtained by differentiating equation (1) with ε replaced by $(L-L_o)/L_0$. This gives

$$\left.\frac{dR}{dL}\right|_{elastic} \cong \frac{2\rho}{A} \tag{3}$$

The above analysis shows that the differing deformation geometry between elastic deformation and slip does not give rise to different resistance increase rate with elongation. We therefore conclude here that the abrupt jump in dR/dL at the yield point (as is shown in Fig.3c) corresponds to an abrupt increase in the resistivity. Next we examine how this resistivity increase brought about by the plastic deformation can be measured.

An important property of plastic deformation is that the deformation is not uniform across the whole length of the specimen, but is localized with slip concentrated in slip bands or Luders bands [8]. As a result, the dislocation density distribution is also non-uniform, leading to non-uniform resistivity increase. It appeared from our literature survey that this important feature was not taken into account in the past measurements of the effect of plastic deformation on the resistivity of materials. These research works all implicitly assumed the plastic deformation to be uniform and thus the resistivity increase was also uniform across the entire specimen. The resistivity increase measured with this assumption underestimated the true resistivity increase of the local plastic sections. A given measured resistance increase, after correction for the dimensional change, could be attributed either to a small increase in the resistivity across the whole specimen, or to a large increase in the resistivity over a short section of the specimen while the rest of the specimen is unchanged. Most plastic deformation process would give rise to the latter situation with plastic strain, and hence dislocations, concentrated locally in slip bands or Luders bands. Based on the uniform deformation assumption, all the past *post mortem* measurements of the effect of plastic deformation gave very small changes in the resistivity, typically by 5%, even in specimens that had been severely deformed [6,7]. The small resistivity increase was interpreted to result from the scattering of conduction electrons from the dislocation cores only [9]. This excludes the contribution to resistivity from the dislocation's elastic distortion field; this appears to be at variance with the long-range scattering of high energy electrons observed as dislocation images in the transmission electron microscopes.

The present method has the advantage that, because the resistance is measured under load and recorded with displacement, the resistivity of the plastically deforming sections can be measured from the rate of resistance increase alone, without the need to know the actual length of the plastic zone. Because the work-hardening rate is low (close to zero for [001] NiAl as is shown in Fig.3c), the overall lengthening of the specimen comes principally from the plastically deforming sections while the rest of the specimen deforms only elastically by negligible amounts because of the low work-hardening rate. In other words, both the resistance increase dR and the lengthening dL come from the plastically deforming sections. This allows the resistivity of the plastically deforming section to be determined from $\rho_p = (A/2) \cdot (dR/dL)$ (derived by differentiating $R = \rho_p L/A$ for constant ρ_p and volume) which does not involve the actual length of the plastic zone. Changes in the cross-sectional area A can be taken into account either by elasticity calculations or by assuming constant volume. In the present experiment, large changes in dR/dL occurred abruptly over a small strain interval during which the change in the cross-sectional area is expected to be very small. This allows the resistivity of the deforming sections to be measured from the initial cross-section area A and the measured dR/dL. Figure 3d shows the resistivity measured with the above method from the same test as Fig.3c. It shows that the resistivity of the deforming section is increased by approximately a factor of 4, from 6 $\mu\Omega$.cm at the start of the test, to 23 $\mu\Omega$.cm at the onset of plasticity. The resistivity settles at about 17 $\mu\Omega$.cm well into the plastic deformation regime. From such large increases in the resistivity we conclude that there is a concomitant large and abrupt increase in the dislocation density at the onset of plastic deformation. The falling of the resistivity with the falling flow stress is particularly telling in that it shows that the resistivity depends sensitively on the stress, but is approximately unchanged by the increasing plastic strain.

The results reported here have suggested the following physical picture as the deformation process in [001] NiAl. The crystal is initially 'dislocation-free' because the special tensile orientation [001] fully suppresses the operation of any <100> dislocations that may have been left after the annealing treatment. The dislocations producing the plastic deformation have either <110> or <111> as the Burgers vector [4,5]. Due to the lack of preexisting sources, the formation of these dislocations requires very high stresses. The resistivity measurement shows that these dislocations are nucleated collectively and abruptly at the onset of plastic deformation with a very high density of dislocations causing the peak in the resistivity at the yield point. Once the dislocations are nucleated, as is the case past the yield point, the crystal deforms by the 'traditional' mechanism in which dislocation multiplication takes place by the Frank-Read type process from the dislocations introduced during the collective formation. The density of the collectively nucleating dislocations depends sensitively on the stress and decreases rapidly with the falling stress as is shown by the falling resistivity with falling flow stress shown in Fig.3d. This picture is consistent with the recent theories on the collective nucleation of dislocations at critical stresses and temperatures [10,11].

ACKNOWLEDGEMENT

We are grateful to Dr. R. Darolia for providing the NiAl single crystal used in this research. Y.Q. Sun acknowledges partial support from the Wright Laboratory under contract number F33615-94-C-5804. P.M. Hazzledine acknowledges AFOSR contract number F 33615-96-C-5253 for support.

References

1. P.B. Hirsch and S.G. Roberts, Phil. Mag., **64A**, 55 (1991).
2. J.R. Rice and R. Thomson, Phil. Mag., **29**, 73 (1974).
3. W.D. Nix, Metall. Trans. A, **20**, 2217 (1989).
4. R.D. Field, D.F. Lahrman and R. Darolia, Acta metall., **39**, 2951 (1991).
5. Y.Q. Sun, G. Taylor, R. Darolia and P.M. Hazzledine, MRS Proceedings, **364**, 261 (1994).
6. F.R.N. Nabarro, *Theory of Crystal Dislocations*, Dover: New York (1967).
7. B.R. Watts, Dislocations in Solids, vol.8, ed. F.R.N. Nabarro, Elsevier Science (1989).
8. R.W.K. Honeycombe, *The Plastic Deformation of Metals*, Edward Arnold (1968).
9. Z.S. Basinski, J.S. Dugdale and A. Howie, Phil. Mag., **8**, 1989 (1963).
10. M. Khantha, D.P. Pope and V. Vitek, Phys. Rev. Lett., **73**, 684 (1994).
11. Y.Q. Sun, P.M. Hazzledine and D.M. Dimiduk, MRS Symposium on Phase Transformations and Systems Driven Far from Equilibrium, 1997 MRS Fall Meeting (to be published).

Part II

Semiconductor and Microelectronic Applications

A NOVEL APPROACH TO SEMICONDUCTOR ELECTRICAL PROPERTIES - THE ADVANCED METHOD OF TRANSIENT MICROWAVE PHOTOCONDUCTIVITY (AMTMP)

S. GRABTCHAK and M. COCIVERA
Guelph-Waterloo Centre for Graduate Work in Chemistry
University of Guelph, Guelph, Ontario, Canada N1G 2W1

ABSTRACT

The advanced method of transient microwave photoconductivity (AMTMP) represents a new method based on cavity perturbation theory, microwave photoconductivity and harmonic oscillator model analysis. AMTMP provides a direct observation of changes to the complex dielectric constant, and free and trapped electron decays can be studied separately. The results obtained for polycrystalline CdSe thin films clearly indicate that a multiple trapping model developed for amorphous materials does not provide a satisfactory description. For SI GaAs the harmonic oscillator model provides a quantitative interpretation. The limitations are discussed for the application of this method to porous Si.

INTRODUCTION

Perturbation theory, developed by Slater [1], relates changes in cavity parameters directly to ε' and ε'', the real and the imaginary parts of the dielectric constant, and it was used for steady-state measurements [2]. Transient microwave photoconductivity (TMP), on the other hand, assumes the signal is proportional to the light induced changes in the conductivity [3, 4]. The interpretation of the kinetics in this case was based on a single measured decay that was assumed equal to $\delta\varepsilon''$ [5], which may not be valid.

To avoid this assumption, we have developed the advanced method of transient microwave photoconductivity (AMTMP) [6-9], which includes three major components (cavity perturbation theory, microwave photoconductivity and harmonic oscillator model analysis). In this method two separate kinetic decays are registered: the bandwidth and frequency shift of the resonance signal. These parameters can be related to $\delta\varepsilon'$ and $\delta\varepsilon''$ by treating the light induced changes as a second perturbation. As was shown earlier [9], the harmonic oscillator model can be used to relate $\delta\varepsilon'$ and $\delta\varepsilon''$ to photogenerated free and trapped electrons quantitatively. In the present paper we outline briefly the major theoretical aspects of AMTMP and present experimental results obtained for various samples (two types of polycrystalline CdSe thin films (CdSe(I) and CdSe(II)), SI (semi-insulating) GaAs, porous Si). The transient behavior is compared with the carrier transport model used for amorphous semiconductors.

THEORY

Second Cavity Perturbation Theory

Insertion of a semiconductor or dielectric material into a microwave cavity (resonator) causes a perturbation, i.e. a change in the resonance frequency, f_0, and the cavity quality factor, Q_L. Cavity perturbation theory, which relates these changes to real and imaginary parts of the complex dielectric constant, $\varepsilon^* = \varepsilon' - j\varepsilon''$, of the material, was developed by Slater [1]. A second perturbation of cavity parameters occurs when a sample already in the cavity is subjected to photons, electric current, temperature, X-rays etc. to produce excess electrons and cause a change in the complex dielectric constant, $\delta\varepsilon^* = \delta\varepsilon' - j\delta\varepsilon''$. This second perturbation is applied to AMTMP.

Although some aspects were considered in photodielectric studies of some semi-conductors in superconducting resonators [10, 11], a detailed treatment has not been developed.

A rigorous treatment of the second perturbation can be found elsewhere [9]. In general, the changes in the complex dielectric constant ε^* can be related to the shift of the resonance frequency, δf_0, and a change in the cavity quality factor, $\delta(1/2Q_L)$, related to the bandwidth:

$$\frac{\delta f_0}{f_0} + j\delta\left(\frac{1}{2Q_L}\right) = -(\varepsilon_2^* - \varepsilon_1^*)G \frac{1}{1 + L_2(\varepsilon_2^* - 1)} \frac{1}{1 + L_1(\varepsilon_1^* - 1)}, \qquad (1)$$

where the subscripts 1 and 2 refer to values for the sample before and after illumination, respectively, G is geometric factor depending on volumes of the cavity and the sample, and L is the depolarization factor determining the electric field inside the sample. Separating the real and the imaginary parts in the right side of Eqn.(1) relates them to δf_0 and $\delta(1/2Q_L)$. The first perturbation can be treated as a special case by using $\varepsilon_1^* = 1$ and $\varepsilon_2^* = \varepsilon' - j\varepsilon''$ to give expressions identical to those derived earlier [12] except that δf_0 was defined to be positive.

Harmonic Oscillator Model

After the expressions for the changes in the real and the imaginary parts of dielectric constant are obtained from the second perturbation theory the subsequent analysis can be done within a framework of the harmonic oscillator model [13, 14]. Benedict and Shockley [15] used this model to determine the effective mass of free electrons in germanium. Recent work [16] showed that this model adequately reproduces the dielectric constant of a semiconductor for excitation levels up to 10^{20} cm^{-3}. Our recent application of the model to experimental data on SI GaAs [9] showed that the harmonic oscillator model can provide both quantitative and qualitative information.

The implications of the generalized model in the context of AMTMP [9] are qualitatively summarized below. The main types of excitations which can be observed in semiconductors including collective effects are free (conduction band) electrons, electrons trapped at subband gap states, plasmons, excitons, and electron-hole droplets. All these excitations sometimes have very distinct contributions to the complex dielectric constant and can be classified as bound and non-bound states on their influence on the real part of the dielectric constant. The condition for the bound electron requires the existence of a restoring force proportional to the electron displacement. The depolarization effect due to the finite dimensions of a confined electron plasma was shown to provide such a restoring force [17]. The exciton has also a characteristic binding energy. Therefore, the non-bound state would include free electrons only and the bound state would incorporate trapped electrons, plasmons, excitons, and electron-hole droplets. The free electrons are known to produce "metal-like" effects in semiconductors **decreasing** the real part of the dielectric constant. The addition of bound states is equivalent to introducing highly polarizable states to the semiconductor, which would **increase** the real part of the dielectric constant. In a former case a **positive** shift of the resonance frequency is observed, and in the later case, a **negative** sign results. The contributions to the imaginary part normally include the ones from free and bound electrons. The later has a clear resonance character and can be omitted at close considerations [9]. The joint analysis of temporal responses of the real and the imaginary parts of dielectric constant sheds a light on thermallization processes. For example, when a thermodynamic equilibrium is established between free and trapped electrons, both kinetics will have the same time dependence. It is understood that because of a relation between a release time and a trap depth, not all trapped electrons will be in equilibrium with the conduction band within a time scale of the experiment. When in traps, electrons will cause changes in the real part of the dielectric constant, and when in the conduction band, the conductivity effects will manifest themselves in the imaginary part. A contribution to the real part from free electrons is normally masked by trapped electrons, if the concentration of trapped electrons is high and the mobility of free electrons is relatively low. Another case when both decays will have the same time dependence would include a scenario when the **same** free electrons produce not only conductivity changes, but a decrease in the real part of the dielectric constant simultaneously. This is peculiar to materials with high mobilities and/or low concentration of shallow traps.

Therefore, operating with two decays would normally provide a clearer picture of possible processes in semiconductors compared to a traditional method of photoconductivity, when the only decay due to conductivity changes is measured. A traditional analysis of kinetics in terms of rate equations can be done on top of the harmonic oscillator analysis.

EXPERIMENT

The experimental setup for AMTMP measurements was described in detail earlier [8]. The sample is located in the microwave TE_{101} cavity in the maximum electric field. The illumination passes into the cavity through a hole covered with a quartz window. The cavity and a piece of waveguide are evacuated for low temperature measurements. The temperature of the cavity was maintained by a water jacket. Power reflected from the cavity is detected by a diode connected directly to an oscilloscope (Tektronix TDS 320) for registration of kinetics, or to a digital voltmeter to measure the resonance curve in the dark. The experiment is computer controlled. Basically, the same microwave elements are used in the traditional microwave conductivity measurement. The main difference is in the microwave source, a synthesized sweep generator (Wiltron 68137 B), and the method of collection. The photoresponse is measured at a number of frequencies within the dark resonance curve (typically 20-30), and the light induced changes in the resonance curve are extracted from the resulting difference curve as discussed earlier [7, 8] to give the time dependence of the cavity quality factor changes and the shift of the resonance.

RESULTS AND DISCUSSION

CdSe (I)

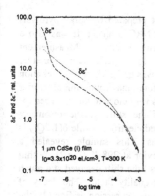

Figure 1. Kinetics of changes in the real and imaginary parts of dielectric constant for CdSe (I).

The results for this sample have been presented elsewhere [6,8], but a short discussion of the main conclusions will provide the necessary background and facilitate discussion of CdSe (II). CdSe (I) was prepared in our laboratory [18] by spray pyrolysis to produce a thin (1 μm) polycrystalline film on a quartz substrate. Measurements were made at 300 K using 10 ns laser pulse at 337 nm. For times longer than 200 ns, Fig.1 illustrates that the $\delta\varepsilon'_{tr}(t)$ and $\delta\varepsilon''_{free}(t)$ transients have nearly identical behavior, consistent with a fast thermal equilibrium between conduction band and shallowly trapped electrons. Deviations from identical time dependence for these two parameters is expected when thermal equilibrium is slow in comparison with the decay processes (i.e. at lower temperature) or when the traps are saturated. The second case may include a direct band-to-band recombination or Auger recombination at saturated traps. This condition was realized in Fig. 1 below 200 ns when the traps were saturated, and the total trapped electron concentration was be determined to be $\sim 10^{18}$ cm^3. In this case, the fast component in $\delta\varepsilon''(t)$, which was absent in $\delta\varepsilon'(t)$, was attributed to the direct band-to-band recombination.

The form of $\delta\varepsilon''(t)$ kinetics was found to be consistent with a dispersive electron transport resulting from the energy distribution of the trap levels within the bandgap [19]. The pertinent equations to describe the time dependence of both parameters over the whole time domain are :

$$\delta\left(\frac{1}{2Q_L}\right) \propto \delta\varepsilon'' \propto \Delta n_{free} \propto \begin{cases} \exp(-t/\tau_{rec}) + t^{\alpha-1}, \ t \leq \tau_{rec} \\ t^{-1}\ln(t/\tau_{rec}), \ t \geq \tau_{rec} \end{cases} \quad (2)$$

$$\delta f_0 \propto \delta\varepsilon' \propto \Delta n_{trap} \propto \begin{cases} t^{\alpha-1}, \ t \leq \tau_{rec} \\ t^{-1}\ln(t/\tau_{rec}), \ t \geq \tau_{rec} \end{cases} \quad (3)$$

in which τ_{rec} is the bimolecular recombination time involving mobile electrons and trapped holes, and the exponential term was not derived from the model and $\alpha=0.69$. The initial power law decay ($t^{\alpha-1}$) in the multiple trapping model is attributed to a redistribution of carriers between different traps, and $t\ln(t/\tau_{rec})$ characterizes the onset of a bimolecular recombination. While the Eqns.(2,3) fit experimental data well, our data show a couple of inconsistencies with the multiple trapping (MT) model [19]: 1) the constant α in our experiments which was found to be nearly temperature independent, but the model predicts $\alpha = T/T_0$ and should exhibit a temperature dependence, 2) our results show that the electrons in the shallow traps exist during 4 orders of magnitude in time, but MT predicts the shallow traps are depleted so fast that there is an accumulation of electrons in deep traps and free electrons can not be in equilibrium with trapped electrons. The expression for Δn_{trap} suggested by the multiple trapping theory and indeed observed in some amorphous semiconductors should be as following [19, 20]:

$$\Delta n_{trap} \propto \frac{1}{1+(t/\tau_{rec})^{\alpha}} \quad (4)$$

Because α in MT is always less than 1 the trapped electron concentration would decay more slowly than the free electron concentration and would approach $t^{-\alpha}$ only at t > τ_{rec}

CdSe(II)

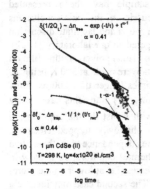

Figure 2. Kinetics of the quality factor changes and the shift of the resonance frequency for CdSe (II).

This sample, which was not prepared by spray pyrolysis was supplied by Dolf Landheer at NRC, Canada. It was also a polycrystalline film with a 1 μm thickness. Measurement conditions were very similar to those for CdSe(I) except a broader temperature range (123 K - 358 K) was used. The laser excitation wavelength was 355 nm with a total maximum initial carrier density of 4×10^{20} el./cm^3, which was very close to the one used for CdSe(I) (see Fig.1). The sign of the shift of the resonance frequency can be attributed to trapped electrons, while $\delta(1/2Q_L)$ is due to conduction band electrons. This sample revealed a completely different transient behavior at 298 K (Fig.2). Within the experimental time scale (5 orders) the shift decayed at a much slower rate than that of the cavity quality factor. The decay of $\delta(1/2Q_L)$ can be described by the same expression as Eqn.(2) (the first part), i.e. the sum of the fast initial exponential decay and a subsequent power law decay ($t^{\alpha-1}$) with a different α (~0.4). We did not observe any changes in a slope at room temperature. The kinetics of the shift of the resonance frequency exhibited a distinct change in slope at ~ 2×10^{-3} s, decaying after that time close to $t^{-1-\alpha}$ with α between 0.4 and 0.7. The uncertainty in this parameter is caused by an increased level of noise in this region. As one can see this behavior is very close to that observed for CdSe(I) after the break point. Fitting a region before the break point to Eqn.(4) gave satisfactory results with $\alpha \sim 0.4$. Therefore, before the break point at room temperature the forms of **both** transients formally coincided with those from MT. In a high temperature range (298 - 358 K) both transients decayed in the same way as discussed above, but α was found to be temperature independent in contradiction to MT. In the low temperature range (298-123 K), $\delta(1/2Q_L)$ decayed also according to Eqn. (2) (the first part) with $\alpha = 0.7$ at 123 K. A sharp change of the slope for $\delta(1/2Q_L)$ was detected at $\sim2\mathrm{x}10^{-3}$ s. While the shift of the resonance frequency exhibited the same change in slope at the break point, the part before that time did not fit Eqn.(4) convincingly. Furthermore in contrast to MT theory, α increased with decreasing temperature.

In summary, a deviation from MT was observed also for this CdSe sample. MT model was developed for **amorphous** semiconductors and provided quantitative results only for an **exponential** distribution of localized states. Although it was believed to apply equally well to time-of-flight (TOF) measurements, other researchers found a **temperature-independent** α for the same samples (a -As$_2$Se$_3$) when measured by TOF [21] and photoabsorption (PA) [20]. From a theoretical point of view it was shown [22] that the change in the slope of the power law decay does not constitute evidence for the presence of a particular exponential energy distribution of states, and analogous behavior can be obtained from the other distributions, i.e. rectangular, linear-tail and Gaussian-type peaked at some energy in the band gap. This raises some concerns about limits of applicability of MT model, even for amorphous semiconductors. For polycrystalline materials the same concern applies. TOF experiments on CdTe polycrystalline films [23] showed that α increased with decreasing temperature. Photocurrent studies of GaN thin films [24] also exhibited the same temperature behavior for α. Therefore, our results for polycrystalline CdSe show the same behavior as these other polycrystalline films, and these results can not be explained satisfactory by the MT model.

SI GaAs

Only a brief discussion of this material is presented here to illustrate the range of effects on AMTMP. Detailed discussion is given elsewhere [9]. This material produced a positive δf_o and with both $\delta(1/2Q_L)$ and δf_o having the same time dependence. According to the harmonic oscillator model, a positive shift indicates that only free electrons contribute to the changes in the real and imaginary parts of the dielectric constant. This response is exhibited by materials having high mobility and low concentration of shallow traps, which would normally dominate in $\delta\varepsilon'$ at higher density. This model indicates the concentration of trapped electrons would not exceed $4\mathrm{x}10^{13}$ cm^{-3} for trap levels ranging between 0.03 and 0.3 eV below the conduction band [9], indicating the harmonic oscillator model was consistent with the data for SI GaAs.

Porous Si

Porous Si samples were kindly supplied by Dr. P. Fauchet, University of Rochester [25]. These films were known for extremely long luminescence decays and we looked for a correlation between photoconductivity and luminescence measurements. Unfortunately, poor response was obtained for all samples except the one having the lowest porosity, which was the least efficient in producing the luminescence. For this sample, both $\delta(1/2Q_L)$ and δf_o exhibited very short transients, comparable to the instrumental time constant. The sign of δf_o was negative indicating trapped electron effects. Therefore, one possible explanation is that the free electrons quickly fill shallow traps, then electrons thermalize very fast down to very deep trap levels which can not be detected by our method. The very long luminescence observed in these samples could support the speculation that it originates from such deep traps.

CONCLUSIONS

We showed that AMTMP can provide a new scope when applied to various semiconductors. The decays proportional to free and trapped electrons can be recorded in a single experiment. This gives an opportunity to check multiple trapping model predictions for polycrystalline semiconductors. It was found that MT could not be used to describe transport of excessive carriers in polycrystalline CdSe which appeals to new theoretical developments in this field. Free electron effects were found to dominant completely in SI GaAs demonstrating the important effect of the mobility on the AMTMP response. The harmonic oscillator model gave an adequate description of the SI GaAs behavior.

ACKNOWLEDGMENTS

The authors are grateful to Dr. D. Landheer (NRC) for CdSe films and to Dr. J. von Behren (University of Rochester) for samples of porous Si. This work was supported in part by a grant (to M. C.) from the Natural Sciences and Engineering Research Council of Canada.

REFERENCES

1. J.C. Slater, Rev. Mod. Phys.,18, p. 441 (1946)
2. R. Hutcheon, M. de Jong, and F.Adams, J. Microw. Pow. Electromagn. En., 27, p. 87 (1992)
3. M. C. Chen, J.Appl. Phys. 64, p. 945 (1988)
4. S. Damaskinos, A. E. Dixon, G. D. Roberts, and I. R. Dagg, J.Appl. Phys. 60, p. 1681 (1986)
5. M. S. Wang and J. M. Borrego, J. Electrochem. Soc., 137, p. 3648 (1990)
6. S.Yu. Grabtchak and M.Cocivera, Phys. Rev. B, 50, p. 18219 (1994)
7. S. Grabtchak and M.Cocivera, Progr. Surf. Sci., 50 (1-4), p. 305 (1995)
8. S.Yu. Grabtchak and M.Cocivera, J. Appl. Phys., 79, p. 786 (1996)
9. S. Grabtchak and M. Cocivera, Phys. Rev. B, 58, iss. 7 (1998)
10. W. H. Hartwig and J. J. Hinds, J. Appl. Phys., 40, p. 2020 (1969)
11. J. J. Hinds and W. H. Hartwig, J. Appl. Phys., 42, p. 170 (1971)
12. L. J. Buravov and I. F. Shchegolev, Prib. Tek. Eksp., 2, p. 171 (1971)
13. H. Ibach and H.Luth, Solid State Physics, 2nd edition, Springer-Verlag, Berlin 1995, p. 292
14. A. K. Jonscher, Dielectric Relaxation in Solids, Chelsea Dielectrics Press, London, 1983, Chapter 4.
15. T.S. Benedict and W.Shockley, Phys. Rev., 89, p.1152 (1953)
16. E. N. Glezer, Y. Siegel, L. Huang, and E. Mazur, Phys. Rev.B, 51, p. 6959 (1995)
17. G. Dresselhaus, A. F. Kip, and C. Kittel, Phys. Rev. 100, p. 618 (1955)
18. S. Weng and M. Cocivera, J. Electrochem. Soc., 139, p.3220 (1992)
19. J. Orenstein, M. A. Kastner, and V.Vaninov, Philos. Mag. B, 46, p.23 (1982)
20. J. Tauc in Semiconductors and Semimetals, ed. by J. I. Pankove, Academic Press, Orlando, 1984, 21 B, Chapter 9.
21. G.Pfister and H. Sher, Phys. Rev. B, 15, p. 2062 (1977)
22. J. M. Marshall, H. Michiel, and G. J. Adriaenssens, Philos. Mag. B, 47, p. 211 (1983)
23. R. Ramirez-Bon, F. Sanchez-Sinencio, G. Gonzales de la Cruz, and O. Zelaya, Phys. Rev. B, 48, p. 2200 (1993)
24. C. H. Qiu, W. Melton, M. W. Leksono, J. I. Pankove, B. P. Keller, and S. P. DenBaars, Appl. Phys. Lett., 69, p. 1282 (1996)
25. J. von Behren, P. M. Fauchet, E. M. Chimowitz and C. T. Lira, Mat. Res. Soc. Proc. 452, p. 565 (1997)

MICROSTRUCTURAL AND ELECTRICAL CHARACTERIZATION OF MISFIT DISLOCATIONS AT THE InAs/GaP HETEROINTERFACE

V.Gopal[1], T.P.Chin[2], A.L.Vasiliev[1], J.M.Woodall[2] and E.P.Kvam[1]

[1] School of Materials Engineering, Purdue University, W.Lafayette, IN.
[2] School of Electrical and Computer Engineering, Purdue University, W.Lafayette, IN.

ABSTRACT

InAs is a narrow band gap semiconductor with potential for such applications as IR detectors, low temperature transistors, etc.. However, the lack of suitable substrates has hampered progress in the development of InAs based devices. In the present study, InAs was grown by Molecular Beam Epitaxy on (001) GaP substrates. Though this system has a high lattice mismatch, (~11%), certain MBE growth conditions result in 80% relaxed InAs layers on GaP with the mismatch accommodated predominantly by 90° pure edge dislocations. Misfit dislocation microstructures were studied using Transmission Electron Microscopy. Electrical characterization using lateral conductance and Hall effect measurements were also performed. Preliminary results indicate the possibility of misfit dislocation related conductivity. The possible correlation between interface structure and electrical properties is discussed.

INTRODUCTION

Epitaxial growth of a lattice mismatched layer results in the formation of strain relieving misfit dislocations [1]. When the mismatch is low (<1%), strained layers of significant thickness can be grown without the formation of misfit dislocations. However, the critical thickness is extremely small at large lattice mismatch (>4%) and pseudomorphic growth is often not possible. In high mismatch systems, it has been reported that the interfacial structure consists chiefly of orthogonal arrays of 90° edge dislocations [2]. 90° dislocations are known to relieve strain more efficiently than 60° dislocations and hence their presence is favored at a high lattice mismatch.

InAs is a narrow gap semiconductor with very high electron mobility. It shows potential for devices such as Infra-Red emitters and detectors, low temperature transistors, thermophotovoltaics, etc.. We have investigated the direct growth of InAs on (001) GaP. This system has the largest lattice mismatch among the III-arsenides and the III-phosphides: ~ 11%. In addition, the large band offsets between InAs (band gap, E_g=0.36 eV) and GaP (E_g=2.26 eV) result in interesting heterojunction properties. Heterojunction InAs/GaP diodes have shown near ideal forward and reverse bias characteristics and reasonable breakdown voltages [3]. The InAs epilayer was found to be 80% relaxed and lattice matched to $In_{0.8}Al_{0.2}As$ and $In_{0.8}Ga_{0.2}As$ [4].

EXPERIMENT

InAs was grown on (001) GaP by Solid Source Molecular Beam Epitaxy (SSMBE) using a Varian GEN-II MBE system. A 1) nm GaP buffer layer was grown followed by the growth of a twenty period superlattice consisting of 5 nm alternating layers of GaP and AlP in order to prevent the out diffusion of sulfur from the substrate. Finally a 200 nm buffer layer of p-GaP (~10^{16} cm^{-3}) was grown at 600°C. An undoped InAs epilayer was grown at 350°C at a growth rate of 0.7 monolayers per second. The first few monolayers were grown under a low V/III beam flux ratio to promote a smoother interface. Finally, a cap layer of undoped $In_{0.8}Al_{0.2}As$ was grown above the InAs layer. The cap layer is lattice matched to the 80% relaxed InAs, and has a wider band gap. Samples with varying thickness of InAs in the range of 5 nm to 30 nm were grown in this manner. A schematic sketch of the multilayer structure grown is shown in Figure 1.

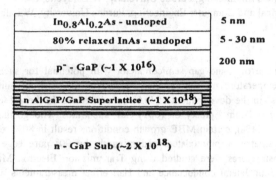

$In_{0.8}Al_{0.2}As$ - undoped	5 nm
80% relaxed InAs - undoped	5 - 30 nm
p^- - GaP ($\sim 1 \times 10^{16}$)	200 nm
n AlGaP/GaP Superlattice ($\sim 1 \times 10^{18}$)	
n - GaP Sub ($\sim 2 \times 10^{18}$)	

Figure 1 : Schematic illustration of the multilayer structure grown by MBE

Cross-sectional and plan view TEM studies were performed using a JEOL 2000 FX microscope with a beam energy of 200 keV. Despite the high lattice mismatch, a perfect epitaxial relationship between the InAs and GaP was observed in all cases, as revealed by electron diffraction patterns. Figure 2 shows a HRTEM micrograph of the InAs/GaP heterointerface. A regularly spaced array (spacing ~ 4 nm) of predominantly 90° misfit dislocations oriented along <110> is clearly visible at the interface. The regions of the interface between the dislocations appears to be atomically smooth and free of distortions. Thus, the interface structure consists of two orthogonal arrays of sessile 90° edge dislocations lying on the (001) plane with mutually perpendicular 1/2<110> Burgers vectors.

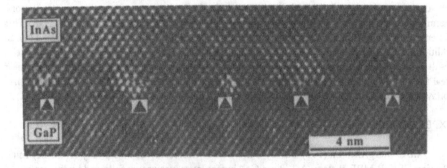

Figure 2 : HRTEM micrograph of InAs/GaP heterostructure displaying regularly spaced 90° misfit dislocations at the interface (arrowed).

The Transfer Length Method (TLM) was used to perform lateral conductivity measurements. TLM patterns were transferred by photolithography onto the samples, followed by Ti/Au metallization and mesa etching to isolate the test structures. TLM patterns were defined parallel to two orthogonal <110> directions - i.e., the two orthogonal directions that the misfit dislocations lie along. These are labeled $<110>_x$ and $<110>_y$ here. The sheet resistance obtained by the TLM technique, R_s, is presented in Table I. A pronounced difference in the values of R_s for the two <110> directions was observed.

Table I : Sheet resistivity data obtained along orthogonal <110> directions

InAs Thickness (nm)	Sheet Resistance (kΩ/$\square$)	
	$<110>_x$	$<110>_y$
5	13	9
20	1.5	1.25

Hall effect measurements were performed at room temperature and at 77K using a van der Pauw geometry. Approximately square specimens were cleaved and In dots were alloyed at the four corners at 300°C in an inert gas atmosphere. The sign of the Hall co-efficient for each of the samples tested indicated that the majority carriers were electrons. Table II lists the Hall mobility and the sheet carrier concentration for samples of differing InAs thickness. No "freeze out" of carriers at liquid N_2 temperature was observed. However, a slight decrease in Hall mobility was observed at 77 K.

Table II : Sheet carrier concentration and Hall mobility data for InAs epilayers of various thickness grown on GaP

Sample #	InAs Thickness (nm)	Room Temperature		77 K	
		Sheet density (cm^{-2})	Hall mobility (cm^2 / V-sec)	Sheet density (cm^{-2})	Hall mobility (cm^2 / V-sec)
1	5	1.5E13	40	1.5E13	25
2	10	1.2E13	500	1.0E13	360
3	15	0.75E13	335	0.75E13	310
4	20	1.0E13	460	0.9E13	430
5	30	0.7E13	795	0.6E13	730

Carrier concentration depth profiling was carried out on a sample with 12.5 nm of InAs and a 0.4 mm capping layer of $In_{0.8}Al_{0.2}As$ using an electrochemical capacitance voltage (ECV) profiler manufactured by Bio-Rad Micromeasurements Inc. A depth resolution of 1 nm can be achieved with this profiler. The resultant semilog plot of carrier concentration vs. depth is shown in Figure 3. A spike in the carrier concentration at the InAs/GaP heterointerface is clearly visible.

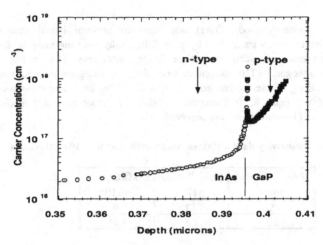

Figure 3 : Electrochemical CV profile showing carrier accumulation at the heterointerface

RESULTS AND DISCUSSION

The results obtained from all the experiments performed for this study indicate that the electrical properties of the InAs/GaP system are dominated by interfacial dislocations. The Hall effect data of Table II has three important features -

i) Given that the InAs layer is undoped, the sheet carrier concentrations are very high,~ 10^{13} cm^{-2}.

ii) The sheet carrier concentration does not vary significantly with InAs layer thickness.

iii) The electron mobilities are very low, nearly three orders of magnitude below the bulk mobility of InAs (33,000 cm^2/V-sec) [5].

All of this indicates that a localized high density sheet of charge exists, and that the carriers are very heavily scattered. Comparing the Hall data to the ECV carrier concentration depth profile shows that this high density of carriers is localized at the InAs/GaP heterointerface. HRTEM images show the presence of an array of misfit dislocations at this interface. Clearly, a linkage can be drawn between the misfit dislocation microstructure and the accumulation of carriers. The dislocations act as scattering centers, hence the low values of Hall mobility.

Further proof that the properties of this system are interface defect dominated comes from the TLM data of Table I. The sheet conductivity, σ_s, is given by :

$$\sigma_s = q.N_s.\mu_e \qquad (1)$$

where N_s is the sheet carrier concentration, μ_e is the electron mobility and q is the electronic charge. Differing values of σ_s in two orthogonal <110> directions indicates that either the mobility or the sheet carrier density varies along different directions. If $q.N_s$ is invariant with direction as expected, then the mobility must vary along the two <110> directions. Since the misfit dislocation microstructure consists of orthogonally arranged arrays, the results of Table I indicate that the scattering potentials of the dislocations in the <110>$_x$ direction differ from those in the <110>$_y$ direction. This anisotropy in scattering potentials probably arises due to differing core structures of the two orthogonal sets of 90° misfit dislocations. Several workers have studied the core structures of a/2<110> type edge dislocations in diamond cubic and zinc blende systems [6,7], but to the best of our knowledge, this is the first indication of different scattering potentials due to these defects.

Free surfaces and defects are known to pin the Fermi level in the conduction band in InAs [8], making the semiconductor degenerate. We propose that misfit dislocations pin the Fermi level at the interface in the conduction band. To test this hypothesis in terms of an energy band model, simulations of InAs/GaP were carried out using the simulator ADEPT. Values for the conduction band offset, ΔE_c, and valence band offset, ΔE_v, were obtained from the work of Chen et al. [3] to be $2/3\Delta E_g$ and $1/3\Delta E_g$ respectively. Figure 4(a) shows the simulated band diagram for this "ideal" n⁻-InAs / p⁻-GaP heterostructure. Undoped InAs, being a narrow gap semiconductor is expected to be n⁻ due to thermal generation. The Fermi level of the InAs is seen to be at midgap. The band bending displayed here indicates that the interface must be depleted of charge and rectifying in nature. Figure 4(b) shows the simulated band diagram with a high density (10^{13} cm⁻²) sheet of electrons inserted at the interface. The Fermi level at the interface is now pinned 0.2 eV above the conduction band edge and the band bending forms a narrow potential well in the InAs, which can confine carriers near the interface.

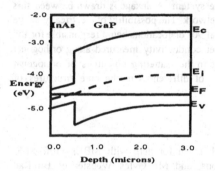

Figure 4(a) : ADEPT simulation of "ideal"
n⁻InAs/p⁻GaP heterostructure.

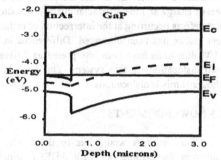

Figure 4(b): ADEPT simulation of InAs/GaP
with electron sheet charge at interface

It appears that the effect of the array of misfit dislocations is to generate electrons, which are confined in a 2 dimensional electron gas (2DEG) near the interface. The specifics of the generation of carriers at the interface due to the dislocations microstructure have yet to be worked out. Several possibilities exist. For instance, Mostoller et al. [9] have proposed that an extended point defect known as the dreidl defect may occur at the intersection of 1/2<110> edge dislocations at the lattice mismatched Ge/Si (001) interface. Similar point defect structures can be envisaged at the InAs/GaP (001) interface. These point defects would form a square lattice with a periodicity of ~ 4 nm. This corresponds to a density of ~ 10^{13} cm⁻²; an almost 1:1 correspondence can be drawn with the sheet carrier concentrations of Table II (~10^{13} cm⁻²). It is possible that the carriers localized at the interface are generated by states introduced by these dreidl-like defects. The dreidl defects themselves could be positively charged centers, accounting for the positive charge seen at the interface in Figure 3.

Residual 60° misfit dislocations could be another possible source for interfacial carrier accumulation. These defects are less likely to occur than 90° misfit dislocations, but a small concentration of 60° dislocations may persist. However, we have not been able to quantify their concentration, and it is also unlikely that this concentration is invariant from one growth run to another. A constant concentration of 60° dislocations is required to account for the invariant sheet carrier density in our samples. We believe sites along 90° misfit dislocations to be an unlikely source for the carriers as well. There are $\sim 10^{14}$ sites along 90° dislocations per cm^2, but the carrier concentration is an order of magnitude lower than this. Fractional ionization of these sites could account for the carriers. However, if that were the case the fraction of ionized sites should drop at lower temperatures. No "freeze out" is observed in this system at 77 K, leading us to believe that 90° misfit dislocations are not responsible for the interfacial carriers.

The carriers would be strongly scattered by the dislocation network, which accounts for the low mobility values. As the temperature is decreased, the confinement would be even stronger, i.e., the mobility would decrease with decreasing temperature, the opposite of the normally expected μ-T relationship. This is consistent with our observations.

CONCLUSIONS

We have demonstrated the existence of a high density sheet of charge at the InAs/GaP heterointerface which dominates the properties of the system. A linkage is drawn between this sheet of charge and the interfacial misfit dislocation network. The possibility of electrically active point defects occurring at the intersection of orthogonal 90° dislocations being responsible for the sheet charge has been discussed. Differences in sheet conductivity measured along orthogonal <110> directions have been interpreted as differences in the scattering potentials of orthogonal misfit dislocations in this system. This is indicative of differences in the core structures of orthogonal misfit dislocations.

ACKNOWLEDGMENTS

The authors would like to thank Mr. E.H. Chen for help with MBE, Prof. M.S. Lundstrom for helping out with ADEPT simulations, and Mr Gyles Webster of Bio-Rad Semiconductor Systems for help in C-V profiling. This work was supported by the National Science Foundation - Materials Research Science and Engineering Center for Technology Enabling Heterostructure Materials through NSF grant 9400415 - DMR.

REFERENCES

1. A. Ourmazd, R. Hull and R.T. Tung in Materials Science and Technology - A Comprehensive Treatment Vol. 4, edited by W. Schroter (VCH Publishers, New York, 1991), pp 402-422.
2. E.P. Kvam, D.M. Maher and C.J.Humphreys, J. Mater. Res. **5**, 1900 (1990).
3. E.H. Chen, T.P. Chin, J.M. Woodall and M.S. Lundstrom, Appl. Phys. Lett. **70**, (1997).
4. J.C.P. Chang, T.P. Chin and J.M. Woodall, Appl. Phys. Lett. **69**, 981 (1996).
5. D.K. Schroder, Semiconductor Device and Material Characterization, (Wiley, New York,1990)
6. J. Hornstra, J.Phys. Chem. Solids **5**, 129 (1958).
7. A.S. Nandedkar and J. Narayan, Phil. Mag. A, **56**, 625 (1987).
8. M. Yazawa, M. Koguchi, A. Muto and K. Hiruma, Adv Mater. **5**, 577 (1993).
9. M. Mostoller, M.F. Chisholm and T. Kaplan, Phys. Rev. Lett **72**, 1494 (1994).

Electrical Measurement of the Bandgap of N^+ and P^+ SiGe Formed by Ge Ion Implantation

Akira Nishiyama, *Osamu Arisumi and Makoto Yoshimi,

Advanced Semiconductor Devices Research Laboratories,

R & D Center, Toshiba Corporation,

*Microelectronics Engineering Laboratories, Toshiba Corporation,

8, Shinsugita-cho, Isogo-ku, Yohohama 235, Japan

Abstract

N^+ and p^+ SiGe layers were formed in the source regions of SOI MOSFETs in order to suppress the floating-body effects by means of high-dose Ge implantation. The bandgaps of the layers were evaluated by measuring the temperature dependence of the base current of the source/channel/drain lateral bipolar transistors. It has been found that the reductions of the bandgaps due to the SiGe formation by the Ge implantation were relatively small, compared to those obtained by the theoretical calculation for heavily doped SiGe. It was also found that the bandgap reduction was larger for n^+ layers than that for p^+ layers.

Introduction

SOI MOSFETs are attractive candidates for future LSI devices in view of their low power consumption and high-speed performance. However, the floating-body effect is a major obstacle which must be overcome before they can be used in practical applications. This effect causes the lowering of the drain breakdown voltage, anomalous decreases in subthreshold swing and kinks in Id-Vd characteristics [1, 2]. These phenomena are caused by the accumulation of excess carriers generated by the impact ionization near the drain region, in the channel region. In order to suppress these phenomena, the authors have proposed the formation of a narrow bandgap material such as SiGe in the source region and confirmed its effectiveness [3]. We used Ge implantation technique in order to form the SiGe layers[4], considering that in the case of the implantation there is no problem about particles which may be formed during SiGe selective CVD deposition[5]. Applicability of the high current ion implanter for the dopant implantation to this process is another advantage of this technique. Therefore, the bandgap of the n^+ and p^+ SiGe layers formed by the Ge implantation is an important factor in this technique.

In this paper, the bandgaps of the SiGe layers in the source regions were evaluated by measuring the temperature dependences of the base current of the source/channel/drain lateral bipolar transistors. The measurement has revealed that the reductions of the bandgaps due to the SiGe formation by the Ge implantation were relatively small, compared to those obtained by the theoretical calculation for heavily doped SiGe[6]. It has also revealed that the bandgap reduction was larger for n^+ layers than that for p^+ layers.

Device Fabrication

LOCOS isolations and gate oxide layers with a thickness of 6nm were formed on SOI wafers with a thickness of 90nm. Phosphorous-doped poly-Si gate layers with a thickness of 300nm were then deposited. After the formation of the poly-Si gate electrodes and subsequent oxidation with a thickness of 5nm, Ge implantation was performed. Ge atoms were implanted perpendicular to the surface at 25keV with dosages of 0.5, 1.0 and 3.0×10^{16} cm^{-2} into source/drain regions. As-implanted Ge profile is shown in ref.[4] for a Ge dosage of 3×10^{16} cm^{-2}. The projected range was about 20nm and the peak Ge concentration was about 4.5, 9, 24 at. % for dosages of 0.5, 1.0 and 3.0×10^{16} cm^{-2}. After the formation of a 20nm thick SiN sidewall, n$^+$ and p$^+$ doping was performed by the implantation of As (30keV, 3×10^{15} cm^{-2}) and BF$_2$(20keV, 3×10^{15} cm^{-2}). Subsequent annealing was performed in a N$_2$ atmosphere at 850°C for 30 minutes in order to recrystallize the SOI layers which suffered implantation damage and activate implanted As and B at the same time. The SiGe crystalline network formation with this annealing was confirmed by Rutherford backscattering spectrometry (RBS) in the previous study [4]. The Ge profiles were hardly changed at this temperature due to a low diffusivity of Ge in Si [7, 8].

Results and Discussion

Figure 1 shows the Gummel plots for npn lateral bipolar transistors in the SOI layers. The measurement setup is depicted as the inset in Fig.1. The SOI MOSFETs with contacts to the channel region were measured as lateral bipolar transistors with source/channel/drain electrodes used as emitter/base/collector electrodes. We applied the gate and substrate voltages of -1V and -3V in order to put the two SiO$_2$/Si interfaces under the accumulation condition and prevent interface states from increasing the recombination current at the source/channel and drain/channel pn junctions. The base width (the gate length) was 0.6μm and the measurement was performed at room temperature. The emitter-base voltage V_{EB} was changed from 0 to -1.0V with the base and collector electrodes connected to the ground. This figure shows that the base current increased as the Ge implantation dosage increased. Figure 2 shows the Gummel plot of pnp lateral bipolar transistors with and without the Ge implantation into the source regions. As in the measurement for npn transistors, the interface states were deactivated by applying appropriate positive gate and substrate biases. The emitter-base voltage V_{EB} was changed from 0 to 1.0V with the base and collector electrodes connected to the ground in this case. Again, as the Ge implantation dosage increased, the base current increased. However the increase was smaller than that for npn transistors.

Under the condition exp(q|V_{EB}|/kT) ≫ 1, and under the assumption that the diffusion length of carriers flowing into the base region is much larger than the base width, the base current can be expressed by the following equation [9]

$$|I_b| = qA \left[\frac{D_E N_{mi}}{L_E} \exp(\frac{q|V_{EB}|}{kT}) \right], \tag{1}$$

where D_E and L_E are the diffusion coefficient and the diffusion length of minority carriers in emitter region, respectively. N_{mi} is the minority carrier concentration in the emitter region in thermal equilibrium, and A is the junction area. The minority carrier concentration can be expressed as follows:

$$N_{mi} = \frac{N_v N_c}{N_{mj}} \exp(-\frac{E_g}{kT}), \qquad (2)$$

where Nv and Nc are the effective densities of states in the valence and conduction bands. N_{mj} is the majority carrier concentration and Eg is the bandgap in the emitter region. The base current can be rewritten using equations (1) and (2) as follows.

$$|I_b| = qA \left[\frac{D_E N_c N_v}{N_{mj} L_E} \exp(\frac{q|V_{EB}| - E_g}{kT}) \right] \qquad (3)$$

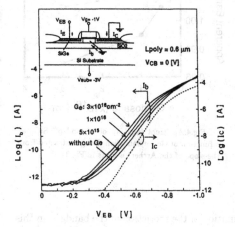

Fig.1:Gummel plots of the npn lateral bipolar transistors with and without the Ge implantation. Bipolar transistor width was 6 μm and base width was 0.6μm. The measurement was performed at room temperature. Measurement setup for the SOI lateral bipolar transistor is depicted as the inset.

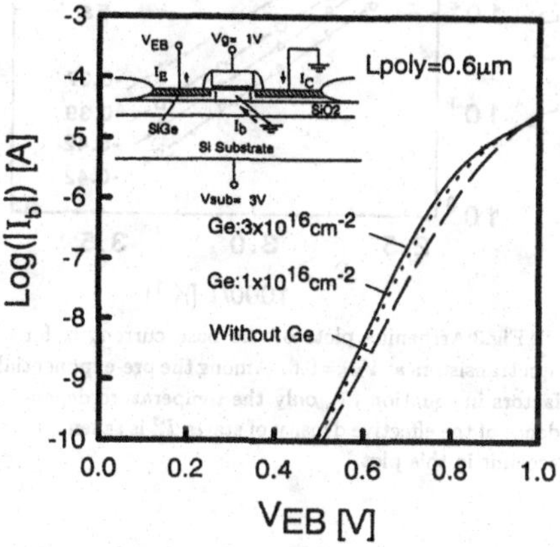

Fig.2:Gummel plots of the pnp lateral bipolar transistors with and without the Ge implantation. Bipolar transistor width was 6 μm and base width was 0.6μm. The measurement was performed at room temperature.

In order to evaluate the bandgaps of the SiGe layers in the source regions, the temperature dependences of the base current of the lateral npn and pnp transistors were measured. Arrhenius plots of the base current I_b for npn transistors at V_{EB}=-0.6V are shown in Fig.3. Among the pre-exponential factors, only the temperature dependence of the effective density of states T^3 [9] is taken into account in these plots [10]. The bandgap in the emitter region extracted from the slope of the I_b/T^3 versus 1000/T relationship is shown in Fig.4 for n$^+$ and p$^+$ cases[†]. The conventional structure gave values of 1.02eV and 1.03eV for n$^+$- and p$^+$-Si, respectively, which are reasonable for Si with As and B doping concentration of around 10^{20} cm^{-3} [11, 12]. As the Ge implantation dosage increased, the bandgap decreased and the bandgap reductions reached 50meV(n$^+$) and 20meV(p$^+$) for the largest implantation dosage

[†]Since the ideality factor 'n' in the emitter/base characteristics for pnp transistors is considerably larger than 1, especially for the reference device (without Ge) as shown in Fig.3, the diffusion current component in the base current was firstly extracted using the procedure mentioned by Sasaki, et al.[10] and is used for the evaluation of the bandgap.

$(3 \times 10^{16}$ cm$^{-2})$. It should be mentioned that the bandgap reductions measured were smaller than those obtained by the theoretical calculation for the same Ge concentration (180mV and 120mV for strained and unstrained n$^+$, 180mV and 100mV for strained and unstrained p$^+$)[6].

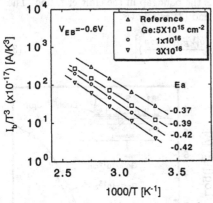 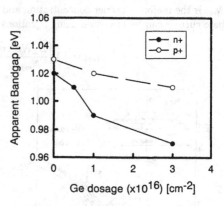

Fig.3:Arrhenius plots of the base current I$_b$ for npn transistors at V_{BE}=0.6. Among the pre-exponential factors in equation (3), only the temperature dependence of the effective density of states T^3 is taken into account in this plot.

Fig.4:Apparent bandgap of n$^+$ and p$^+$ SiGe as a function of Ge implantation dosage extracted from the slope of the Arrhenius plots in Fig.3.

There are possible reasons for the underestimation of the reduction of the bandgap in this experiment. Firstly, the SiGe layer occupied only a part of the SOI layer in the source region as shown in the inset of Fig.1. Figure 5 shows the influence of the ratio of SiGe thickness to the whole SOI thickness. In this calculation, the base current was considered to consist of two components I$_{b,Si}$ and I$_{b,SiGe}$. Since I$_{b,Si}$ was the base current without Ge implantation, I$_{b,SiGe}$ could be extracted using the following equation,

$$I_{b,SiGe} = I_b - (1 - r)I_{b,Si} \qquad (4)$$

and the slope of the Arrhenius plot of the I$_{b,SiGe}$ was measured. In the equation (4), r is the ratio of SiGe thickness to the whole SOI thickness. Figure 5 shows that the bandgap reduction increases from the value obtained under the assumption that the whole thickness (90nm) is occupied by the SiGe layer. The SIMS measurement[4] showed that the thickness of the Ge peak region, in which the Ge concentration is almost uniform, extends from the surface to a depth of about 30nm and the tail region width is about 30nm. As only the peak region is considered to work effectively for SiGe, it can be concluded that the bandgap reductions are larger than those in Fig.4 by 5meV for the Ge dosage of 0.5×10^{16} cm^{-2} and by 10meV for the Ge dosages of 1 and 3×10^{16} cm^{-2} for n$^+$ case. The same procedure also gave further 10meV reduction in the bandgap for the Ge dosages of 1 and 3×10^{16} cm^{-2} for p$^+$ case.

We must also consider that the SiGe layer was located away from the source/channel pn junction as shown in the inset of Fig.1.The continuity of quasi-Fermi level and current at the SiGe/Si interface in the source region gives the following equations.

$$I_b = \frac{qAD_{Si}}{\Delta x}\left[N_{mi}\exp(\frac{qV}{kT}) - N_{int}\right] = \frac{qAD_{SiGe}}{L_{SiGe}}(N_{int} - N_{mi})\exp(\frac{\Delta Eg}{kT}) \qquad (5)$$

where D_{Si} and D_{SiGe} are diffusion coefficients of the minority carriers in Si and SiGe. L_{SiGe} is the diffusion length of minority carriers in SiGe, Δx is the distance between the SiGe/Si interface and pn junction. ΔEg is the amount of the bandgap reduction. N_{int} and N_{mi} are the minority carrier concentration at the Si side of the interface, and that in Si in thermal equilibrium, respectively. From these equations the following equation was extracted.

$$I_b = I_{b,Si}\frac{D_{SiGe}L_{Si}}{D_{Si}L_{SiGe}\exp(-\frac{\Delta Eg}{kT}) + D_{SiGe}\Delta x} \qquad (6)$$

Figure 6 shows the influence of Δx on the relationship between $I_b/I_{b,Si}$ and ΔEg. Since the pre-exponential factor in equation (3) did not change much as mentioned in the previous work[14], we assumed $D_{Si}/D_{SiGe}=1$ and $L_{SiGe}=L_{Si}=0.7\ \mu m$[11, 13] in this calculation. The measured Δx between pn junctions and the Ge peak regions was about 10nm both for n and p channel MOSFETs. Figure 6 shows that if the apparent bandgap reduction is less than about 0.07eV, the influence of Δx is negligible in this case.

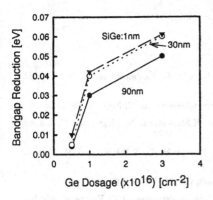

Fig.5:Influence of the ratio of SiGe thickness to the SOI thickness on the measured bandgap of the SiGe layers. This result is for the npn transistor.

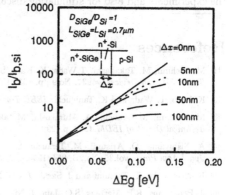

Fig.6:Influence of distance Δx between SiGe/Si interface and pn junction on the I_b increase as a function of bandgap reduction. Δx is taken as a parameter.

As a result, we could conclude that the reductions of the bandgaps of heavily doped SiGe formed by the Ge implantation are smaller than those obtained by the theoretical calculations, especially for p+. The influence of Ge implantation and heavy doping on the film strain can affect the bandgap of these films. Moreover, Ge atoms implanted into Si may not be fully incorporated with the presence of As or B. Further investigation is necessary to clarify the reason for this result.

Conclusion

N$^+$ and p$^+$ SiGe layers were formed in the source regions of SOI MOSFETs in order to suppress the floating-body effects by means of high-dose Ge implantation. The bandgaps of the layers were evaluated by measuring the temperature dependences of the base current of the source/channel/drain lateral bipolar transistors. The measurement has revealed that the reduction of the apparent bandgap by the Ge implantation with a dosage of 3×10^{16} cm^{-2} was 50meV for n$^+$ and 20meV for p$^+$, respectively. The fact that the SiGe occupies only about 30% of the SOI thickness in the source region gave 60meV and 30meV as the estimation of the bandgap of the SiGe layers. However, the bandgap reduction was still small compared to that in the theoretical calculation for heavily doped SiGe, especially in the case of p$^+$. Film strain and the segregation of Ge atoms from the lattice must be investigated for explaining why the bandgap reductions are smaller for these SiGe layers.

Acknowledgments

The authors would like to thank A.Murakoshi and K.Suguro for their cooperation in the experiments and also for stimulating discussion.

References

[1] M. Yoshimi, M. Takahashi, T.Wada, K. Kato, S. Kambayashi, M. Kemmochi and K. Natori, *IEEE Trans. Electron Devices*, vol.37, No.9 , p. 2015,1990.

[2] K. Kato, T. Wada and K. Taniguchi, *IEEE Trans. Electron Devices*, vol.37, No.2 , p.458, 1985.

[3] M. Yoshimi, M.Terauchi, A. Murakoshi, M.Takahashi, K.Matsuzawa, N. Shigyo and Y. Ushiku, *Technical Digest of IEDM 1994*, p.429.

[4] A. Nishiyama, O. Arisumi, M. Terauchi, S. Takeno, K. Suzuki, C. Takakuwa and M. Yoshimi, *Jpn.J.Appl.Phys.*, vol.35, No.2B, 954, 1996.

[5] K. Aketagawa, T. Tatsumi and J. Sakai, *J. Cryst. Growth*, vol.111, No.1-4, 860, 1991.

[6] J. Poortmans, R.P. Mertens, S.C. Jain, J. Nijs and R. van Overstraeten, 19th European Solid State Devices Research Conf., 807, 1989.

[7] G.F.A. van de Walle, L.J. van Ijzendoorn, A.A. van Gorkum, R.A. van den Heuve, A.M. Theunissen and D.J. Gravesteijn, *Thin Solid Films*, vol.183, No.1, 183, 1989.

[8] R.B. Borg and G.J. Dones, *An Introduction to Solid State Diffusion*, Academic Press, New York, 1990.

[9] S.M. Sze, *Physics of Semiconductor devices (2nd edition)*, A Wiley-Interscience Publication, 1981.

[10] K. Sasaki, T. Miyajima, Y. Kubota and S. Furukawa, *IEEE Trans. Electron Devices*, vol.39, No.9. 2132, 1992.

[11] J.A. del Amalo, R.M. Swanson, *IEEE Tran. Electron Devices*, vol.34, No.7, 1580, 1989.

[12] S.E. Swirhun, Y. Kwark and R.M. Swanson, *Technical Digest of IEDM 1986*, p.24.

[13] H.T. Weaver and R.D. Nasby, *IEEE Trans. Electron Devices*, vol.28, No.5, 465, 1981.

[14] A. Nishiyama, O.Arisumi and M. Yoshimi, *IEEE Trans. Electron Devices* vol.44, No.12, 2187, 1997.

INVESTIGATION OF THE DOPANT DISTRIBUTION IN THIN EPITAXIAL SILICON LAYERS BY MEANS OF SPREADING RESISTANCE PROBE AND SECONDARY ION MASS SPECTROMETRY

Ilya Karpov*, Catherine Hartford***, Greg Moran*, Subramania Krishnakumar*, Ron Choma**, and Jack Linn**
*Mitsubishi Silicon America, Salem, OR
**Harris Semiconductor Sector, Palm Bay, FL
***Solid State Measurements, Inc., Pittsburgh, PA

ABSTRACT

In this paper, we examine the dopant distributions in 1.8 to 4 micron-thick boron- and phosphorus-doped epitaxial silicon layers. These layers were grown by chemical vapor deposition (CVD) on arsenic-, antimony-, or boron-doped (100)- and (111)-oriented substrates. We performed doping profile studies by means of local resistivity measurements using a spreading resistance probe (SRP). Chemical profiles of the dopants were also obtained using secondary ion mass spectrometry (SIMS).

INTRODUCTION

Epitaxial silicon layers with tight doping profile tolerances are required for bipolar, CMOS, and BiCMOS applications in microelectronics. The thickness of these layers ranges from over 100 μm for some power discrete devices to those under 2 μm used in CMOS DRAM integrated circuits. Depending upon the application, the epitaxial layers can have uniform, step-like, as well as graded doping profiles. Spreading resistance probe (SRP), an electrically based technique to probe local resistivity, has been widely used to investigate the electrically active dopant distribution in the epitaxial layers [1-3]. Secondary ion mass spectrometry (SIMS), a chemical analysis technique was traditionally used to study ion-implanted and diffused profiles [2,4. We compare SRP and SIMS profiles and discuss the correlation between the two techniques. We also discuss potential limitations for each technique and show examples where the techniques provide complementary information.

EXPERIMENT

Boron- and phosphorous-doped epitaxial silicon layers were grown on (100) and (111)-oriented silicon substrates doped with As, B, and Sb. 1.8 to 4 μm-thick layers were grown in a commercial epitaxial rector by the conventional trichlorosilane technology [2].

SRP measurements were performed using an SSM-150 spreading resistance probe. For these experiments lightweight probe arms, a 5g probe load, and a special conditioning procedure were used to reduce pressure effects and to improve the depth resolution [5]. The samples were beveled to a 1°9' angle with the 0.05 μm grade diamond paste using a specially textured glass plate to produce a uniform bevel surface and to minimize the surface damage. The profiles were acquired using a 2.5 μm lateral step which corresponds to a 5 nm in-depth step. A Nomarski microscope was used to find probe marks on the sample surface and to eliminate the points found above the bevel edge from the SRP profile. The accuracy of the depth measurement was determined to be within 10%, mostly due to the uncertainty in the bevel angle measurement. Choo's variational multilayer correction procedure was used to correct the local resistivity profile for the field effects of the resistivity variations and underlying pn junctions [6]. The local resistivity profile was converted to the carrier density profile using Thurber's equations [7].

SIMS analyses were performed with a Perkin-Elmer PHI 6600 instrument with a quadrupole mass spectrometer. Boron profiles were acquired with a 7 KeV O_2^+ ion beam. Phosphorous, arsenic, and antimony profiles were acquired with a 7 KeV Cs^+ ion beam. Typical raster areas were 350 μm x 350 μm with a secondary ion collection area of 15 to 20%. Primary ion beam currents were set to provide sputtering rates of 2 to 3 nm/s. No charge neutralization was employed. The concentration axis was determined by profiling standard wafers with known implant doses of the elements of interest using the same analysis conditions as those implemented for the samples. The depth axis was established by calculating a sputter rate from the same standard wafers. This was accomplished by measuring the sputter time required to reach the known center point of the implant peak.

RESULTS

Fig. 1 shows an example of SIMS/SRP profiles acquired for a staircase-doped epitaxial layer. As the doping concentration decreases from 10^{19} to 10^{15} cm^{-3}, the noise in the SIMS data increases. A signal-to-noise ratio, defined as the signal standard deviation over the signal mean, decreases from 100 to 0.4. The signal-to-noise ratio was over 20 for the SRP profile and only slightly dependent on the doping level. Though Fig. 1 indicates a good agreement between the values for the doping concentration derived from the SIMS and SRP data, we note that the deviation between the absolute values of the concentration increases with the doping level.

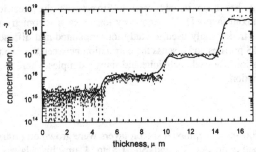

Fig. 1. SRP (solid line) and SIMS boron (dashed line) profiles of a staircase boron-doped epitaxial layer grown on boron-doped (100) substrate at 1120 °C.

The dopant distribution between an epitaxial layer and a substrate has been the focus of the extensive research [2]. The transition region has been found to be determined by the solid state diffusion and autodoping phenomena [2]. Fig. 2 shows SIMS/SRP profiles for the phosphorus-doped epitaxial layer grown at 1150 °C on the arsenic-doped (111) substrate. Two parts in the transition region can be distinguished. The doping profile near the substrate is steep and is determined by the solid state diffusion of arsenic from the substrate into the growing epitaxial layer. The arsenic evaporation from the substrate and reincorporation of arsenic through the gas phase into the growing layer contributes to the tailing portion of the transition region. Since the epitaxial layer was doped with phosphorous, the SRP profile of the electrically active dopant within the layer is different from the As doping profile by SIMS.

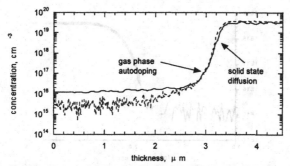

Fig. 2. SRP (solid line) and SIMS arsenic (dashed line) profiles of a phosphorous-doped epitaxial layer grown on a arsenic-doped (111) substrate at 1150°C

For epitaxial layers thinner than 2 μm, we observed differences between SIMS and SRP profiles in the transition region. Fig. 3 shows an example of a SIMS/SRP profile of a 2 μm-thick boron-doped epitaxial layer grown at 1100 °C on a boron-doped substrate. Fig. 3 indicates that the SIMS profile exhibits a sharper slope at the interface. It was found that the deviation between the dopant and the carrier density profiles is a result of the carrier diffusion effects generally occurring over a rather small distance on the order of several tenths of a micron, but extending to as much as one micron in very lightly doped layers [8, 9]. The carrier diffusion phenomenon results in the deviation between the SR profile in Fig. 3, which follows a net on-bevel carrier density, with the SIMS profile which follows the doping concentration.

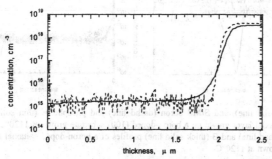

Fig. 3. SRP (solid line) and SIMS boron (dashed line) profiles of a boron-doped epitaxial layer grown on a boron-doped (100) substrate at 1100°C

To test whether the SIMS profile in the transition region near the substrate is consistent with a solid-state diffusion mechanism, we fit the SIMS profile using an error function model developed in Ref. 10. The results are shown in Fig. 4 for the same structures shown in Fig. 3. A good agreement between the experimental data and the fitting model was obtained. The diffusion coefficient extracted from the model was $5.6 \cdot 10^{-13}$ at 1100 ^{0}C for B in agreement with a diffusion coefficient $5 \cdot 10^{-13}$ obtained from the extrinsic diffusion data for B [11].

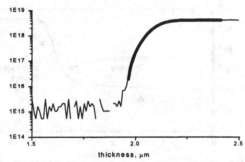

Fig.4. SIMS profiles for boron (thin solid line) was fitted with a function y=a+b·erf(cx-d) (thick solid line) to test the solid state diffusion mechanism at the epitaxial layer-substrate interface and to extract a diffusion coefficient for boron.

When both SRP and SIMS were used to profile ion-implanted structures, the SRP profile, in contrast to Fig. 3, was found to have a sharper slope than that obtained by SIMS [3,4,12]. An extended tail was observed for the SIMS profile. We suggest that an atom mixing/knock-on effect [4], which deteriorates the SIMS depth resolution when profiling a region with a sharp negative doping slope, does not affect the depth resolution when profiling a region with a sharp positive doping slope.

The difference between the SRP and SIMS profiles becomes significant when a doping type

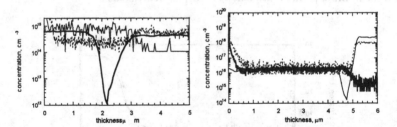

Fig. 5 (a) SRP (thick solid line), and SIMS boron (dashd line) and phosphorus (thin solid line) profiles of a phosphorous-doped epitaxial layer grown on a boron-doped (100) substrate grown at 1120°C, (b) SRP (thin solid line), and SIMS Sb (dashed line) and B (thick solid line) profiles of a boron-doped epitaxial layer grown on a Sb-doped (100) substrate grown at 1120°C,

for an epitaxial layer is opposite to that of the substrate. Fig. 5 (a) and (b) show examples of SIMS/SRP profiles acquired for epitaxial layers which contain a pn junction. The area of reduced carrier density in Fig. 5 (a) and (b) corresponds to a space charge region at the pn junction. At thermal equilibrium, the depletion layer width in Si is equal to 0.3 µm and 2 µm for doping levels of 10^{16} and $4 \cdot 10^{14}$ cm^{-3} [13], respectively. This is in agreement with the width of the negative peaks, 0.25 µm and 1.5 µm, in the carrier density profiles shown in Fig. 5 (b) and (a), respectively.

Fig. 5 (a) and (b) show a shift between the location of the metallurgical pn junctions derived from the SIMS profile (where $N_A=N_D$) and the position of the electrical junction derived from the SRP profile (where n=p). In Fig. 5(a), the electrical junction appears to be much more shallow at 2.2 µm, than the metallurgical junction at 3 µm. The effect was studied in Ref. [5,8] and was attributed to the removal of the material above the pn junction by beveling. The calculated junction ratio [6], i.e. the ratio of the electrical junction depth to that of a metallurgical junction, was found to increase as the doping concentration in the epitaxial layer decreases in agreement with Fig. 5 (a) and (b).

Fig. 5 indicates that the net on-bevel carrier density profile derived from the spreading resistance data does not provide the accurate information about the dopant distribution in the space charge region. Fig. 6 shows an example of the SRP2 algorithm developed in Ref. 13 to improve the accuracy of the dopant distribution analysis with SRP. SRP2 is a forward modeling analysis using a Poisson solver. The procedure begins with the development of a trial set of the dopant profiles derived from a conventional SRP profile. To calculate a resistivity profile, the Poisson equation and the charge density equations are solved simultaneously with boundary conditions of a zero surface charge and a zero electric field in the uniformly doped substrate. The process is repeated until the satisfactory agreement is obtained between the calculated and measured spreading resistance profiles. Fig. 6 indicates that the position of the metallurgical junction (where $N_A=N_D$) determined by SRP2 is within 10% of that determined by SIMS. We note that B concentration measured at the n-side, as well as P concentration measured at the p-side of the pn junction, are at the noise level for the quadrupole SIMS.

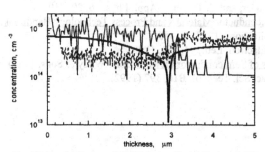

Fig. 6. SRP2 doping profile (thick solid line), and SIMS boron (dashd line) and (thin solid line) profiles of a phosphorous-doped epitaxial layer grown on a boron-doped (100) substrate grown at 1120°C.

CONCLUSION

We examined the dopant distributions in 1.8 to 4 micron-thick boron- and phosphorus-doped epitaxial silicon layers by SRP and SIMS. Our results indicate that SIMS profiling can follow very sharp transitions at the layer/substrate interface, while SRP profiles are very sensitive to concentration variations in high resistivity epitaxial layers. We found that probe conditioning, the implementation of Choo's multilayer correction procedure, as well as the forward modeling algorithm (SRP2), significantly enhance the depth resolution of the SRP technique.

ACKNOWLEDGMENTS

The authors gratefully thank Robert Garza (MSA) for his invaluable help with SRP measurements, and Alex Demkov (Materials Research and Strategic Technologies at Motorola) for reviewing the manuscript.

REFERENCES

1. D. K. Schroder, Semiconductor Material and Device Characterization, (John Wiley & Sons, 1990).
2. H. M. Liaw and J. W. Rose, in Epitaxial Silicon Technology, edited by B. J. Baliga (Academic Press, 1986).
3. W.M. Bullis, D.G. Seiler, A.C. Diebold, Eds., Semiconductor Characterization. Present Status and Future Needs, (American Institute of Physics, 1996).
4. A. Benninghoven et al, Secondary Ion Mass Spectrometry SIMS VII (John Wiley & Sons, 1990).
5. A. Casel and H. Jorke, Appl. Phys. Lett. 50, 989 (1987).
6. S. Choo, M. S. Leong, J. H. Sim, Solid State Electron. 26, 723 (1983).
7. ASTM F723-88, Standard Practice for Conversion between Resistivity and Dopant Density for Boron-Doped and Phosphorous-Doped Silicon, ASTM Annual Standards, v10.05 (1996).
8. S.M. Hu, J. Appl. Phys. 53, 1499 (1982).
9. J. Albers, in Emerging Semiconductor Technology, ASTM STP 960, edited by D. Gupta and P. H. Langer (ASTM, 1986), p.480.
10. A. S. Grove, A. Roder, and C. T. Sah, J. Appl. Phys. 36, 802 (1965).
11. R. B. Fair in Semiconductor Materials and Process Technology Handbook, edited by G. E. McGuire (Noyes Publications, 1988), p. 455.
12. S. Clayton et al, Electron. Lett., 24, 881, 1988.
13. S. M. Sze. Phisics of Semiconductor Devices (John Wiley & Sons, 1981), p. 78.
14. R.G. Mazur, J. Vac. Sci. Technol. B10, 397 (1992).

EVALUATION OF GAP STATES IN HYDROGEN-TERMINATED SILICON SURFACES AND ULTRATHIN SiO₂/Si INTERFACES BY USING PHOTOELECTRON YIELD SPECTROSCOPY

S. Miyazaki, T. Tamura, T. Maruyama, H. Murakami, A. Kohno and M. Hirose
Department of Electrical Engineering, Hiroshima University, Higasi-Hiroshima 739, JAPAN
miyazaki@sxsys.hiroshima-u.ac.jp

ABSTRACT

The energy distributions of gap state densities for H-terminated Si surfaces and thermally-grown SiO₂/c-Si interfaces have been evaluated by total photoelectron yield spectroscopy (PYS) with a dynamic range of eight orders of magnitude which is sufficient to detect the density of states as low as $10^{10}cm^{-2}eV^{-1}$. It is confirmed from the threshold energies for direct and indirect photo-excitations that, for monohydride-terminated Si(111) surfaces prepared by an NH_4F treatment, no significant band-bending is observable. For H-terminated n-type sample, the gap state densities of the order of $10^{11}cm^{-2}eV^{-1}$ were estimated in the region within 0.4 eV from the valence band edge, which may be attributable to a very little oxidation in the sample preparation. It is also found that, for as-grown 2.5nm-thick SiO₂/n⁺ Si, there exist interface states around midgap with densities as high as $\sim 10^{12} cm^{-2}eV^{-1}$.

INTRODUCTION

A clear insight into chemical and electronic structures of Si surfaces and SiO₂/Si interfaces has become increasingly important with the continuing shrinkage of MOS device dimensions. To gain a better understanding of the nature of the surface and interface states, and to control them, precision measurements for the gap state distributions in the surfaces or the interfaces are thought to be crucial. It is well known that hydrogen-terminated, unreconstructed Si surfaces (1x1), which are prepared by a wet-chemical cleaning in a dilute HF or pH-controlled HF solution [1, 2], exhibit high chemical stability against oxidation [3] and slow surface recombination [4], in contrast to the hydrogen-free, reconstructed Si surfaces (for example, 7x7 or 2x1). These observations indicate that the gap state density in the H-terminated surfaces is fairly small. However, the perfectness of the surface passivation or the energy distribution of gap states for H-terminated surfaces has not been well-characterized yet. Even for an ultrathin SiO₂/Si system, the determination of the energy distribution of interface states is still a matter for research because conventional C-V characterizations are no longer useful under high tunneling current.

In this work, we have investigated the energy distributions of gap states for H-terminated Si(100) and (111) surfaces and thermally-grown ultrathin SiO₂/c-Si interfaces by means of total photoelectron yield spectroscopy (PYS).

EXPERIMENTAL

Czochralski grown Si(100) and Si(111) wafers with boron or phosphorous concentrations of 1.3×10^{15} to $2.0 \times 10^{19} cm^{-3}$ were precleaned by conventional wet-chemical cleaning steps. H-terminated Si(100) and atomically-flat, monohydride-terminated Si(111) surfaces were prepared by 4.5%HF and 40%NH$_4$F treatment, respectively, followed by pure water rinse for 3sec to remove residual fluorine atoms from the surface. Hydrogen bonding features on the surfaces so prepared were monitored by Fourier-transform infrared attenuated total reflection (FT-IR-ATR) and ultraviolet (HeII) excited photoelectron spectroscopy (UPS). The thermal oxidation was carried out in a furnace at a temperature of 1000°C in dry O$_2$. The energy distributions of filled gap states for these samples were evaluated from total photoelectron yield spectra which were measured as a function of incident photon energy in the region from 4 to 6eV, with an energy resolution of ~20meV, and normalized by the incident photon flux.

RESULTS AND DISCUSSION

The typical photoelectron yield spectrum for monohydride-terminated p$^+$ Si(111) with a boron concentration of ~5x10^{18}cm^{-3} is shown in Fig. 1. The spectrum shows a signal dynamic range of about eight orders of magnitude. Obviously, an extremely low yield for photons with energies below 4.8eV is obtained. The surface Fermi-level position determined by a Kelvin probe method is located at 4.7eV below the vacuum level, namely shifted toward midgap by at least 0.25eV from the bulk Fermi-level. This Fermi-energy shift is mainly attributed to hydrogen-induced passivation of acceptors in the near surface region during the wet-chemical process [5,6]. In fact, after 5min annealing at 350°C the surface Fermi-level shifts close to the bulk position without any change in not only the yield spectrum but also surface hydrogen termination as confirmed by UPS measurements [5]. In addition to this, for both as-prepared and annealed samples, no photovoltage under illumination conditions for the photoemissions was detected within a detection limit of ~1mV, and the surface concentrations for ionic contaminants such as Na$^+$ were proved by X-ray photoelectron spectroscopy to be at least an order of magnitude lower than needed to explain the Fermi-level position for as-prepared sample. The yield between the Fermi-level positions before and after annealing at 350°C indicates that a few gap states still exist on the surface. The yield spectra from differently-doped samples, which were prepared with the same cleaning procedure, were also measured and compared as represented in Fig.2. A fairly big difference in the yield among the samples is observed. However, this is not so surprising because the yield spectrum reflects the integrated energy distribution of filled states. In order to evaluate the band bending in the surface, or the tailing of the valence band edge for heavily-doped samples, the observed yield

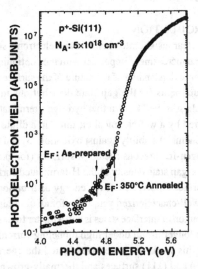

Fig. 1 Total photoelectron yield spectrum for as-prepared or annealed Si(111) with a boron concentration of ~5x10^{18} cm^{-3} of which the surface was passivated with monohydrides. The thermal annealing was carried out at 350°C for 5min. The Fermi level positions before and after annealing are indicated by short solid lines.

spectra were characterized by the power low dependence [7, 8] predicted for photoemission processes as indicated in Figs. 3 and 4. Figure 3 shows a linear plot of each yield spectrum shown in Fig. 2 as a function of photon energy to determine the threshold energy for direct excitations at a higher photon-energy region. It is evident that, for p and n-type samples, the same threshold energy is obtained at 5.56eV, which agrees well with the reported value for samples with flat band which were cleaved in vacuum [8]. On the other hand, the threshold energy of p^+ sample is decreased by 0.13eV. At a photon energy lower than the threshold, photoelectrons can emit through an indirect excitation process, and then the yield approximately follows a cubic power law [8, 9]. Figure 4 represents a cubic root plot for each yield by which the threshold energy for indirect transitions, namely the valence band edge, can be determined. As in the case of the linear plot, p- and n-type samples show the same threshold energy (5.11eV), which is identical to the reported energy separation between the valence band edge with nearly parabolic density of states and the vacuum level [9]. In contrast, for the p^+ sample, again the threshold energy decreases by 0.13eV, and the fairly low yield below the indirect excitation threshold implies a very few surface states in the fundamental gap. Therefore, the observed red-shift in the direct and indirect thresholds might be attributable mainly to the band edge tailing and partially to the surface band bending [10]. When the yield is fitted with the 5/2 power law which has been theoretically predicted for indirect transition processes [7], the indirect threshold energy for each case becomes higher by 40meV than the value determined from the cubic power law. Thus, the photoemission yield from gap states can be distinguished from the contribution of photoemissions from the valence band. As the next step, a crude estimation of the gap state density was made from the observed PYS spectra by considering that the XPS valence band data, to a first approximation, provides the real density-of-states distribution in the valence

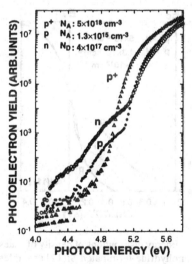

Fig. 2 PYS spectra for differently-doped Si(111) samples of which the surfaces were passivated with monohydrides.

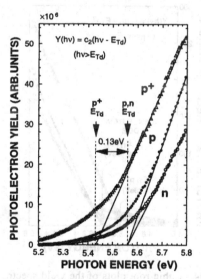

Fig. 3 Linear plots of the yield spectra shown in Fig. 2. E_{Td} for each plot denotes the threshold energy for the direct excitation processes.

band. Assuming that the photoelectron escape depth is constant at 2.5nm [8], we first converted the photoelectron yield at an energy above the direct transition threshold (typically around 5.8eV) into the density of states, and then calculated the first derivative of the yield with respect to energy so as to determine the energy distribution of gap states as shown in Fig. 5. It is found that, around 0.2eV above the valence band edge, there exist occupied surface states with a density of $\sim 3 \times 10^{11}$ $cm^{-2}eV^{-1}$ for n-type and $\sim 3 \times 10^{10} cm^{-2}eV^{-1}$ for p-type sample. Taking into account the fact that the native oxidation rate in air or in deionized water depends on the Fermi level position and is basically the order $n^+ > n > p > p^+$ [11], a small amount of oxidation, which could occur until the samples were loaded in UHV chamber ($<10^{-9}$ mbar), might be responsible for the observed surface states. Furthermore, for as-prepared H-terminated p-Si(111), the yield spectrum measured after 12hr storage in 10^{-8} mbar shows an upward shift of the valence band edge by 70meV as determined from the cubic root plot, and an increase in the gap states up to $\sim 5 \times 10^{11}$ $cm^{-2}eV^{-1}$ around 0.2eV above the valence band edge. This implies that undesirable surface reactions with residual gases, such as hydrocarbons and water molecules, during the storage in 10^{-8} mbar induce the gap states.

We also evaluated the interface states for an ultrathin SiO_2/Si system from the total photoelectron yield spectrum. Figure 6 represents the photoelectron yield spectrum for n^+-type Si(100) through 2.5 nm-thick SiO_2 grown at 1000°C. The photoelectron yield spectrum taken after oxide removal by dilute-HF etching is also shown as a reference. Because the energy band gap of SiO_2 is around 8.9eV as determined from the onset of energy loss signals for O1s photoelectrons [12], the valence electrons of the SiO_2 layer can not contribute to the yield for the SiO_2/Si system. For photons below 4.8eV, an enhanced yield for the SiO_2/Si sample is originated from photoemissions from the

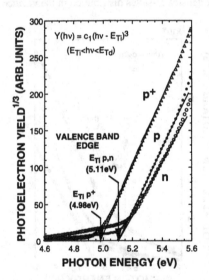

Fig. 4   Cubic root plots of the yield spectra represented in Fig. 2. E_{Ti} for each plot denotes the threshold energy for the indirect exitation processes, namely the valence band edge.

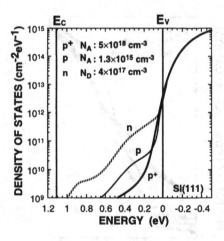

Fig. 5   Energy distributions of density of states for monohydride-terminated Si(111) samples which were estimated from the PYS spectra shown in Fig. 2. E_C and E_V denote the conduction and valence band edges, respectively.

interface states. In the higher energy region, the yield for the SiO_2/Si sample becomes significantly smaller than that for H-terminated Si(100). The scattering for photoelectrons to path through the SiO_2 layer is partially responsible for this reduction. The threshold energies for direct and indirect excitations, which were determined from the linear plot and the cubic root plot of each yield spectrum as indicated in Fig. 7, respectively, indicate that the band edge for the oxide sample is shifted toward the higher photon energy side by 40meV from that for the H-terminated Si. Taking into account both this energy shift and the oxide thickness, the escape depth for photoelectrons with an energy around 5.7eV is estimated to be 2.1nm from the yield reduction. Figure 8 shows the energy distribution of gap states estimated from the yield spectra shown in Fig. 7 by the same manner as mentioned above. It is found that, for an as-grown 2.5nm-thick SiO_2/n^+ Si, the interface state

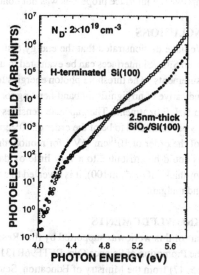

Fig. 6   PYS spectrum for as-grown SiO_2 (2.5nm)/n^+- Si(111) and H-terminated n^+- Si(111). The oxidation temperature was 1000°C.

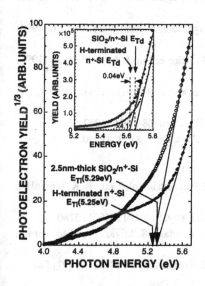

Fig. 7   Cubic root plots of the PYS spectra represented in Fig. 6. Inset shows linear plots for a higher energy region.

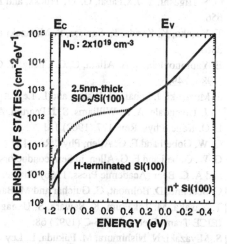

Fig. 8   Energy distributions of density of states for as-grown SiO_2(2.5nm)/n^+- Si(111) and H-terminated n^+- Si(111) which were estimated from the PYS spectra shown in Fig. 6.

density around midgap is as high as $\sim 10^{12}$ cm^{-2}eV^{-1}. This is presumably because the thermal annealing to improve the interface properties was not done.

CONCLUSIONS

We have demonstrated that the energy distributions of gap state densities for Si surfaces and ultrathin SiO_2/Si interfaces can be evaluated to the values as low as 10^{10} cm^{-2}eV^{-1} by using the PYS measurements. The threshold photon energies for direct and indirect excitations in the PYS spectra are indicative of no significant band-bending for monohydride-terminated Si(111) surfaces prepared by an NH_4F treatment. The gap state densities in the region within 0.4 eV from the valence band edge were estimated to be of the order of 10^{11}cm^{-2}eV^{-1} for monohydride-terminated n-type Si(111) and of the order of 10^{10}cm^{-2}eV^{-1} for monohydride-terminated p-type Si(111). The observed gap states could be attributed to a very little oxidation in the sample preparation. For an as-grown 2.5nm-thick SiO_2/n$^+$ Si(100), it is revealed that interface states as high as $\sim 10^{12}$cm^{-2}eV^{-1} distribute around midgap.

ACKNOWLEDGMENTS

Part of this work was supported by the "Research for the Future" Program in the Japan Society for the Promotion of Science (No. RFTF96R13101) and a Grant-in-Aid for Scientific Research (No. 08455147) from the Ministry of Education, Science, Sports and Culture.

REFERENCES

[1] T. Takahagi, I. Nagami, A. Ishitani, H. Kuroda and Y. Nagasawa, J. Appl. Phys. 64 (1988) 3516.

[2] G. S. Higashi, Y. J. Chabal, G. W. Trucks, and K. Raghavachari, Appl. Phys. Lett. 56 (1990) 656.

[3] T. Yasaka, K. Kanda, K. Sawara, S. Miyazaki and M. Hirose, Jpn. J. Appl. Phys. 30 (1991) 3567.

[4] E. Yablonovich, D. A. Allara, C. C. Chang, T. Gmitter and T. B. Bright, Phys. Rev. Lett, 57 (1986) 249.

[5] S. Miyazaki, J. Schäfer J. Ristein and L. Ley, Appl. Phys. Lett. 68 (1996) 1247.

[6] A. J. Tavendale, A. A. Williams, S. J. Pearton, Appl. Phys. Lett. 48 (1988) 590.

[7] E. O. Kane, Phys. Rev. 127 (1962) 131.

[8] G. W. Gobeli and F. G. Allen, Phys. Rev. 127 (1962) 141.

[9] G. W. Gobeli and F. G. Allen, in Semiconductors and Semimetals Vol. 2, eds. R. K. Willardson and A. C. Beer (Academic Press, 1966) Chapter 11.

[10] C. Sebenne, D. Bolmont, G. Guichar and M. Balkanski, Phy. Rev. B 12 (1975) 3280.

[11] T. Yasaka, M. Takakura, K. Sawara, S. Uenaga, H. Yasutake, S. Miyazaki and M. Hirose, IEICE Trans. Electron, E75-C (1992) 68.

[12] S. Miyazaki, N. Nishimura, M. Fukuda, L. Ley and J. Ristein, Appl. Surf. Sci. (1997) 585.

Conductance Transients Study of Slow Traps in Al/SiN$_x$:H/Si and Al/SiN$_x$:H/InP Metal-Insulator-Semiconductor Structures

S.Dueñas,* R.Peláez*, E.Castán*, J.Barbolla*, I.Mártil** and G.Gonzalez-Diaz**

*Dept. Electricidad y Electrónica, Facultad de Ciencias, Universidad de Valladolid,

47011 Valladolid, SPAIN, sduenas@ele.uva.es

** Dept. Electricidad y Electrónica, Facultad de Física, Universidad Complutense de Madrid, 28040 Madrid, SPAIN

ABSTRACT

We have obtained Al/SiN$_x$:H/Si and Al/SiN$_x$:H/InP Metal-Insulator-Semiconductor devices by directly depositing silicon nitride thin films on silicon and indium phosphide wafers by the Electron Cyclotron Resonance Plasma method at 200°C. The electrical properties of the structures were first analyzed by Capacitance-Voltage measurements and Deep-Level Transient Spectroscopy (DLTS). Some discrepancies in the absolute value of the interface trap densities were found. Later on, Admittance measurements were carried out and room and low temperature conductance transients in the silicon nitride/semiconductor interfaces were found. The shape of the conductance transients varied with the frequency and temperature at which they were obtained. This behavior, as well as the previously mentioned discrepancies, are explained in terms of a disorder-induced gap-state continuum model for the interfacial defects. A perfect agreement between experiment and theory is obtained proving the validity of the model.

INTRODUCTION

Present by, ultrathin silicon dioxide gates (30-40 Å) are required as a consequence of the reduction in the ultralarge-scale-integration (ULSI) silicon device dimensions. On the other hand, silicon nitride, Si$_3$N$_4$, has been successfully used as an insulator with different III-V semiconductors. Two important properties of silicon nitride, Si$_3$N$_4$, with respect to silicon dioxide, SiO$_2$, make it a candidate for ultrathin dielectric structures: a higher dielectric constant and a better performance as a diffusion barrier. However, the interface between Si$_3$N$_4$ and Si is not as well known as the SiO$_2$/Si interface and significantly higher densities of interface states are always displayed by the silicon nitride/silicon structure.

As for the deposition technique, there is a growing interest in the deposition of insulator thin films by the Electron Cyclotron Resonance Plasma method (ECR) to fabricate MISFET devices. This process provides insulator films with good physical properties at low substrate temperature (≤200 °C). Device quality films of SiN$_x$ [1], SiO$_2$ [2] and SiO$_x$N$_y$[3] have been obtained with this technique. This good behavior is attributed to the "soft" character of the ECR plasma method. Usually, an Interface Control Layer (ICL) is deposited before the ECR film deposition [1,4] to passivate the surface of the semiconductor. Recently, we reported studies of films deposited on Si [5] and InP [6] by ECR in which the ICL layer was not deposited first.

In this work we summarize the results of the electrical characterization of these films. We used Capacitance-Voltage and DLTS to measure the interface-state densities present in the MIS structures made with these films. Some discrepancies appear that can be explained in terms of a spatially distributed interface states model. This assumption is confirmed by the existence of transients in the conductance of the structure when Admittance characterization is attempted.

EXPERIMENT

Sample Preparation

The devices were obtained as follows: Substrates used were <100> n-type silicon wafers with a resistivity of 5 Ωcm, and unintentionally doped n-type InP wafers (5×10^{15} cm^{-3}) respectively. Prior to the insulator deposition, back electrodes were deposited by thermal evaporation of aluminum on silicon and an AuGe/Au alloy on InP wafers. After this, the wafers were cleaned with organic solvents and, then, the native oxides were stripped in a solution of H_2O:HF, followed by a 3 min. rinse in deionized water and drying in N_2. Then they were transferred to the insulator deposition chamber: a vacuum chamber of our design attached to an ECR reactor (Astex 4500). Substrate temperature (200 °C) and total pressure (0.6 mTorr) were kept constant in all the experiments. Microwave power was varied in the range 50-200 W for silicon wafers and was kept constant (100 W) for the InP samples. Gases used were N_2 to generate the plasma and pure SiH_4. The value of the gases flux ratio [N_2]/[SiH_4] was varied from 1 (x=0.91) to 9 (x=1.49) for the two types of semiconductor wafers. The deposition time was scaled to obtain insulator film thickness of 500 Å. Finally, Al dots (1.2×10^{-3} cm^2) were thermally evaporated with a mechanical mask and post metallization annealing was conducted in inert atmosphere.

Electrical Characterization

The electrical characteristics of the semiconductor/insulator interfaces are similar to those using passivation layers. In the case of films deposited over silicon, we observe that better values of the electrical properties are obtained for near stoichiometric films and poor values were obtained for both Si-rich and N-rich films. We used the C-V technique to determine the existence of traps in the insulator-semiconductor interface. The interface state density existing in these structures was kept in the range (3-5)$\times10^{11}$ cm^{-2} eV^{-1} for films with [N]/[Si] ratio lower than 1.38. For nitrogen rich films, the trap density suddenly increases following the same trend as the concentration of N-H bonds in the SiN_x:H film . This result is explained on the basis of the model reported by Lucovsky et al. [7] for Oxide-Nitride-Oxide/Si structures. The model is additionally supported by DLTS measurements (Fig.1) that show the presence of silicon dangling bonds at the insulator/semiconductor interface (the so called P_{bN0} centers). The concentration of these centers depends on the insulator composition in a similar way as the interface trap density and, i.e., as the N-H bond concentration. Moreover, this result supports the assumption that the N-H bonds located at the interface act as precursor of •Si≡Si$_3$ type defects, that is, of P_{bN0} centers. In conclusion, a close relation between interface trap density, P_{bN0} centers and N-H bond density can be established. As for the films deposited on InP, the C-V and DLTS results (Fig.2) suggest an inverse correlation between the insulator composition and the density of interface traps. The minimum trap density concentration (2×10^{12} cm^{-2} eV^{-1} at 0.32 eV above the midgap) was obtained for the films with the maximum [N]/[Si] ratio. To explain this behavior we suggest substitution mechanisms consisting of N atoms coming from the insulator that occupy phosphorous vacancies, V_p, giving place to N_{Vp} configurations. The values obtained for the interface state densities are similar to those obtained by other authors for devices in which chemical and physical passivation processes of the InP surface have been made previously to the deposition of the insulator.

Although C-V and DLTS experiments show similar trends with composition, some differences can be pointed out in the experimental curves. We observed that the shape of the distributions is very different. C-V measurements give U-shaped distributions, whereas DLTS results show a different dependence with the energy. These differences have been already observed in devices like SiO_2/InP [8] and SiN_x/InGaAs [9]. Hasegawa et al. [10] observed that DLTS measurements tend to underestimate the interface trap density, whereas C-V method overestimates the density values. Furthermore, C-V plots exhibit hysteresis phenomena. All these phenomena can be explained assuming that interface sates are not located just in the interface but distributed both in energy and in space. This distribution is called disorder-induced gap-state (DIGS) continuum.

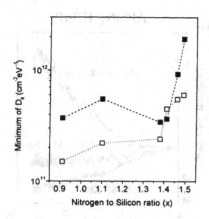

Figure 1.- Minimum Value of D_{it} as a function of x deduced from C-V (■) and DLTS (□), for silicon wafers

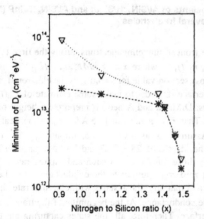

Figure 2.- Minimum Value of D_{it} as a function of x deduced from C-V (▽) and DLTS (*), for InP wafers.

Emission and capture of free electrons by states located far from the interface can occur by means of tunneling mechanisms. When the bias varies from inversion to accumulation, electrons in the semiconductor conduction band are captured by emptied interface states, whereas when moving in the opposite direction electrons are emitted from filled interface states to the conduction band of the semiconductor. Since these processes are tunneling assisted, they are slow and nonsymmetrical. That causes the experimental capacitance to depend both on the direction and on the speed of the voltage variation and, therefore, hysteresis effects are observed.

CONDUCTANCE TRANSIENTS

In a recent work [11], we reported the existence of room temperature conductance transients in the MIS silicon structures fabricated as described before. In the present paper we describe that these transients also appear in the case of the InP samples. These transients occur when we apply a positive pulse bias that drives MIS structures from deep to weak inversion. Subsequently, capture processes take place in which the empty DIGS states trap electrons coming from the conduction band. This process is assisted by tunneling and is time consuming. In consequence, electrons are captured first by states near the interface than by those located farther away in the dielectric bulk. In Figure 3(a) we show several conductance transients at three frequencies for Al/SiN$_x$:H/Si wafers. The most noticeable point is that the transient shape varies significantly with the frequency. For the lowest frequency, the conductance remains constant at the beginning of the transient and it increases at times longer than about 1 second. Conversely, at high frequency we obtained decreasing transients. And at intermediate values the transients have a mixed behavior: first they increase, then reach a maximum value and, after that, they decrease towards a stationary value. Similar trends are also obtained in the case of Al/SiN$_x$:H/InP samples (Fig.3(b)). Conductance transients can be explained in terms of the spatial distribution of the interface states [12]. When the states are spatially distributed the electron emission and capture processes involve both thermal excitation and tunneling. Figure 4(a) is a schematic of the insulator-semiconductor (IS) structure during a transient. E_F and E_F' are the relative locations of the Fermi level with respect to the interface states before and after the pulse. In this plot we also draw lines corresponding to the states having the same electron emission rate [8].

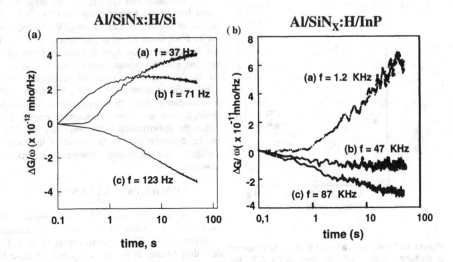

Figure 3.- Room Temperature Conductance Transients of Al/SiN$_x$:H/Si (a) and Al/SiN$_x$:H/InP (b) MIS structures at several frequencies

The point $x_c(t)$ is the distance covered by the front of tunneling electrons during the time t. This distance is given [13] by $x_{cn}(t) = x_{on} \ln(\sigma_{on} v_{thn} n_s t)$, where $x_{on} = h/4\pi(2m_{eff} H_{eff})^{1/2}$ is the tunneling decay length, σ_{on} is the electron capture cross section value for x=0, v_{thn} is the electron thermal velocity and n_s is the free electron density at the interface. In order to make some estimate of $x_{cn}(t)$ we used the following values of the different parameters: DiMaria and Arnett [14] reported a height of 1.9 eV for the silicon-nitride/silicon energy barrier, H_{eff}. That gives a value of 1.25 Å for x_{on}. Empirical data for σ_{on} are typically in the range 0.5-5×10^{-14} cm^2. Assuming weak inversion conditions ($n_s \approx 10^{17}$ cm^{-3}) we obtained that the front of tunneling electrons reaches depths of 28.8, 31.7 and 34.3 Å after times of 1, 10 and 100 s, respectively. It is well known that only those traps with emission and capture rates of the same order of magnitude as the frequency have non zero contributions to the conductance [15]. Let us assume in Figure 4(a) an experimental frequency ω_b and that only those states with emission rates in the range $\omega_b \pm \Delta\omega$ have non negligible contribution to the conductance. At the beginning of the transient the front of tunneling electrons is very close to the interface. Therefore, all the states capturing electrons have emission rates very much higher than the frequency and do not produce any change in the conductance of the structure. When $x_c(t)$ reaches the point A, the states with emission rate equal to ω_b and with energy slightly higher than E_F give some contribution to the conductance signal. After that, more and more states are subsequently incorporated. When $x_c(t)$ reaches the point B, the range $\omega_b \pm \Delta\omega$ is completed. In summary, conductance signal increases during the time employed in going from A to B. From then, the states contributing to the conductance at a time t′ are those indicated by the segment PQ. Once $x_c(t)$ reaches the point C the states with emission rates in the measurable range have now energies above the Fermi level, E_F', and remain emptied. Afterwards, no contribution of the interfacial states can be measured and the conductance reaches a stationary value. As the profile of states usually decreases with x, the shape of the conductance transient is like that plotted in Figure 4(b).

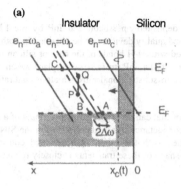

(a)

Insulator   Silicon

(b)

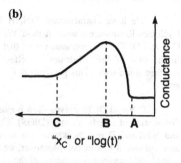

Figure 4.- (a) Schematic band diagram of an I-S interface illustrating the capture of electrons by DIGS continuum states during a conductance transient. (b) General shape of the conductance transient.

Finally, the transient corresponding to the frequency ω_d is similar to that for ω_b but with an additional delay. This model explains our experimental transients of Figure 3. Curves (b) and (c) correspond to the cases of ω_b and ω_c in Figure 4(a). In the case of the lowest frequency (37 Hz) the transient is ever increasing. The 50 s record seems not to be long enough to complete the measurable emission rate range for this frequency because of the fact that the time needed by electrons to reach deeper positions in the insulator increases exponentially with the distance.

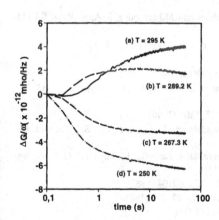

Figure 5.- Conductance Transients of Al/SiN$_x$:H/Si at f = 37 Hz and several temperatures.

In addition, we have made measurements keeping the constant frequency and varying the temperature (Fig.5). We have detected conductance transients at temperatures as low as 200 K. This fact supports the assumption of a tunneling assisted (i.e., temperature independent) capture process. Besides, as temperature decreases the transients are modified in a similar way as when frequency is increased at constant temperature: the lower the temperature is, the faster the transients become. That is easily explained taking into account that emission rate is a function of the temperature: as temperature decreases the emission rates of all interface states decrease exponentially and the equi-emission lines shift approaching the interface and shorter distances have to be covered by the front of tunneling electrons.

CONCLUSIONS

We have characterized SiN_x:H films directly deposited on silicon and InP by the Electron Cyclotron Resonance plasma method. We proved that good quality devices can be obtained in both cases. C-V and DLTS measurements reveal that traps are related with N-H bonds in the case of silicon wafers. In the case of InP substrates, an effective passivation of the InP surface takes place when N-rich conditions are used. This passivation is explained as due to substitutional nitrogen atoms saturating V_p sites.

C-V and DLTS curves show important differences that can be explained with a model of a spatial distribution of interface states (DIGS). The anomalous conductance transients observed in the SiN_X:H/Si and SiN_X:H/InP interfaces are also very well explained with the DIGS approach and confirm the assumptions of this model. Moreover, we show that the shape of these transients is closely related to the spatial and energetic distribution of the interface states.

ACKNOWLEDGMENTS

This work has been partially supported by the Castilla y León Government, under grant VA 35/96.

REFERENCES

1. S.García, J.M. Martín, I. Mártil, M.Fernández and G.González-Díaz, Phil. Mag. **B73**, 487 (1996).
2. K. Semo, S. Hayashi, S. Wickramanayaka and Y. Hatanake, Thin Solid Films **281-282**, 397 (1996).
3. R.I. Hegde, P.J. Tobin, K.G. Reid, B. Maiti and S.A. Ajuria, Appl. Phys. Lett. **66**, 2882 (1995).
4. K. Vaccaro, H.M. Dauplaise, A.Davis, S.M. Spaziani and J.P. Lorenzo, Appl. Phys. Lett. **67**, 527 (1995)
5. S.García, I. Mártil, G.González-Díaz, E.Castán, S.Dueñas and M.Fernández, J. Appl. Phys. **83 (1)**, 332 (1998).
6. S.García, I. Mártil, G.González-Díaz, E.Castán, S.Dueñas and M.Fernández, . J. Appl. Phys. **83 (1)**, 600 (1998)
7. G. Lucovsky, Z. Ying and D.R. Lee, J. Vac. Sci. Technol. **B13**, 1613 (1995)
8. T. Hashizume, H.Hasegawa, R. Riemenscheneider and H.L. Hartnagel, Jpn. J. Appl. Phys. **33**, 727 (1994)
9. P.J.M. Permiter and J.G. Swanson, J. Electron. Mater. **25**, 1506 (1996).
10. H. Hasegawa, M. Akazawa, H. Ishii, A. Uraie, K. Iwadate and H. Ohno, J. Vac. Sci. Technol. **B8**, 867 (1990).
11. S.Dueñas, R.Peláez, E.Castán, R. Pinacho, L. Quintanilla, J. Barbolla, I. Mártil and G. González-Díaz, Appl. Phys. Lett. **71(6)**, 826-8, (1997).
12. P. van Staa, H. Rombach, and R. Kassing, J. Appl. Phys. **54**, 40104 (1983).
13. L. He, H. Hasegawa, T. Sawada, and H.Ohno, J. Appl. Phys. **27**, 512 (1988).
14. D.J. DiMaria and P.C. Arnett, Appl. Phys. Lett. **26**, 711 (1975).
15. J. Barbolla, S. Dueñas, and L. Bailón, Solid-State Electron. **35**, 285 (1992).

THE INFLUENCE OF IONIC ACTIVITY ON THE ELECTRICAL PROPERTIES OF PECVD (TEOS) SILICON DIOXIDE

A. ROMANELLI CARDOSO*, M.L. PEREIRA da SILVA* , J.J. SANTIAGO-AVILES**
*LSI/PEE/EPUSP, University of Sao Paulo, Sao Paulo, Brazil
**Dept. of EE, University of Pennsylvania, Philadelphia, PA 19104

ABSTRACT

A Plasma enhanced CVD system was modified to place a potential screen between a plasma and a silicon wafer. Silicon dioxide from an organometallic precursor (TEOS) was deposited onto silicon wafers. In this way, any alteration of the screen potential resulted in a modified ion speed, or the removal of the ion flux incident on the wafer. The oxides films produced in this manner were analyzed by Raman spectroscopy, and both C-V and I-V techniques. The characterization results suggest an important role for Oxygen ion bombardment on the TEOS oxidation process, such as the removal of carbon compounds from the film as TEOS oxidizes. We have evidence that ion bombardment decreases the dielectric constant and increases the hysteresis of the SiO_x films. A qualitative model to explain the experimental results was developed.

INTRODUCTION

From experimental results one may construct a qualitative model of deposition processes. For depositions enhanced by plasmas, one may like to consider surface reactions as well as those occurring inside the plasma. For the surface case, the adsorption of the reagents and its energetic is of fundamental importance. For the present case, the surface species oxidation seems to be one of the most important factors in determining the quality of the film. The oxidation can be enhanced by increasing the O_2/TEOS ratio, which seems to be the alternative favored for low temperature deposition processes.

Bunchah [1] points out the importance of ion bombardment on surface processes. They can be altered in many ways, such as dissociation of adsorbed molecules, creation of active sites, or the removal of heteroatoms that may influence chemisorption, micro-structure, and physical properties. Hey [2] showed that the etching rate is related to the energy density transferred to the film. As described by Raupp [3], this may suggest a mechanism based simultaneously on ion bombardment and the reactivity of oxygen species.

The importance of ions on thin film deposition by PECVD systems has been confirmed. For example in a microwave plasma, the deposition of Si from $SiCl_4$ follows an ion \ molecule mechanism [4]. The O_2^+ ion strongly influences the quality of a film deposited by electron cyclotron resonance [5]. With SiH_4, Si_nH_m oligomers are formed and are the main species for depositions at pressures below 100 mTorr [6]. For the TEOS system, the reader can refer to the study by Campostrini [7] de Silva [8] and Voronkov [10]. Common in most of these studies is the observation that the positive ions generated by the TEOS decomposition may produce ionized species with three or more silicon atoms, possibly via gas phase polymerization.

EXPERIMENTAL RESULTS

To facilitate the understanding of the physico-chemical phenomena involved in PECVD, in particular ions generated by TEOS decomposition or by oxygen, a metal screen was placed as shown in figure 1, this allowed separation of charged species from neutrals in the deposition process.

The experiments were performed in two different runs, using 10-20 Ω.cm, (100), p-type Si wafers. The common deposition conditions were a pressure of one Torr, RF power of 200 W, wafer temperature of 360°C, electrode distance from substrate of 22 mm, and TEOS and O_2 fluxes of 5 and 300 sccm, respectively.

During the first run (samples A1 to A8), we biased the screen from -125 to 125 V. For the second run (samples B1 to B17) we biased the screen from -500 to 300V and increased the O_2 volumetric flow rate. After deposition, the films were characterized by ellipsometry (for refractive index and thickness), by Raman scattering (to determine homogeneity and carbon content), and by electrical means, such as 1KHz to 10 MHz C-V and I-V. For the electrical measurements, we fabricated MOSCAP structures with 1mm diam. circular aluminum gates.

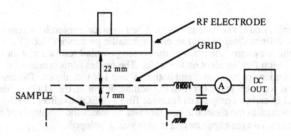

Figure 1. Experimental arrangement utilized for the acceleration or suppression of ionic species in a PECVD system.

The screen current measured by the ammeter (figure 1) never exceeded 0.5 µA in either experimental run. The refractive index seldom showed a value different from 1.44.

With a floating grid (no potential applied) the deposition rate was higher, and the refractive index marginally higher. Under this condition we observed no plasma between grid and sample.

For the first run the deposition rate slightly diminishes with the magnitude of the applied voltage. For the second run one can observe that the deposition rate tends to a unique value when the applied voltage goes to – 500 V (see figure 2). This suggests that physical processes such as sputtering may be important under these conditions.

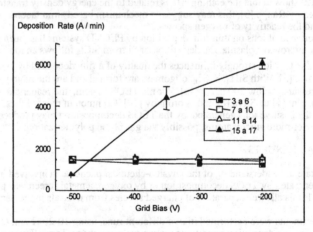

Figure 2. Deposition rate as a function of bias applied to grid. For samples B3-B14.

For samples B3-B6, an increase in voltage resulted in a slight increase in the deposition rate. For this group of samples, the ion flux increment translated into an increase in surface reactions. Here most probable the ions are participating or otherwise enhancing the reaction rate with little or no effect on species removal. For samples B7-B10, where the bombardment is by ions originating from oxygen molecules, the deposition rate remains constant. For samples B11-B17 increasingly negative voltages diminishes the deposition rate. The strong reduction of the

reaction rate for films formed with a high TEOS concentration (samples B15-B17) is due to enhanced sputtering. Raman intensities and line profiles suggest poor film formation.

In general one can infer that for high negative voltages, physical processes (sputtering) dominate. For positive voltages or less negative ones, chemical processes are favored. This was an expected result, but placing the biasing screen in the deposition chamber allowed us to favor one process or the other.

The lack of variation in the deposition rate for voltages between -300 to -500V (samples from B7-B10) suggest an energy efficient process, that is an enhanced condition of available energy per TEOS molecule is reached. Therefore for low TEOS concentration we expect the best oxide films a matter already corroborated by previous studies [11]

The film appearance under the Raman scattering microscope (80 μm spot size) is a heterogeneous surface. The heterogeneity's correspond to a segregated phase rich in carbon on a silicate background. For the first run samples we noticed a great deal of uniformity in their Raman spectra. Some of these samples received an interrupted TEOS flux, and those samples showed a more homogeneous film deposition. This is an indication of the relative importance of solid phase reactions, where a diminution on the TEOS concentration allowed a cleaner film formation (less carbonatious residues). The Raman spectra of the second run samples showed a correlation between the extent of phase segregation and the screen potential. At high negative voltages the oxidation process of the deposited film compounds seem more efficient.

We noticed that increasing the potential is effective up to about -300V for samples B3-B6 but diminishes continuously for samples B11-B14. Consider that the only difference between these groups of samples is that in the second group, only the O_2 ions are accelerated. These are the most efficient ions in promoting oxidation of the film precursor carbonatious compounds.

From the C-V profiles we obtained the flat band voltages, effective charges, hysteresis and dielectric constants. From the I-V profiles, beside the breakdown voltages, we obtained the oxide leakage currents.

For samples in the first run, we noticed a different behavior between the two different grid biasing during deposition (constant biasing, first group, samples A1-A5 and interrupted TEOS and biasing for the second group of first run, samples A6-A8). The C-V profiles for the first group were similar for all samples showing strong room temperature hysteresis and the presence of leakage currents. The electrical parameters showed a tendency to improve with increasing flux of positive ions (increasing negative bias). Extrapolating the experimental dielectric constant as a function of applied voltage, and assuming a linear relationship, one must supply a negative bias of at least -1kV to obtain a value of 4, the dielectric constant typical of silicon dioxide. The electrical characteristics of the samples in the second group showed greater variation. For the samples with the largest impinging O_2 flux, we noticed the best values of electrical parameters. Perhaps this implied the best removal of carbonaceous heterogeneity's.

In the second run, it is interesting to note the high value of the flat band voltage, V_{fb} obtained after bombardment with a relatively small flux of positive ions (i.e. sample B2, V_{grid} = 300V). With the exception of the previous sample, the other noticeable case is that of the sample with the most negative grid bias (sample B17, V_{grid} = -500V). The Raman data shows sample B17 to posses the most heterogeneous surface, but that same sample, on the average do posses the best characteristics as a dielectric film, since it posses the refractive index nearest in value to a typical thermal oxide, the lowest dielectric constant (although still high) and its deposition rate was the lowest (which correlate strongly with film quality [12]). The I-V profiles were converted into I-E (current - electric field) to allow easy comparison between films of different thickness (figure 3). Note that the highest leakage current correspond to the sample with the highest deposition rate. In the second group of the first run one can point to sample A6 which also showed a noticeable leakage current. This sample suffered the lowest intensity of positive ion bombardment, suggesting the importance of this factor in quality film formation.

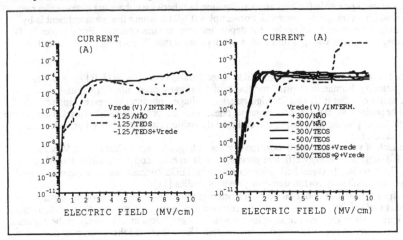

Figure 3. I-E profiles for first run, second group samples (left) and second run samples (right).

CONCLUSIONS

The energetic of solid phase reactions can be altered by two distinct means, by increasing the substrate temperatures or by increasing the ion flux incident on the substrate. The first method may be constrained by the impositions of multi-level metallization schemes, so the study of ion bombardment enhanced film depositions may be of relative importance. In this context, the most efficient process may be the bombardment with reactive species, such as oxygen neutrals or ions.

The plasma in this system, besides providing reactive ions, also provides reactive neutrals. Segregated phases rich in carbonaceous species must be avoided if one desires the best possible film structure and composition, TEOS may provide the ionic and neutral species capable of plasma phase polymerization and therefore, carbon rich compounds from these sources may be incorporated into the solid film.

REFERENCES

1. Bunshah, R.F. Handbook of Deposition for Films and Coatings 2nd Ed. Noyes Pub. New Jersey, 1994.
2. Hey, H.P.W., Shijik, B.G. and Hemmes, D.G. Solid State Technology, 139 (1990).
3. Raupp, G.B., Cale, T.S. and Hey, H.P.W. Chemical Perspectives in Microelectronics Materials 2nd Symp. 495 (1991).
4. Avni, R. Symp.Proc.- 6th.Int Symp Plasma Chem. 2, 522 (1983).
5. Fukuda T. Suzuki, K Mochizuchi, Y. Ohue, M. Momma, N. Sonobe, T. 20th Int Conf. on Solid State Devices and Materials Extended abstracts 65 (1988).
6. Perrin, J. Int. Congress Phenomena Ionized Gases, 54 (1987).
7. Campostrini, R. Carturan,G. Belli, B and Traldy, P. J. of Non-Cryst.Solids 108,143 (1989).
8. Silva, M.L.P, Riveros, J.M. J. Mass Spectrom. 30, 733 (1995).
9. Holtgrave, J. Riehl, K. Abner, D. Haaland, P.D. Chem. Phys. Letters, 548 (1993).
10. Morimoto, N.I. et al. Anais do VII SBMicro (Brazil) 565 (1992).
11. Cardosos, A.R. PhD Thesis, University of Sao Paulo, Brazil (1997)
12. Deshmuckh, C.S. Aydil, E.S. Appl. Phys. Lett. 65, (25) 3185 (1994).
13. D.A.Hackenberg,J.J. Linn,J.H, J. Electrochem.Soc 143 (3) 1079 (1996).

ANALYSIS OF TEOS SILICON DIOXIDE: THE IDENTIFICATION OF CARBONATIOUS CONTAMINANTS

M. L. PEREIRA da SILVA*, A. ROMANELLI CARDOSO* , J.J.SANTIAGO-AVILES**
* LSI/ PEE/ EPUSP, University of Sao Paulo, Sao Paulo, Brazil
** Dept. of EE, University of Pennsylvania, Philadelphia, PA, 19104

ABSTRACT

This work presents the analysis performed on a SiO_2 film deposited from organometallic precursors with the aim of correlating their physico-chemical properties including electrical characteristics with processing variables. The characterization tools used in this study included SEM for film homogeneity; SIMS for the determination of total carbon content; FT-MS and Raman scattering spectroscopy for surface characterization. GC-MS was used to understand the electrochemical reactions taking place while performing I-V characteristics measurements.

The use of these multiple characterization techniques pointed out to deposited films with reasonable deposition characteristics but poor electrical ones. Phase segregated heterogeneity's rich in carbonatious residues influenced the degradation of the electrical characteristics.

INTRODUCTION

The deposition of oxides from organometallic precursors although frequently utilized in microelectronics, the details of the reaction mechanisms are not clearly understood. In this work we are interested in the potential correlation's between physico-chemical parameters, electrical characteristics and processing conditions for the plasma enhanced CVD deposited oxide films utilizing TEOS and O_2 as precursors. Films obtained this way often showed physical and chemical characteristics (refractive index, IR spectra) similar to those of thermally obtained oxides but with poor electrical characteristics. A first examination of these films showed them to be microscopically heterogeneous. The origin of these segregated phases most likely are carbonatious compounds from the incomplete oxidation of the TEOS molecule. Kulisch [1] believe that the processing of TEOS at low temperature should produce a film free of segregated organic residues. Emesh [2] claims that the carbonatious residues should not exceed 1 atomic %, leaving some room for the possibility of carbonatious contaminants. According to Dobkin [3], in a PECVD utilizing TEOS precursors, the resultant ratio of C / Si = 0.01% (about 100 ppm). So the task is clear, to characterize the deposited films and try to identify the carbonatious residues.

EXPERIMENTAL RESULTS

The silicon dioxide films were deposited from the reaction of $Si(TEO)_4 + O_2$ on a home built multichambered PECVD system. The silicon substrates were (100), p-type, 10-20 Ω-cm , and the deposition parameters, pressure = 1.5 Torr, RF power = 200 W, temperature of 360ºC , electrode - substrate distance of 1.5 cm , and volumetric flow rates of O_2 and TEOS of 100 and 40 sccm respectively. The initial characterization showed an oxide thickness of 0.125 microns, an ellipsometrically obtained index of refraction with a value of 1.434, and a IR spectra with the characteristic bands for Si-O, and Si - OH stretching modes. No evidence of carbon contamination within the detectivity limits of the instrument.

Previous Experiments

While attempting to perform C-V and I-V measurements on MOSCAP structures using the deposited oxide as the dielectric layer we noticed the formation of bubbles under the aluminum gates upon biasing the MOSCAP. We wanted to be certain that the gas originating from the biasing was the result of a chemical reaction we performed some simple experiments. Detail description of the I-V characteristics are presented in a companion article [4].
1.   By applying positive voltage steps to the gates we could accelerate the formation of the gas bubbles, suggesting a field dependence for the gas release.

2. When we ramped the voltage, less gas was released, perhaps an indication of enhanced diffusion.

3. We never observed molten aluminum, so local heating and metal evaporation was not the origin of the gas.

4. A sample was heated to 200°C for 15 min to outgass, later, it was subjected to a voltage bias and a bubble was formed, eliminating the possibility of desorption as the origin of the gas.

5. The moderate frequency C-V characteristics show a time dependence for the appearance of the first bubble (see figure 1).

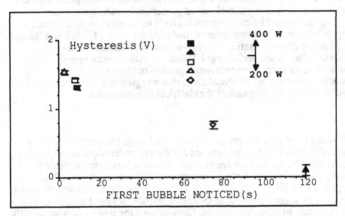

Figure 1. Extent of Hysteresis vs time for occurrence of the first noticeable bubble on the Al gate.

The above facts suggest an electrochemical reaction as the possible origin of the bubbles and the gas causing their formation. At this time we decided to divide the characterization effort in two parts, one effort to elucidate the nature of the gas phase, and a second one for the solid phase.

Analysis of the Solid Phase

We utilized SEM pictures to seek evidence of heterogeneity's which may spatially correlate with the formation of bubbles. No evidence was found to the magnification limit of the instrument.

We decided next on SIMS for its low detection limit, we used a beam of O_2^+, at 13 KeV on a sample with a 150 nm Al gate. The mass spectra was mostly what we expected, except for a high value of C^+ (m/z=12) which points to a carbon concentration superior to ten times (200 ppm) what is expected in a thermal oxide, as shown in figure 2. A calibrated SIMS run, yield a more reasonable amount of carbon (0.0098% or about 100 ppm as experienced by Dobkin [3]). Since Raman scattering spectrometry yield a detection limit superior to IR, we decided to utilize a Raman microscope to look at the surface. The Raman data show that the samples surface are not homogenous, presenting segregated phases with diameters between 1-2 μm randomly scattered through the film surface. We noticed that the poorer the electrical characteristics the higher the density of those heterogeneity's. When looking at the heterogeneity's, noticeable bands at 1300 and 1580 cm^{-1}, and associated with C=C (as in graphite [5]) appear prominently (see figure3).

Analysis of the Gaseous Phase

     For the analysis of the gaseous evolution from the MOSCAP structures we used FT/MS and gas chromatography combined with mass spectrometry (GC / MS). To retain the gas in the bubbles for subsequent analysis, we biased the MOSCAP with a moderate electric field (5 MV/cm) for a short time. This way we managed to create bubbles without damaging the gates, and therefore trapping the gas. Once in the instrument chamber, we applied a pulse from a laser, perforating the gates and liberating the gas. After all these efforts in elucidating the nature of the gas, only a small band attributed to ethylene ($C_2H_3^+$) was detected and this at the detection limit of the instrument. At this time we decided to explore if some of the evolved gas residues remained adsorbed in the solid. Raman scattering was used again confirming the existence of the C=C graphitic bonded carbon.

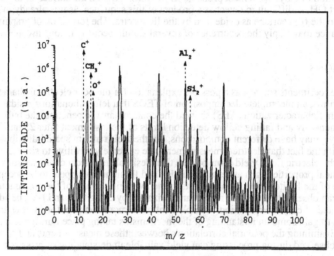

Figure 2. SIMS spectra of the TEOS oxide.

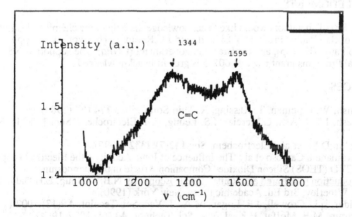

Figure 3 . Raman scattering spectra of a typical heterogeneity, showing the bands associated with "graphitic" carbon.

99

The Electrochemical Reaction

Using the Raman scattering results as a lead, we thought that the most likely electrochemical reaction occurring must be the oxidation - reduction of the TEOS fractions on the surface and in the bulk of the film. The CG results of elylene molecules in the gas is interesting, as the formation of C_2H_4 has been proposed [6,7,8] to explain the pyrolysis of TEOS over a silica surface. The proposed reactions are:

$$(SiO)Si(TEO)_3 \text{-----}> (SiO)_2Si\,(TEO)_2 + C_2H_4$$
$$(SiO)_2Si(TEO)_2 \text{------}>(SiO)_2Si(TEO)(OH) + C_2H_4$$
$$(SiO)_2Si(TEO)(OH) \text{-----}> (SiO)_2Si(OH)_2 + C_2H_4$$

So the formation of ethylene is most probable as we have TEO on the surface of the oxide. The other product OH, is difficult to detect as a product of this sequence, as it is already in great abundance on the film surface as evidenced by the IR spectra. The formation of graphitic nuclei in some abundance may imply the occurrence of several simultaneous reactions involving the TEOS fractions.

CONCLUSIONS

The experiments suggest as a possible explanation for the poor electrical characteristics of these oxide films, an incomplete decomposition of TEOS that left carbonatious residues on the film surface. Further characterization (AES) showed the contaminants to penetrate the bulk decreasing almost exponentially and falling bellow detection limits of the instrument after 25 nm. These heterogeneity's may have different compositions, but the their random spatial distribution, combined with the fact that electrical measurements are extremely sensitive to singularities that concentrate the electric field, yield films of less than desirable electrical characteristics. The fact that the physical properties such as refractive index and IR signature appear to be acceptable is a consequence of the limited instrument resolution and averaging nature of these characteristics.

We can observe that the electrical characterization by using C-V and I-V , besides inexpensive and of easy interpretation, complement other more complex forms of characterization and allow us to gain some understanding of the mechanisms taking place during the oxide film deposition. Examining the potential correlation's between these measurements and simultaneous phenomena observed during processing can yield valuable understanding of process control and optimization.

ACKNOWLEDGMENTS

The Brazilian authors would like to acknowledge the help of the chemistry institute of the University of Sao Paulo (Profs. H.V.Linnert, and J.C.Rubim). This work was supported by the state of Sao Paulo (Fapesp), and the Brazilian government (CNPq). The support of NSF International Programs grant DMI 94 00775 is gratefully acknowledged.

REFERENCES

1. Kulish, W. Lippman, T. Kassing, R. Thin Solid Films **174** (57 (1989).
2. Emesh, I.T. D'Asti, G.Mercier, J.S. Leung, P. J. Electrochem. Soc. **136**(11), 3404 (1989).
3. Dobkin,D.M. et al J.Electrochem. Soc.**142**(7),2332 (1995).
4. A. Romanelli Cardoso et al. "The Influence of Ionic Activity o the Electrical Properties of PECVD (TEOS) Silicon Dioxide" Companion Article in this symposium.
5. Introduction to Infrared and Raman Spectroscopy by N.B. Colthup, L.H.Daly, and S.E.Wiberley, 3rd Ed. Academic Press, New York (1990).
6. Tedder, L.L. Crowell, J.E. Logan, M.A. J. Vac. Sci. Techlol. **A9**(13),1002 (1991).
7. Bartram, M.E. Moffat, H.K. J. Vac. Sci. Technol. **A12**(4),1027(1994).
8. Tedder, L.L. Lu, G. and Crowell, J.E. J. Appl. Phys. **69**,7037 (1991).

ELECTRICAL PROPERTIES OF INTEGRATED Ta$_2$O$_5$ METAL-INSULATOR-METAL CAPACITORS

B. C. MARTIN*, C. BASCERI†, S. K. STREIFFER†, and A. I. KINGON†

* Materials Research and Strategic Technologies, Semiconductor Products Sector, Motorola, Mesa, Arizona 85202.
†Department of Materials Science and Engineering, North Carolina State University, Raleigh NC 27695-7907.

ABSTRACT

We demonstrate the feasibility of a Ta$_2$O$_5$-based metal-insulator-metal (MIM) capacitor module which is integrated into the backend-of-the-line of a 0.5 μm CMOS process flow. The demonstration utilizes 6-inch wafers, sputtered Ta$_2$O$_5$ films with TaN and TiN electrodes, reactive ion etch (RIE) processes for defining the capacitor stack, and a metallization scheme which uses hot Al(Cu) and W plugs and a standard forming gas anneal (N$_2$-5% H$_2$). Acceptable electrical properties have been achieved within these processing constraints for the integrated MIM capacitor module, including specific capacitance (C/A) and leakage current density (J) of ≥ 5 fF/μm^2 and ≤ 1 pA/pF-V, respectively. In addition, we compare the fundamental properties of the Ta$_2$O$_5$ dielectric with literature reports and point out that leakage mechanisms must be analyzed with care due to significant dielectric relaxation in the films under certain processing/measurement conditions. For the baseline integrated films, we confirm that leakage is most consistent with a Poole-Frenkel mechanism.

INTRODUCTION

Increasing semiconductor device performance requirements make it desirable to integrate capacitors, such as bypass capacitors, directly on-chip. The value of the on-chip silicon real estate makes it important to minimize the capacitor area. As the thickness of SiO$_2$ or oxide-nitride-oxide (ONO) reaches an effective minimum imposed by charge tunneling and reliability constraints, it becomes necessary to consider a dielectric with a larger permittivity. Ta$_2$O$_5$ is being considered for these on-chip applications, as well as for DRAMs. While Ta$_2$O$_5$ displays a lower permittivity than alternatives such as (Ba,Sr)TiO$_3$ (BST), it has the advantage of having only two components, and therefore being significantly simpler to deposit.

A development program was undertaken to demonstrate the feasibility of the integration of a Ta$_2$O$_5$ MIM capacitor module in the backend of an existing 0.5μm CMOS process flow. The final process is described below, and it is the first such description to our knowledge. We then compare the properties obtained from the capacitors to the literature. In doing so we have emphasized the issues of leakage currents and time dependent polarization. We have shown previously for the BST system that it is necessary to take into account the dielectric relaxation, which corresponds to a time dependent polarization over a large time domain, when one considers the capacitor performance. In particular, the time dependent polarization can result in a voltage drop under open circuit conditions after pulse charging [1,2]. Similarly, we have shown that these polarization currents can be mistaken for leakage currents when undertaking an analysis of leakage mechanisms. In this study we have therefore investigated both the polarization and leakage behavior for the Ta$_2$O$_5$ capacitors. While there have been two brief reports of "transient currents" in Ta$_2$O$_5$ films [3,4], we suspect that the origin of the behavior has not been correctly assigned.

FABRICATION AND INTEGRATION OF THE CAPACITOR MODULE

A. Capacitor structure and deposition methods

The typical MIM stack capacitor employed in our studies comprised 2000 Å TaN bottom electrode, 400 Å Ta_2O_5 dielectric, and 2000 Å TiN top electrode. The thin amorphous Ta_2O_5 films were reactively sputter deposited from a pure tantalum target at a rate of 1.7 Å/s in a mixture of O_2 and Ar to a nominal thickness of 400 Å. The ratio of O_2:Ar in the gas mixture was carefully controlled to produce stoichiometric Ta_2O_5 films, for it has been reported in the literature and observed in practice that sub-stoichiometric films have higher leakage currents. The TiN and TaN electrode layers were sputter deposited in the same sputter deposition system from Ti and Ta targets, respectively, at 300 °C in a mixture of 50:50 Ar:N_2. Both TiN and TaN have a columnar microstructure.

B. Integration in a 0.5 µm CMOS process flow

The planar capacitor module was integrated in an existing 0.5µm CMOS process flow which incorporated W plugs and hot Al(Cu) metallization. Figure 1 is a schematic of the MIM module integrated in a 0.5µm short flow between Metal 1 and Metal 2. Blanket layers of the bottom electrode, dielectric, and top electrode were sputter deposited onto thermally oxidized silicon substrates. Two additional masking layers defined the capacitor: the TiN / Ta_2O_5 was etched in one step, followed by a TaN pattern etch. The capacitor stack was then covered with a plasma-enhanced chemical vapor deposited tetraethylorthosilicate (PETEOS) dielectric. The existing 0.5µm process flow defined the contact window, contact metal, and passivation windows. Via photo and etch defined the contact openings and W plugs. Tungsten deposition and Al(Cu) metallization provided contacts to the top and bottom electrodes. All wafers were annealed in forming gas (N_2-5% H_2) at 390 °C for 30 minutes prior to automated Keithley and hand probing. The process steps critical to the successful integration of the MIM capacitor module were identified to be the deposition and etch processes.

Integrated Ta2O5 MIM Capacitor

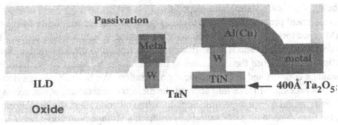

Figure 1: Schematic of a planar MIM capacitor module integrated in a 0.5 µm CMOS flow with W plugs and hot Al(Cu) metallization.

Integration of the capacitor module required 2 masking layers and the following 3 RIE steps, all using standard RIE equipment and chemistries:

1. Definition of the TiN top electrode and Ta_2O_5 dielectric using CF_4-Cl_2 chemistry: The etch was visually endpointed and stopped on the TaN (a thin residual layer of Ta_2O_5 may remain). A typical vertical etch profile is shown in Figure 2.
2. Definition of the TaN bottom electrode using CF_4-Cl_2 chemistry: the etch stopped on the ILD1 PETEOS. The typical sidewall profile was vertical.
3. Anisotropic via etch: Any residual Ta_2O_5 from Step 1 would be removed from the TaN bottom electrode during this step.

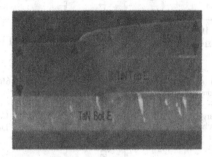

Figure 2: Cross-sectional SEM micrograph of the MIM capacitor edge after the RIE step to define the top electrode and dielectric layer.

ELECTRICAL CHARACTERIZATION

DC currents were measured between room temperature and 125°C using a Keithley 617 programmable electrometer with a built-in voltage source. A heating block was used to vary the temperature of the samples. Capacitance-voltage analyses were performed at 1 kHz using an HP4192A impedance analyzer. A value of 0.1 V was chosen for the ac oscillation level, and it was verified that this was not so large as to effect the measurements.

RESULTS AND DISCUSSION

Figure 3 (a-d) shows current-time curves at -3.5V for a series of films of varying Ta_2O_5 thickness. If one considers first the 1600 Å Ta_2O_5 capacitor, it can be seen that the charge and discharge curves are almost superimposed, especially at shorter times. This is direct confirmation that the observed currents correspond to polarization, rather than leakage, currents [1,2]. As one moves to thinner dielectrics at the same applied voltage, and thus to larger applied fields, it is clear that the charging current densities are considerably larger than the discharge, especially at longer times. The implication is simply that the polarization currents are only weakly (near linearly) field dependent, while the leakage currents increase very strongly as one increases the applied field, directly analogous to the BST case [5]. Furthermore, as in the case of BST capacitors, a major

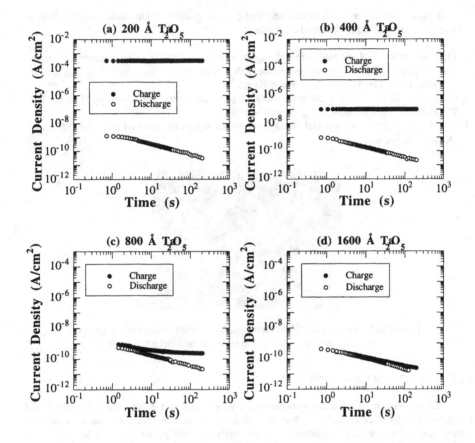

Figure 3: Current density versus time curves at 25°C for capacitor modules with four different dielectric thicknesses. (Data shown are all for negative 3.5 V applied to the top electrode, but the currents under positive 3.5 V are virtually identical).

cause of any voltage drop of the capacitor under open circuit conditions is the time dependent polarization, in addition to the leakage currents [1,2].

There are two previous reports of such "transient" currents in Ta_2O_5 films in the literature. Sundaram *et al.* observed the phenomenon in MIS, but not MIM structures [3]. The transient currents observed in their MIS structure were explained by interfacial charging at the Ta_2O_5/SiO_2 interface. This behavior was reported as a non-intrinsic dielectric relaxation of Ta_2O_5 films. Pignolet *et al.* [4] did not consider dielectric relaxation as an explanation. Short time currents were attributed to electrode polarization. In neither study were the defining measurements undertaken, namely the charge-discharge behavior.

The magnitudes of the currents should also be noted. The implications are that at 3.5V the charging current densities are acceptable for all films except the 200 Å sample (Fig. 4.a-b).

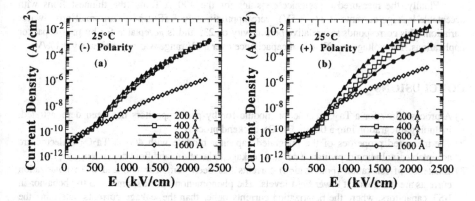

Figure 4: Current - electric field characteristics of different film thicknesses at (a) negative and (b) positive voltage polarities.

However, the actual operating fields will likely be considerably lower than the measured condition for the 200 Å film. The current densities under negative voltage polarity are similar for different film thicknesses, except for the 1600 Å film. This may indicate that thicker films are microstructurally different from the thinner films. Under positive voltage polarity, the current densities exhibit a thickness dependence which may arise from the differences in the electrode/film interfaces for the different film thicknesses, and from nonuniformities in the film thickness direction. More work is needed to establish the precise cause.

A number of research groups have studied the leakage behavior of Ta_2O_5 films, but the conduction mechanism in the field range of interest for many integrated circuit applications is still under debate [6-14]. This can be attributed to large variations in defects/trap densities generated in different processing techniques. Also, in many cases, only the slope of the field dependent leakage currents have been used to distinguish the possible conduction mechanisms. However, it has been shown that the currents in Ta_2O_5 films with relatively few traps may have similar field-dependent slopes when analyzed in terms of both the Schottky emission and Poole-Frenkel effects; this raises concerns about the utility of such analyses [10]. The temperature dependence of the leakage currents must also be investigated, since the parameters necessary for describing the leakage behavior can be obtained from such measurements in conjunction with the field dependent data.

Additionally, it is very useful to analyze the effect of different electrodes on the field and temperature dependencies of the leakage behavior [11]. Careful analyses of the true leakage currents, specifically excluding polarization currents, as a function of applied electric field, temperature, film thickness and different electrode materials indicate that the Poole-Frenkel effect is likely to describe the conduction mechanism occurring in our films. Trap levels were calculated to be around 1.2 eV. This is consistent with zero-bias thermally stimulated current measurements on amorphous Ta_2O_5 films by Devine et al. [12], although other workers have deduced traps to be closer to the conduction band. We have also analyzed the leakage behavior in terms of a Schottky barrier model. However, the calculated barrier heights do not show a clear polarity dependence with the nitride electrodes, suggesting that a bulk-limited conduction mechanism (Poole-Frenkel effect) is more likely. In the case of Pt electrodes where the work function is larger, Schottky barrier heights do not show a correlation with metal work function.

Finally, the measured capacitance densities for the 400 Å films (the thinnest films with acceptable leakage currents at +/− 3.5V) were approximately 5 fF/μm^2, with small cross-wafer variation. This corresponds to a relative permittivity of 23, and is acceptable for by-pass capacitor applications. The voltage coefficient of capacitance over a 5V range was approximately 0.1%/V.

CONCLUSIONS

1) A process flow for a Ta$_2$O$_5$ capacitor module for by-pass capacitors has been demonstrated, including integration into a 0.5μm CMOS backend process.
2) The measured properties of the integrated capacitor module with 400 Å Ta$_2$O$_5$ dielectric are acceptable, including capacitance density, leakage, relaxation, and field dependence.
3) Existence of relaxation (polarization) currents has been demonstrated, and these polarization currents are dominant at lower field levels. The phenomenon appears similar to the behavior in BST capacitors, where the polarization currents rather than the leakage currents determine the charge storage performance under normal operating conditions.
4) Careful analyses of the true leakage currents, specifically excluding polarization currents, indicate that Poole-Frenkel effect is likely to describe the conduction mechanism in our films.

REFERENCES

[1]. M. Schumacher, G. W. Dietz, and R. Waser, Integr. Ferroelectr. **10**, 231 (1995); G. Dietz and R. Waser, Integr. Ferroelectr. **9**, 317 (1995).
[2] S. K. Streiffer, C. Basceri, A. I. Kingon, S. Lipa, S. Bilodeau, R. Carl, and P. C. van Buskirk, Mat. Res. Symp. Proc. **415**, 219 (1996).
[3] K. Sundaram, W. K. Choi, and C. H. Ling, Thin Solid Films **230**, 95 (1993).
[4] A. Pignolet, G. M. Rao, and S. B. Krupanidhi, Thin Solid Films **258**, 230 (1995).
[5] G. Dietz, M. Schumacher, R. Waser, S. K. Streiffer, C. Basceri, and A. I. Kingon, J. Appl. Phys. **82**, 2359 (1997).
[6] C. A. Mead, Phys. Rev. **128**, 2088 (1962).
[7] H. Seki, Phys. Rev. B **2**, 4877 (1970); S. Seki, T. Unagami, and O. Kogure, J. Electrochem. Soc. **132**, 3054 (1985); S. Seki, T. Unagami, O. Kogure, and B. Tjusiyama, J. Vac. Sci. Technol. A **5**, 1771 (1987).
[8] P. L. Young, J. Appl. Phys. **47**, 235 (1976); P. L. Young, J. Appl. Phys. **47**, 242 (1976).
[9] G. S. Oehrlein and A. Reisman, J. Appl. Phys. **54**, 6502 (1983); G. S. Oehrlein, F. M. d'Heurle, and A. Reisman, J. Appl. Phys. **55**, 3715 (1984).
[10] S. Banerjee, B. Shen, I. Chen, J. Bohlman, G. Brown, and R. Doering, J. Appl. Phys. **65**, 1140 (1989).
[11] H. Matsuhashi and S. Nishikawa, Jpn. J. Appl. Phys. **33**, 1293 (1994).
[12] R. A. B. Devine, L. Vallier, J. L. Autran, P. Paillet, and J. L. Leray, Appl. Phys. Lett. **68**, 1775 (1996).
[13] F. C. Chiu, J. J. Wang, J. Y. Lee, and S. C. Wu, J. Appl. Phys. **81**, 6911 (1997).
[14] W. S. Lau, L. Zhong, A. Lee, C. H. See, T. H. n, N. P. Sandler, and T. C. Chong, Appl. Phys. Lett. **71**, 500 (1997).

EFFECTS OF SC OR TB ADDITION ON THE MICROSTRUCTURES AND RESISTIVITIES OF Al THIN FILMS

SHINJI TAKAYAMA
Hosei University, Dept. of System and Control Engineering, 3-7-2, Kajino-cho, Koganei, Tokyo, 184-8584, Japan

I. ABSTRACT

The resistivities of Al thin films with added Sc or Tb largely decrease at over 350 °C. They reach to 5 –7 $\mu\Omega$cm after annealing at 450 °C, together with the segregation of fine Al_3RE (RE = Sc or Tb) metallic compounds. However, the temperatures at which resistivity starts to decrease largely are much lower for Al-SC alloy films (150 °C) than for Al-Tb ones (250 °C). Furthermore, thermal defects of hillocks or whiskers start to appear on the film surface after annealing at 200 °C and 450 °C for Al-Sc and Al-Tb alloy films, respectively. It was revealed that the further addition of Zr to these binary alloy films largely retards a large decrease of resistivities on annealing and enhances the formation of hillocks or whiskers. On the contrary, the addition of Cu to Al-Tb or Al-Sc films significantly suppresses the formation of thermal defects and shows relatively low resistivities after annealing at 350 °C.

II. INTRODUCTION

To achieve larger and higher-resolution thin-film-transistor liquid-crystal displays (TFT-LCDs), Al alloy thin films with low resistivity have recently received much attention [1 - 6]. It has been reported by the present authors that Al-rare-earth alloy systems in particular show very low resistivities without the formation of hillocks and whiskers, even after annealing at high temperatures (350 °C - 450 °C) [3-6]. In the course of this alloy investigation, we study here in more detail the effects of adding Sc and Tb to Al thin films on the microstructures and resistivities after annealing. Furthermore, the effects of addition of the third elements, Zr and Cu to Al-Sc or Al-Tb binary alloy films were also investigated and discussed in terms of a chemical interaction between constituent elements.

III. EXPERIMENT

Al-rare-earth alloy films about 400 - 600 nm thick were deposited on a 7059 glass substrate by using a DC magnetron sputtering apparatus at an Ar pressure of 0.7 Pa. The composite target consisting of 100 mm Al disks (4N purity) and 5 x 5 x 1 mm^3 rare-earth-metal element chips (3N purity) was used. The film composition (atomic %) was determined by inductively coupled plasma (ICP) spectroscopy. The film's electrical resistance was measured by using a conventional four-point probe at room temperature. The film structures were examined by using a X-ray diffractometer operated at 50 kV and 200 mA, and a transmission electron microscope (TEM) operated at 400 kV.

IV. RESULTS AND DISCUSSION

X-ray diffraction measurements showed that in all the alloy systems studied here, the growth of a highly oriented Al (111) plane was strongly suppressed by adding only a small amount of the Sc or Tb to pure Al. However, the diffraction peaks continued to increase or decrease with a further addition of Tb or Sc, respectively. This is shown representatively in Fig. 1 for Al-Tb alloy samples. The above facts indicate that thin film structures tend to lose their initial highly oriented crystal structure and to acquire a more randomly oriented polycrystalline structure with the addition of present foreign atoms to Al. Though the solid solubility of these rare-earth metal elements is very low (almost negligible) at a room

107

temperature, the noticeable diffraction peaks associated with Al alloy metallic compounds are barely observable, as the figures show. TEM observation revealed that these as-made Al- rare-earth alloy films (containing about 3 to 4 at% of added elements) have very fine grains about 100 nm in diameter (see Figs. 2(a) and 2(b)). Note that the feature of the observed grains of both the as-made alloy films is relatively clear,

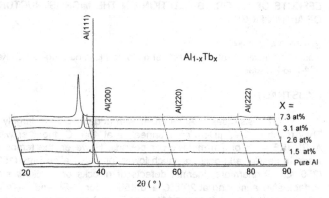

Fig. 1 X-ray diffraction profiles of Al-Tb alloy thin films as a function of the content of added Tb elements

unlike other Al-rare-earth alloy thin films (Al-RE, RE = Y,La,Pr,Nd,Sm,Gd and Dy)(3-6). Selective area diffraction analysis of the samples in Fig. 2 indicates that their structures consist mainly of highly supersaturated solid solutions of the Al phase. The corresponding energy dispersive X-ray analysis (EDX) of the Al-Tb samples in Fig. 2 revealed that the added Tb element was homogeneously distributed in the Al matrix, though this is not presented here.

All of the as-made Al-rare-earth alloy film samples in Fig. 1 were isochronally annealed in a vacuum (less than 10^{-4} Pa.) at various temperatures for 30 min. The changes in their resistivities are shown as functions of both the annealing temperature and the composition of each added element in Fig. 3. The results for pure Al films are also included for comparison. The solid and dotted lines in the figure represent the curves obtained for the Al-Tb and Al-Sc alloy systems, respectively. The at% values shown on the right of the figure indicate the amounts of added elements to Al. Note that the general features of the resistivity change in Al-Sc alloy system are quite different from those of Al-Tb alloys. The resistivities of former alloy samples decrease significantly at relatively low temperature 150 °C, and then more sharply at 200 °C. On the other hand, the latter Al-Tb alloys start

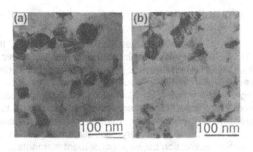

Fig. 2 Bright field images of as-deposited (a) $Al_{97}Sc_3$ and (b) $Al_{96.9}Tb_{3.1}$ alloy thin films.

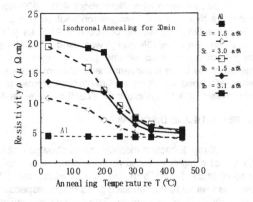

Fig. 3 Change in resistivity ρ of Al-X alloy films (X=Sc or Tb) as a function of the isochronal annealing (for 30 min) and composition.

to decrease sharply at 250 °C, and then approach the values of pure Al above 350 °C. Note here that the resistivity change of pure Al thin films on isochronal annealing is negligibly small (see the bottom part of Fig. 3). This fact most likely implies that the resistivity changes due to grain growth, internal stress relief, and annihilation of point or line defects are expected to be small in the present sputtered Al films within the accuracy of the present measurement. Thus, we can conclude that the large resistivity change in Al alloy films studied here results mainly from (1) removal of impurities from the Al matrix and (2) precipitation of metallic compounds. A detailed discussion of these subjects was made in reference 5.

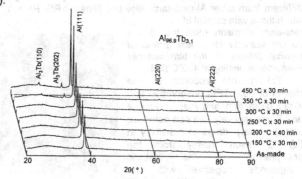

Fig. 4   X-ray diffraction profiles of the $Al_{96.9}Tb_{3.1}$ alloy thin films in Fig. 3

The representative X-ray diffraction profiles of $Al_{96.9}Tb_{3.1}$ are shown in Fig. 4. The salient diffraction peaks associated with the segregation of metallic compounds start to appear after annealing above 300 °C. This is also true for $Al_{97}Sc_3$ samples though not presented here. Some of these diffraction peaks were indexed, from available data, as belonging to an Al_3Tb metallic compound, as indicated in the figure.

TEM observations of samples annealed at 350 °C for 80 minutes are shown in Figs. 5(a) and 5(b) for $Al_{97}Sc_3$ and $Al_{96.9}Tb_{3.1}$ alloy thin films, respectively. Note that in comparison with the results of the as-made samples in Fig.2, the grain size of both the annealed samples are slightly larger, but not as much as those of other Al-rare-earth alloy films reported earlier (3-6). Both selective area diffraction and EDX analyses indicated that most of the relatively dark image

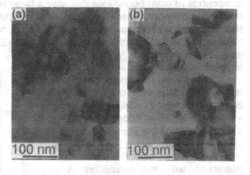

Fig.5   Bright field images and diffraction patterns of (a) $Al_{97}Sc_3$ and (b) $Al_{96.9}Tb_{3.1}$ thin films annealed at 350 °C for 80 min.

grains in these photos were more likely Al_3RE (RE = Sc or Tb) metallic compounds, while relatively large and bright gray image grains corresponded to pure Al matrix. However, EDX analysis showed that the added Tb elements after annealing were not markedly segregated in the Al matrix, as they were in the case of other annealed Al-rare-earth alloy films (3-6). This means that the removal of impurities from the Al matrix after annealing is more likely less in both Al-Sc and Al-Tb alloy films than in the other Al-RE (RE = Y, La, Pr, Nd, Sm, Gd and Dy) ones.

Figure 6 shows representative observations, obtained by using a polarized optical microscope, of the above (a) $Al_{97}Sc_3$ and (b) $Al_{96.9}Tb_{3.1}$ film surfaces after isochronal annealing at 350 °C and 450 °C, respectively in Fig. 3. The white spots in the photos indicate hillocks or whiskers. In the case of Al-Sc alloy samples, these hillocks and whiskers started to appear at 200°C, whereas in Al-Tb alloys, these were observed at 450°C. These facts indicate that the addition of Tb shows much higher resistance against growth of hillocks on annealing than that of Sc. However, this tendency for the growth of thermal defects is quite different from other Al-rare-earth alloy thin films (Al-RE, RE = Y, La,Pr, Nd, Sm, Gd and Dy), in which those with content of rare-earth elements more than 3 at% do not scarcely show any growth of thermal defects on the film surfaces even after annealing at 450°C (3-6).

Taking into account of the results of both TEM observation and X-ray diffraction analyses in Figs 4 and 5, respectively, we can conclude that the sharp drops in resistivity in Fig. 3 result mainly from the precipitation of the metallic compounds, together with the removal of impurities from the Al matrix, as explained above. However, the precipitation of metallic compounds at grain boundaries is less clear in annealed Al-Sc alloys

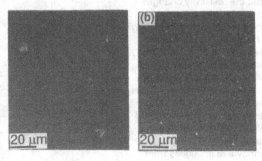

Fig. 6 Polarized optical micrographs of Al alloy thin film surfaces annealed at 350 °C for 80 min: (a) $Al_{97}Sc_3$ and (b) $Al_{96.9}Tb_{3.1}$.

than in Al-Tb ones as seen in Fig. 5. Thus the growth rate of metallic compounds during annealing is expected to be slower in Al-Sc alloy films than in Al-Tb alloy systems. Since the growth of thermal defects mainly results from grain boundary diffusion, precipitation of metallic compounds at grain boundaries can play a large role in preventing grain boundary diffusion, resulting in strong suppression of growth of hillocks and whiskers on the film surfaces. Thus both the slow precipitation rate of metallic compound at grain boundaries and slow rate of removal of solutes from the Al matrix , like in Al-Sc alloy systems as mentioned above, can lead to a lot of chances to form hillocks and whiskers on the film surfaces.

To investigate in more detail the different effects of the growth of thermal defects between added alloying elements, we further added Zr or Cu as a third element to the above binary alloy systems. Note that Zr or Cu shows repulsive or attractive chemical interaction with the constituent rare-earth-elements (such as Sc and Tb) in Al alloy films, respectively. Fig. 7 representatively shows a change in resistivity of the Al-Tb-TM (TM = Cu or Zr) alloy films as a function of temperature when they were isochronally annealed in a vacuum at each temperatures for 30 min. Note that

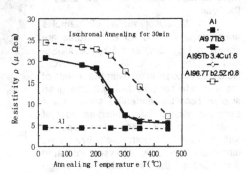

Fig. 7 Change in resistivity ρ of Al-Tb-TM (TM = Cu or Zr) alloy films as a function of the isochronal annealing (for 30 min).

in comparison with the binary Al-Tb alloys, Zr addition to Al-Tb alloy films largely increase resistivity over the whole experimental temperature range, and that the resistivity does not sharply drop above 350 °C. Conversely, in the ternary Al-Tb-Cu alloy, the general trend of resistivity change is the same as that of the binary Al-Tb alloy films. They largely decrease above 250 °C and then approach the value for pure Al above 350 °C. Furthermore, the striking difference between these two transition metal additions is that the Cu addition to Al-Tb alloys can largely suppress the growth of hillock or whiskers even at 450 °C, whereas the Zr addition strongly induce these thermal defects above 350 °C annealing. This is clearly seen in Fig.8. The figure shows representative optical microscopic observations, obtained by using a polarized optical microscope, of the above (a) $Al_{96.7}Tb_{2.5}Zr_{0.8}$ and (b) $Al_{95}Tb_{3.4}Cu_{1.6}$ after annealing at 350 °C. The dark spots in the photos indicate hillocks or whiskers grown on the film surfaces. These tendencies of resistivity change and formation of hillocks on annealing are also true for

Al-Sc-TM (TM = Cu or Zr) though this is not presented here.

The representative X-ray diffraction profiles of (a) $Al_{96.7}Tb_{2.5}Zr_{0.8}$ and (b) $Al_{95}Tb_{3.4}Cu_{1.6}$ samples in Fig. 7 are shown in Figs. 9(a) and 9(b), respectively. The salient diffraction peaks associated with the precipitation of metallic compounds start to appear at 250 °C for $Al_{95}Tb_{3.4}Cu_{1.6}$, and at 300 °C for $Al_{96.7}Tb_{2.5}Zr_{0.8}$. Some of these diffraction peaks were tentatively indexed, from available data, as indicated in the figure. It

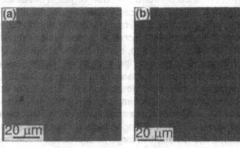

Fig. 8 Polarized optical micrographs of Al alloy thin film surfaces annealed at 350 °C for 80 min. (a)$Al_{96.7}Tb_{2.5}Zr_{0.8}$ and (b) $Al_{95}Tb_{3.4}Cu_{1.6}$.

should be noticed here that in the Al-Tb-Zr samples, only Al-Tb compounds were observed after annealing, while in the Al-Tb-Cu samples, Cu-Tb, and Al-Cu and/or Al-Tb compounds were formed. Now, we recall here that precipitation of metallic compounds in the present alloy films takes place mostly at grain boundaries, and that the impurities of added elements are removed from the Al matrix on annealing, giving rise to decrease of resistivities. These facts more likely indicate that added alloying elements, at first, tend to segregate at grain boundaries and then form metallic compounds with other constituent elements if they have a strong compound-forming tendency with each other.

Segregation of impurities to grain

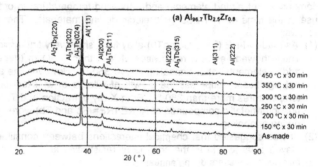

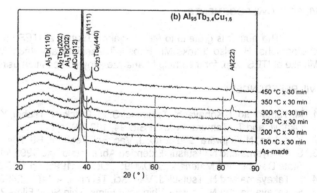

Fig. 9 X-ray diffraction profiles of the (a) $Al_{96.7}Tb_{2.5}Zr_{0.8}$ and (b) $Al_{95}Tb_{3.4}Cu_{1.6}$ alloy thin films in Fig. 7, isochronally annealed up to 450 °C.

111

boundaries in multicomponent alloy has been discussed by M. Guttmann in temper embrittlement steels [7]. He thermodynamically derived general equations of segregation to free surfaces and showed that balance between all the chemical interactions between constituent atoms in alloys determined the tendency of the segregation of impurities to the grain boundaries. It shows that a relative attraction or repulsion between the solutes with respect to the solvent enhance or retard the cosegregation of solute atoms to the grain boundaries, respectively. It should be noticed in the present result of Fig. 7 that Zr which has a repulsive interaction with the constituent rare-earth-elements (such as Sc and Tb) largely increase resistivities and retards the decrease of resistivities on annealing, whereas Cu which has an attractive interaction with them, does not, and shows nearly the same resistivity change with the Al-Tb binary alloy films during annealing. Therefore, by applying Guttamann analysis of surface segregation to the present results in Figs. 7 and 9, the addition of Zr to the Al-Tb binary alloy films causes the desegregation of the Tb or Zr elements to grain boundaries, resulting in remaining most of impurities in Al solvent after annealing. Conversely, the addition of Cu enhances the cosegregation with Tb at grain boundaries, toward the formation of metallic compounds, resulting in large decrease of resistivities during annealing. Thus, it turns out that the Cu addition largely suppresses the formation of hillocks or whiskers at the grain boundaries, whereas the Zr addition does not, as seen in Fig. 9.

V. CONCLUSIONS

Al-RE-TM alloy films (RE = Sc or Tb, TM = Cu or Zr) were studied as functions of the composition of added elements and annealing temperature, in order to assess the potential use of the films as TFT-LCD gate electrode line materials. The results are summarized as follows:
(1) As-made Al-RE (RE = Sc or Tb) alloy films show very fine grain less than 100 nm in size. They showed very low resistivities of less than 6 $\mu\Omega$cm (depending on their content of added elements) after annealing above 350 °C. However, the addition of Sc and Tb does not significantly suppress the growth of hillocks or whiskers on the film surfaces at high temperatures more than 350 °C. Both X-ray diffraction and TEM observation showed that fine intermetallic compounds of Al_3RE (RE= Sc and Tb) were precipitated in the Al matrix during annealing.
(2) It was suggested that chemical interactions between constituent elements in alloy films played a large role to the change of resistivities and the suppression of the growth of hillocks or whiskers during annealing.

VI. ACKNOWLEDGEMENTS

The author is grateful to Mr . Naganori Tsutsui at ITES corp. for his help taking TEM photographs. He also thanks Mr. Hiroshi Takashima at Hitachi Metal Corp. and Mr. Shinji Miyake at ITES corp. for their help to analyze alloy composition using ICP method.

VII. REFERENCES

1. M. Yamamoto, I. Kobayashi, T. Hirose, S. M. Bruck, N. Tsuboi, Y. Mino, M. Okafuji, and T. Tamura, Proc. of SID 94, 142 (1994).
2. Y. K. Lee, N. Fujiwara, and T. Ito, J, Vac. Sci. Technol. B9, 2542 (1991).
3. S. Takayama and N. Tsutsui, Extended Abstracts of the 1995 Int. Conf. on Solid State Devices and Materials (SSDM '95), Osaka, 1995, pp. 318-320.
4. S. Takayama and N. Tsutsui, J. Vac. Sci. Technol. A 14(4), 2499 (1996).
5. S. Takayama and N. Tsutsui, Thin solid Films, Thin Solid Films 289, 289 (1996).
6. S. Takayama and N. Tsutsui, J. Vac. Sci. Technol. B 14(5), 3257 (1996).
7. M. Guttmann, Surface Sci., 53, 213 (1975).

ELECTRICAL PROPERTIES OF NOVEL ANODIC FILMS FORMED IN NONAQUEOUS ELECTROLYTE SOLUTIONS

F. Mizutani*, S. Takeuchi, T. Nishiwaki, N. Sato, and M. Ue
Tsukuba Research Center, Mitsubishi Chemical Corporation,
8-3-1, Chuo, Ami, Inashiki, Ibaraki 300-0397, JAPAN.
* 3803960@cc.m-kagaku.co.jp

ABSTRACT

An aluminum alloy (Al - 1.0% Si - 0.5% Cu) was anodized in nonaqueous electrolyte solutions containing propylene carbonate (PC), ethylene glycol (EG), or γ-butyrolactone (GBL) as a solvent. MIM (metal-insulator-metal) elements using these anodic films were made to evaluate their electrical properties. Leakage current of the anodic films formed in the nonaqueous solutions was lower than those formed in aqueous electrolyte solutions. These novel films are useful as a high insulating film used for a gate insulator in thin film transistor liquid crystal displays (TFT-LCD).

INTRODUCTION

Recently, aluminum metal or alloys covered with barrier-type anodic oxide films have been considered as attractive candidates for metallizaton in micro device applications such as TFT-LCD. The main advantage to using such anodic oxide films is the suppression of hillock formation on aluminum, which is a promising wiring metal due to its low resistivity [1, 2]. Demands for high insulating anodic films are increasing, because leakage current of the anodic films is not sufficiently low.

We have reported anodic oxidation of pure aluminum in nonaqueous electrolyte solutions and characterization of the resultant anodic films [3-6]. In this study, insulating properties of anodic films formed in nonaqueous electrolyte solutions were examined and compared with the results obtained for aqueous counterparts.

EXPERIMENTAL

Substrate

Non-alkaline glass substrates (Corning Inc., #7059) were thoroughly washed with ultra pure water. An aluminum alloy (Al - 1.0% Si - 0.5% Cu) was deposited onto the glass substrates by a DC magnetron sputtering machine (ULVAC, MLX-3000). Residual gas pressure was below 2.0×10^{-5} Pa and the substrates were kept at 100 ℃ during sputtering. Thickness of the films was estimated to be about 400 nm from changes in weight.

Mat. Res. Soc. Symp. Proc. Vol. 500 © 1998 Materials Research Society

Film formation

Sputtered aluminum alloy films were anodized by a dc power source (Keithley, 2400 Digital SourceMeter). Anodization was carried out at a constant current (current density: 0.1 mA cm^{-2}) up to 100 V, followed by a constant voltage (voltage: 100 V) for 120 min. Electrolyte solutions used are listed in Table 1. After anodization, these films were annealed at 350 ℃ in a nitrogen atmosphere for 1 hr using a clean oven　(Koyo Lindberg, INH-9CD).

Table 1.　Electrolyte solutions

Solvent	Solute	Water Content / wt.%	Electrolytic Conductivity / mS cm^{-1}
Propylene Carbonate (PC)	0.4 mol dm^{-3} Triethylmethylammonium Salicylate	1.0	5.8
Ethylene Glycol (EG)	0.4 mol dm^{-3} Triethylmethylammonium Salicylate	1.0	1.1
γ-Butyrolactone (GBL)	0.4 mol dm^{-3} Triethylmethylammonium Salicylate	1.0	7.2
Water (W)	0.4 mol dm^{-3} Triethylmethylammonium Salicylate		12.0

Electrical Measurements

Aluminum for counter electrodes was deposited onto the annealed films by sputtering. Photolithography was used to form ，an array of dots 1mm in diameter and 400 nm thick. The electrical contact to the MIM elements was established through pressure probes (Signatone, H100). Using a source-measure unit (Keithley, 237), a stair wave form voltage was applied from 0 V to 25 V with 1 V step and 1 s delay after each measurement.

Surface Analysis

Annealed films were characterized by scanning electron microscopy (SEM; Hitachi, S-900), transmission electron microscopy (TEM; Hitachi, H-600), and auger electron spectroscopy (AES; JEOL, JAMP-7800).

RESULTS AND DISCUSSION

Anodization behavior

A typical anodization curve is shown in Fig. 1. During the constant current process, voltage rises linearly to 100 V and current decreases rapidly in the constant voltage process. Q_{cc} and Q_{cv} are the electric charges used in constant current and constant voltage process, respectively. Residual current is measured at the end of the anodization process, and t_{cc} is the time of the constant current process.

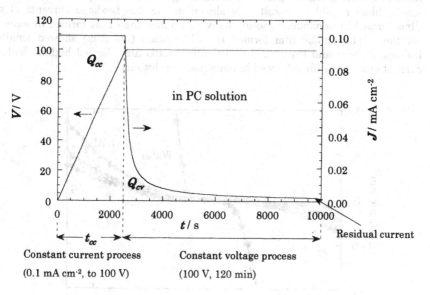

Fig. 1. Anodization behavior in propylene carbonate solution

Table 2. Comparison in anodization behavior

Solvent	t_{cc} / min	Q_{cc} / C cm^{-2}	Q_{cv} / C cm^{-2}	$\dfrac{Q_{cc}}{Q_{cc}+Q_{cv}}$ / %	Residual Current / μA cm^{-2}
PC	43	0.26	0.05	84	2.2
EG	49	0.29	0.06	83	2.8
GBL	46	0.27	0.09	75	2.3
Water	120	0.72	0.34	68	39.6

115

Comparison in anodization among nonaqueous solutions and an aqueous solution is shown in Table 2. Anodization behavior in the nonaqueous solutions was different from that in the aqueous solution. In nonaqueous solutions, anodization speeds were faster and the ratios of Q_{cc} to $Q_{cc} + Q_{cv}$ were higher. Low residual current can indicate high electrical resistance of the films.

Current - voltage characteristics

Current - voltage characteristics were measured after annealing because most TFT manufacturing processes include high temperature process to make the anodic films reliable. Results are shown in Fig. 2. Leakage currents of the films formed in nonaqueous solutions were smaller than that formed in aqueous solution. The anodic film formed in EG solution (EG film) showed smallest leakage current and the next was PC film. GBL film showed highest leakage current among the films formed in nonaqueous solutions.

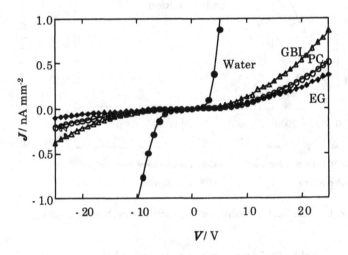

Fig. 2. Leakage currents of anodic films

Surface Analysis

By SEM observation, every film was barrier-type film. TEM observation revealed the film thicknesses of the anodic films are similar to those formed in aqueous solutions (PC film; 140 nm, EG film; 140 nm). AES depth profiles of the films formed in nonaqueous electrolyte solutions are shown in Fig. 3. EG film contained highest level of carbon, and the order in carbon level is the same as the order in low leakage current.

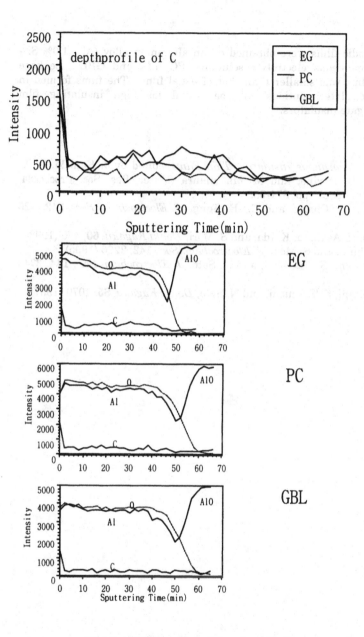

Fig. 3. AES depth profiles of anodic films formed
in nonaqueous electrolyte solutions

117

CONCLUSIONS

Novel anodic films were obtained on an aluminum alloy (Al - 1.0% Si - 0.5% Cu) in nonaqueous electrolyte solutions; PC, EG, and GBL. Leakage current of the films was smaller than that of usual film. The films formed in nonaqueous electrolyte solutions will be useful for high insulating film applications like gate insulators.

REFERENCES

1. T. Tsukada in *Amorphous Insulating Thin Films*, edited by J. Kanicki, W. L. Wallen, R. A. B. Devine, and M. Matsumura (Mater. Res. Soc. Proc. 284, Warrendale, PA, 1993), pp. 371-382.
2. R. -L. Chiu, P. -H. Chang, and C. -H. Tung, *J. Electrochem. Soc.*, **142**, 525 (1995).
3. M. Ue, T. Sato, H. Asahina, K. Ida, and S. Mori, *Denki Kagaku*, **60**, 480 (1992).
4. M. Ue, H. Asahina, and S. Mori, *J. Electrochem. Soc.*, **142**, 2266 (1995).
5. M. Ue, F. Mizutani, S. Takeuchi, and N. Sato, *J. Electrochem. Soc.*, **144**, 3743 (1997).
6. M. Ue, F. Mizutani, S. Takeuchi, and N. Sato, *Denki Kagaku*, **65**, 1070 (1997).

ANODIC OXIDATION OF NITROGEN-ADDED AL-BASED ALLOYS FOR THIN-FILM TRANSISTORS

Toshiaki Arai and Hideo Iiyori
IBM Yamato Laboratory, 1623-14 Shimo-tsuruma, Yamato-shi, Kanagawa 242, Japan

ABSTRACT

Novel anodized films of nitrogen-added aluminum-based alloys were proposed for use in the fabrication of gate insulators for thin-film transistors, and the effect of nitrogen addition on the anodized aluminum-based alloys was investigated. Gadolinium and neodymium were employed as alternative alloy components. The film thickness, the dielectric constant, and the roughness average of the anodized films decreased as the nitrogen content increased, and the nitrogen content was required to be lower than 20 at.%. The most improved values of the breakdown electric fields of anodized aluminum-gadolinium and aluminum-neodymium alloy were 10.1 MV/cm with 6.0 at.% nitrogen content and 9.9 MV/cm with 4.0 at.% nitrogen content, respectively. The leakage currents of the anodized films under a negative bias, which could not be suppressed by high-temperature annealing, were adequately suppressed by nitrogen addition, especially in anodized aluminum-gadolinium alloy. The current leakage of the anodized aluminum-gadolinium alloy with 6.0 at.% nitrogen content became -8E-13 A at -10 V and 150℃. This value is nearly equal to that of chemical-vapor-deposited (CVD) films.

INTRODUCTION

Bottom-gate thin-film transistors (TFTs) have been widely investigated for use in active-matrix liquid crystal displays (AMLCDs). In recent years, the panel size and resolution of AMLCDs have increased dramatically, and aluminum (Al) and its alloys have been widely studied for their potential use in the low-resistivity gate bus lines [1-2]. Aluminum is susceptible to stress migrations such as hillocks and whiskers under thermal stress, and therefore, the surface of the aluminum is often anodized. This anodized film functions as a protective layer and effectively suppresses the stress migrations. The electrical properties of the anodized film are important because it also functions as a gate insulator. The anodic oxidation of aluminum has been investigated, and the effects of the electrolyte, current density, and cathode position have been reported [3-5]. These oxidation conditions improved the electrical properties of the anodized aluminum film, but its leakage current was 1-3 orders of magnitude larger than that of films formed by chemical vapor deposition (CVD). We previously clarified the mechanism of the current leakage and reported that annealing can decrease the current leakage [6]. However, annealing did not adequately suppress the current leakage under a negative bias. In the work described in this paper, we added nitrogen to the aluminum oxide to decrease the current leakage, and used Al-based alloys for anodic oxidation, to increase the thermal resistance. Rare-earth metals such as gadolinium (Gd) and neodymium (Nd) have been studied as additional elements to increase the thermal resistance of aluminum [7-8]. However, the thermal resistances of these alloys are not sufficient for the fabrication of large, high-resolution AMLCDs, and therefore, we used anodic oxidation for these alloys.

The film thickness, surface morphology, nanostructure, and electrical properties of anodized nitrogen added Al-based alloys are described, and the effects of nitrogen addition are investigated. We plan to publish a separate paper on nitrogen-added Al-based alloys.

Mat. Res. Soc. Symp. Proc. Vol. 500 © 1998 Materials Research Society

EXPERIMENT

Aluminum alloys were deposited on 5-inch-square LCD-grade glass substrates by means of a dc magnetron sputtering apparatus at a substrate temperature of 150℃. Gadolinium (Gd) and neodymium (Nd) were used as alloy components to increase the thermal resistance of the aluminum. Al-Gd alloy with a composition of 1 at.% was sputtered from a composite target, while Al-Nd alloy with a composition of 2 at.% was sputtered from an alloy target. Argon and nitrogen were used as the sputtering gases, and the pressure was controlled at 0.4 Pa. The flow ratios of the nitrogen were from 0% to 20% in the Al-Gd alloy, and from 0% to 15% in the Al-Nd alloy. The nitrogen contents were determined by electron spectroscopy for chemical analysis (ESCA: SSI-M-probe ESCA). The thicknesses of the sputtered aluminum alloys were measured by a surface profiler (Tencor-P1) at 290-390 nm in the Al-Gd alloy, and 170-300 nm in the Al-Nd alloy.

These alloys were anodized in an electrolyte consisting of a mixture of 3 wt% tartaric acid solution and ethylene glycol, whose pH was kept at 6.5-7.5 by addition of an ammonium solution. The cathode was positioned on the rear of the anode [3]. A positive bias, applied to the anode with a current density of 1.5 mA/cm^2, was increased until it had reached 150 V, and was then maintained at this level until the anodizing reaction had ended.

These samples were annealed at 350℃ in a nitrogen atmosphere for 1.0 hour for the Al-Gd alloy and 2.0 hours for Al-Nd alloy, to decrease the leakage current [6]. This temperature was determined according to the maximum temperature of the AMLCDs' fabrication.

The surface morphology was observed by means of a tapping-mode atomic force microscope (AFM: Digital Instrument-Nanoscope-3). A cross-sectional view was observed with transmission electron microscope (TEM) after a sample had been prepared by focused ion beam (FIB) etching. The thickness of Al_2O_3 was estimated from cross-sectional scanning electron microscope (SEM: Hitachi-S800) and TEM views. The electrical properties of the anodized films were measured after the fabrication of 1-mmφ molybdenum (Mo)/Al_2O_3/Al capacitors. The dielectric constant was calculated from the capacitance measured at a frequency of 10 kHz. The breakdown electric field was measured by means of a curve tracer (Tektron-K213). The rate of voltage increase was 100 V/sec. The leakage current was measured by a triangular voltage sweep (TVS) method [9], using a computer-controlled HP-4140B system. In this measurement, the samples were heated to 150℃, and the bias applied to the aluminum electrode was swept at the rate of +/- 0.1 V/sec.

RESULTS AND DISCUSSION

Fig. 1 shows the thickness of the anodized Al-based alloys. In each alloy, an approximately 2000-Å-thick aluminum oxide was formed by anodic oxidation with an applied voltage of 150 V, when the nitrogen content was lower than 18 at.%. However, the thickness suddenly decreased when the nitrogen content exceeded approximately 20 at.%. There may be better conditions for the nitric aluminum alloys, and it may be possible to extend the limit, because the oxidation condition used in this paper is fixed to a single-condition optimum for a pure aluminum.

Fig. 2 shows the dielectric constant. This decreased as the nitrogen content increased. For the gate insulator of the TFT, a high dielectric constant is required in order to obtain a high drain current (Id). Therefore the nitrogen content should be lower than approximately 20 at.%.

Fig. 3 shows the average roughness (Ra) of as-deposited and anodized Al-based alloys calculated from AFM observation. In each alloy, the Ra had almost the same tendency in relation to the nitrogen content: the Ra decreased as the nitrogen content increased. The anodized Al-Gd alloy presented a smoother surface than the anodized Al-Nd alloy. The surface of the anodized Al-Nd alloy appeared to take essentially the same shape as the surface of the as-deposited alloy, and therefore the latter needed to be smoothed.

Fig. 4 shows the breakdown electric field (E_{BD}). Films with a nitrogen content of more than 20 at.% have a large E_{BD}, but they cannot be used as gate insulators in TFTs because of their small dielectric constant. The most improved values of the E_{BD} of anodized Al-Gd alloy and Al-Nd alloy were 10.1 MV/cm with 6.0 at.% nitrogen content and 9.9 MV/cm with 4.0 at.% nitrogen content.

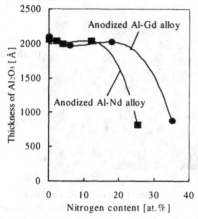

Fig. 1 Thickness of anodized films vs. nitrogen content.

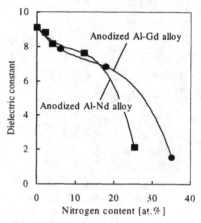

Fig. 2 Dielectric constant of anodized films vs. nitrogen constant.

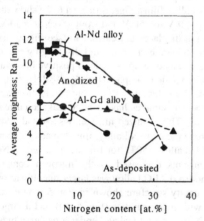

Fig. 3 Average roughness (Ra) of as-deposited and anodized films vs. nitrogen content.

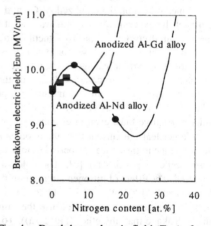

Fig. 4 Breakdown electric field (E_{BD}) of anodized films vs. nitrogen content.

Fig. 5 shows cross-sectional TEM views of these anodized films. The grain growth of Al was suppressed by nitrogen addition, and segregation of Gd and Nd was observed as a result of annealing after anodic oxidation. In the case of the anodized Al-Nd alloy, the film with a nitrogen content of 4.0 at.% showed a clear cavity structure, although the nitrogen content was lower than that of the anodized Al-Gd alloy. This may decrease the E_{BD}. The cavity structure is thought to originate in nitrogen included in the anodized film, and might be caused by outgassing of the nitrogen from the specimen. It may be possible to suppress its formation by optimizing the anodic oxidation condition for nitric Al-based alloys, but additional examination is needed to reach a fuller interpretation.

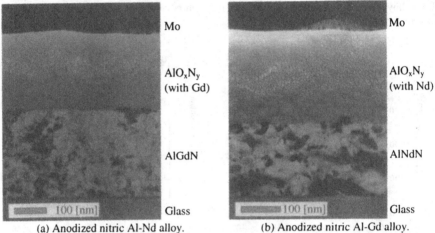

(a) Anodized nitric Al-Nd alloy.　　　　　　(b) Anodized nitric Al-Gd alloy.

Fig. 5　　TEM cross-sectional view of anodized Al-Gd and Al-Nd alloys. The nitrogen contents were 6.0 at.% in Al-Gd alloy, and 4.0 at.% in Al-Nd alloy.

To obtain detailed information about current leakage, we used a triangular voltage sweep (TVS) method [9] on the anodized Al-Gd and Al-Nd alloy films. Figs. 6-(a) and 6-(b) show the results of the experiment. The bias was swept from +50 V to –50 V and from –50 V to +50 V. The measured current in the TVS method includes a displacement current Ic (see Fig. 6-(a)) associated with the capacitive component, resulting from the charging and discharging of the capacitor. The current leakage of the nitrogen-less, annealed samples under a positive bias was sufficiently low, as compared with that of the non-annealed samples (Fig. 6-(a)-(Ref), 6-(b)-(Ref)), because the current leakage under a positive bias dominated by ionic conduction was reduced by 350℃ annealing [6]. However, the annealed samples have a large leakage current under a negative bias, even after annealing at 350℃. The anodized nitrogen-less Al-Gd alloy has a larger leakage current than the anodized nitrogen-less Al-Nd alloy; this is mainly due to the difference in the annealing period (anodized Al-Gd alloy: 350℃ for 1 hour; anodized Al-Nd alloy: 350℃ for 2 hours) [6]. These leakage currents decreased as the nitrogen content increased. The effect of nitrogen addition on the anodized Al-Gd alloy was larger than it was on the anodized Al-Nd alloy. It is supposed that the cavity structure shown in Fig. 5 would not suppress the current leakage, because the sample with many cavities (Fig. 5-(b)) had a larger current leakage than the other (Fig. 5-(a)). To study the current leakage under a negative bias, we focused on the negative bias region. Fig. 7-(a) and 7-(b) show the leakage current under a negative bias, and the displacement current Ic was deducted from the measurement. The dashed additional lines in Fig. 7 show that the current leakage is practically proportional to V^2, indicating that the current is dominated by a space-charge-limited current. The patterned region in Fig. 7-(a) indicates the current decrease caused by electron trapping.

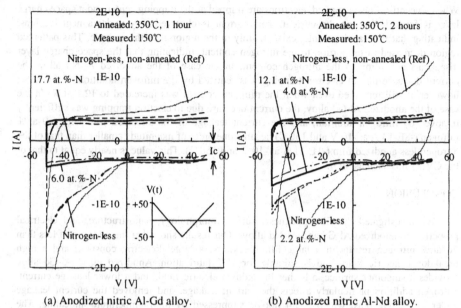

(a) Anodized nitric Al-Gd alloy.　　　　(b) Anodized nitric Al-Nd alloy.

Fig. 6　Leakage current measured by the TVS method. The bias was swept from +50 V to -50 V and from -50 V to +50 V. The measurement temperature was 150℃.

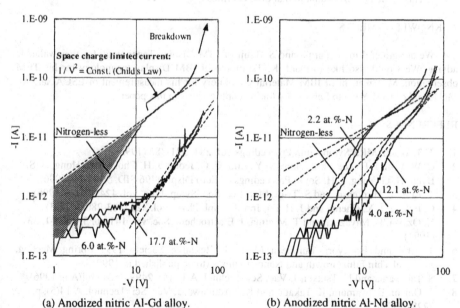

(a) Anodized nitric Al-Gd alloy.　　　　(b) Anodized nitric Al-Nd alloy.

Fig. 7　Leakage current measured by the TVS method. The bias was swept from 0 V to between -50 V and -100 V. The measurement temperature was 150℃. The displacement current Ic associated with the capacitive component was deducted from the measurement.

When a negative bias is applied, electrons are trapped at the trapping center and a space charge layer is formed, limiting the current. This current is observed only under a negative bias, indicating that the trapping center exists mainly in the region near the cathode. This patterned region decreased in proportion to the nitrogen content, indicating that the space charge layer disappears according to the nitrogen content. In the case of the anodized Al-Nd alloy, the current decrease due to electron trapping was reduced by the nitrogen addition of 4.0 at.%; however, it still remained even when the nitrogen content was increased to 12.1 at.%. In the case of the anodized Al-Gd alloy, the current decrease due to electron trapping was sufficiently removed by nitrogen addition, and the current leakage of the anodized film with 6.0 at.% nitrogen (-8E-13 A at -10 V and 150°C) became one order of magnitude smaller than that of the nitrogen-less anodized film (-1E-11 A at -10 V and 150°C). This value is nearly equal to that of the CVD films such as SiO_x.

CONCLUSION

We investigated the film thickness, surface morphology, nanostructure, and electrical properties of anodized Al-Gd and Al-Nd alloys. Our results show that the addition of less than 20 at.% nitrogen results in a large film thickness, a large dielectric constant, and a high breakdown electric field, and is suitable for TFT fabrication. Anodized nitric Al-Gd alloy provides a smoother surface, a higher breakdown electric field, and a lower leakage current. Nitrogen addition particularly affects the current leakage, and removed the current leakage under a negative bias, which could not be suppressed by high-temperature annealing. The current leakage of the anodized Al-Gd alloy with 6.0 at.% nitrogen was -8E-13 A at -10 V and 150°C. This value is nearly equal to that of CVD films.

ACKNOWLEDGMENTS

We are grateful to K. Furuta and S. Tsuji of IBM Display Technology for their valuable advice. We would also like to thank K. Tsujimoto of IBM Display technology for his TEM observation, M. Sakauchi of IBM Material Laboratory for his measurement of ESCA, and M. McDonald of IBM Yamato Laboratory for his contributions to this paper.

REFERENCES

1. T. Tsukada, MRS Symposium Proceedings vol. 284, 371-382 (1993).
2. C. W. Kim, J. H. Lee, H. R. Nam, S. Y. Kim, C. O. Jeong, J. H. Choi, M. P. Hong, H. S. Byun, H. G. Yang, J. H. Souk, Proceedings of Euro Display '96 SID, 591-594 (1996).
3. T. Arai, Y. Hiromasu, and S. Tsuji, Mat. Res. Soc. Symp. Proc. vol. 424, 37-42 (1996).
4. C. F. Yeh, J. Y. Cheng, and J. H. Lu, Jpn. J. Appl. Phys., vol. 32, 2803-2808 (1993).
5. K. Ozawa, K. Miyazaki, and T. Majima, J. Electrochem. Soc. vol. 141 no. 5, 1325-1333 (1994).
6. T. Arai and H. Iiyori, European Materials Research Society 1997 Spring Meeting, Epitaxial Thin Film Growth and Nanostructures (to be published in 1997).
7. S. Takayama and N. Tsutsui, J. Vac. Sci. Technol. A 14 (4), 2499-2504 (Jul/Aug 1996).
8. T. Onishi, E. Iwamura, K. Takagi, and K. Yoshikawa, J. Vac. Sci. Technol. A 14(5), p. 2728 (1996).
9. N. J. Chou, J. Electrochem. Soc. vol. 118 no. 4, 601-609 (April 1971).

NUCLEATION AND GROWTH AT REACTIVE INTERFACES
FOLLOWED BY IMPEDANCE MEASUREMENTS

F. VOIRON *, M. IGNAT *, T. MARIEB **, H. FUJIMOTO **
* LTPCM, INP Grenoble, 1130 Rue de la piscine 38402 Saint Martin d'Heres, fvoiron@ltpcm.inpg.fr
** INTEL, Components Research, Santa Clara

ABSTRACT

To evaluate the thickness of different constitutive layers of a TiAl multilayer, two different experiments, based on impedance measurements on multilayers, have been performed. In the first one, the thickness of the layer is deduced from resistivity measurements, applied to a parallel resistance model. In the second one, the thickness of the layer is deduced from comparing of calculated and measured values of surface potentials.

INTRODUCTION

To study the process of interfacial reactions in a multilayer, one of our major concern is to follow the thickness evolution in multilayered samples during the reaction. Usually, thickness determinations of thin films, intended for microelectronics devices are performed by different experimental techniques. For example, for metallic layers, the square resistance method can be used [1]. This method provides fast information about the thickness of a monolayer. However the model used for the thickness interpretation take into account a single experimental measurement, therefore additional hypothesis are necessary to measure thickness of multilayer samples (number of equations should be at least equal to number of unknown). Furthermore, because of symmetry of the measurement device, classical square resistance method is not sufficiently sensitive to give precise information about the earlier stage of a new phase growth.

In the first part, we present an attempt to improve the determination of the thickness or conductivity in multilayers with and without interfacial reaction, from electrical potential determinations and numerical fitting. This method allows accurate thickness determination in one step even in the case of multilayer sample, it has been called « Potential Method » (PM).

In the second part we compare the results obtained by the potential method to the results obtained by the classical square resistance measurements method (SRA).

The potential method gives the thickness of the different layers by taking into account the electrical perturbations induced by the other layers of a sample. The method includes two main steps which are :

- The numerical calculation of the electrical potential on the surface of the sample (direct method).

- An iterative procedure which compares the measured values, to the calculated ones, through an optimization algorithm (inverse method), which tends to produce a unique solution for a given structure .

Schematically the procedure is shown in Figure 1.

We may recall that a similar procedure is used in Geophysics to determine the conductivity and the thickness of underground layers. This method is called the electrical drilling [2], [3].

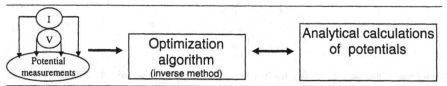

Figure 1 : Schematic representation of the procedure followed to determine different parameters of a multilayer, by applying the potential method.

POTENTIAL METHOD THICKNESS MEASUREMENT

Physical background

This approach considers a n-stratified media, with a cylindrical symmetry, and an infinite boundary. The n layers are supposed to be deposited on an infinitly thick substrate. We use for calculation the standard cylindrical frame (o, r, z). An electrical current I is injected at the system origin $(r=0, z=0)$.

For this structure, the electrical potential is described in each layer by the Laplace equation :

$$\frac{d^2v_i}{dr^2} + \frac{1}{r}\frac{dv_i}{dr} + \frac{d^2v_i}{dz^2} = 0 \qquad (1)$$

v_i is then the potential in the ith layer, of thickness h_i, r and z correspond to cylindrical coordinates. The solution for v_i will present a vanishing wave form :

$$v_i = U_i(r)e^{\pm kz} \qquad (2)$$

By replacing (2) into (1) we obtain the Bessel equation of the second order.

$$\frac{d^2U_i}{dr^2} + \frac{1}{r}\frac{dU_i}{dr} + k^2U = 0 \qquad (3)$$

The Bessel function J_0 is a possible solution for U_i.
At this point we can make the following remarks :
- the surface potential is divided in two contributions :
    - a fundamental one due to the first layer,
    - a perturbation due to the buried layers
- the fundamental potential diverges at the point $(r=0,z)$ and strongly decreases when r increases. Consequently, a possible solution for the fundamental contribution is :

$$v_i^{Fund.} = \frac{\rho_1 I}{2\pi r} \qquad (4)$$

where ρ_1 represents the surface resistivity, and I the injected current intensity.
From relation (2) and relation (4) the electrical potential corresponding to a layer i, can be expressed in each layer, by taking into account both contributions :

$$v_i = \frac{\rho_1 I}{2\pi}\left(\frac{1}{r} + A_i(k)J_0(kr)e^{-kz} + B_i(k)J_0(kr)e^{+kz}\right) \qquad (5)$$

From the conditions of the potential continuity through the interfaces, the following boundary conditions are obtained :

$$v_i(r,h_i) = v_{i+1}(r,h_i)$$
$$\left(\frac{dv_i}{dz}\right)_{h_i} = \left(\frac{dv_{i+1}}{dz}\right)_{h_i} \qquad (6)$$

If we consider that the upper surface $(z=0)$ is in contact with air, we can write the following boundary condition at the surface :

$$\left(\frac{dv_1}{dz}\right)_{z=0} = 0 \qquad (6bis)$$

As an example, in the following section we develop the analytical approach, for a monolayer deposited on substrate. The same formalism is used for calculating the electrical potentials in multilayers and for stratified media in Geophysics [4].
For simplicity of the calculation, and by using the Weber formula, relation (5) can be rewritten :

$$v_i = \frac{\rho_1 I}{2\pi}\left(\int_0^\infty J_0(kr)e^{-kr}dk + \int_0^\infty f_i(k)J_0(kr)e^{-kr}dk + g_i(k)J_0(kr)e^{+kr}dk\right) \tag{7}$$

We can notice then, that the substrate's potential is null when z tend to ∞ (vanishing wave) ; this gives $g_{sub}(k)=0$. By using the boundary conditions (6) and (6bis) and the relation (7), the harmonic functions $f_1(k)$, $g_1(k)$ and $f_{sub}(k)$ are solutions of the following system (8) :

$$\begin{cases} f_1(k) = g_1(k) \\ f_1(k)\left(e^{-kh_1} + e^{kh_1}\right) = f_{sub}(k)e^{-kh_1} \\ -e^{-kh_1} + f_1(k)\left(-e^{-kh_1} + e^{kh_1}\right) = \dfrac{\rho_1}{\rho_{sub}}\left(-e^{-kh_1} - g_2(k)e^{-kh_1}\right) \end{cases} \tag{8}$$

ρ_1 and ρ_{sub} being the resistivities of the surface layer and of the substrate. Finally we obtain a solution for $f_1(k)$:

$$f_1(k) = \frac{1 + \beta e^{-2kh_1}}{1 - \beta e^{-2kh_1}} \tag{9}$$

with β the ratio among resistivities of the system :

$$\beta = (\rho_{sub} - \rho_1)/(\rho_{sub} + \rho_1) \tag{10}$$

As we are working with an insulating substrate β is equal to 1.
By using (8), (9) and the superposition principle, and by setting the curent injection at a distance d from the extracting point, we obtain the potential at each point of the surface :

$$v_{surf} = \frac{\rho_1 I}{2\pi r}\left(\frac{\alpha - 1}{\alpha} + 2\int_0^\infty \frac{1 + e^{-\frac{2uh_1}{r}}}{1 - e^{-\frac{2uh_1}{r}}}\left(J_0(u) - J_0(\alpha u)\right)du\right) \tag{11}$$

u is an adimensional integration variable corresponding to $u=kr$ and $\alpha=(d-r)/r$.
This expression had a finite value for each point of the surface and can be evaluated by different numerical approximations, as for example a quadrature or a filter method [5], [6]. We apply in our case the Gauss quadrature method which provides a sufficient precision and a short convergence time .

Inversion method : parameters optimization

We apply the inversion method to the calculated potentials, for determining the values of a set of parameters (h_i, ρ_i) which gives the shortest gap between an experimental set of measured potentials and a calculated one. Theoretically a unique set of parameters (in this case h_i, ρ_i) should be the solution, when applying the inversion method. However, we may point out that the probability to obtain a solution, which will be close to the exact one, is directly relied to the number of the treated experimental measurements. In our case the number of measurements was increased by measuring potentials with three different spaces among the current points and measurements points (distance d-2r).The gap between measured and calculated potentials is therefore fixed by the variance evolution. To avoid the problems induced by local minimums when performing the inversion, and also to reduce the calculation time, we used two different minimization methods :

• First, a fast global Monte-Carlo method, which gives a general idea of the low variance zones, in regard of the investigated parameters (Figure 3a) ;
• Second, a Hook &Jeeves [7] method, which avoids local minimum problems and provides a fine convergence. Besides, this method allows the preconditioning of the parameters. Because this method is expensive in terms of calculation time, we used it only on a small scale (Figure 3b).

127

EXPERIMENTAL RESULTS

In the following, we will present the results obtained by the potential method, for a bilayer, partially and totally transformed by the interdiffusion among the consisting layers. Before annealing, the system is constituted of Al (~3950 Å) and Ti(~1600 Å) deposited on a silicon substrate having a thermally-grown SiO_2 (~1000 Å) on the top. The transformations were obtained by annealing the bilayers under nitrogen atmosphere. In one case, partial transformation is obtained by annealing the sample during 11 hours at 430°C, in the other case the annealing during 15 hours at 500°C totally transforms the bilayer into a single layer of $TiAl_3$. Transformed fractions have been checked by XRD phase detection [8].

To perform the potential determinations we have used a distance of *12mm, 16mm* and *24 mm* between the current input and output points. Furthermore we adopted a particular arrangement for these points known as the Wenner arrangement. It provides thickness values which are representative to the center of symmetry of the electrical system. As a result of the described formalism, a comparison among calculated and measured potentials is shown on Figure 2.

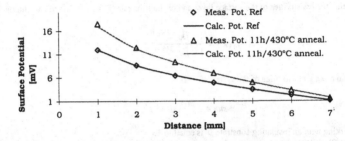

Figure 2 : Comparison between measured and calculated potentials on an Al/Ti bilayer before and after thermal treatment with different electrodes spacing *r*.

The results correspond to a bilayer with and without a converted fractions of $TiAl_3$. On Figure 3 we illustrate the two methods of inversion, which are applied to our results

- figure 3a corresponds to a fast Monte-Carlo inversion, which permits to identify a zone of low variance near $h_{Al}=0.4\mu m$ on an unreacted sample
- figure 3b presents a fine optimization which permits to deduce Al and Ti thickness, following the Hook&Jeeves method [7].

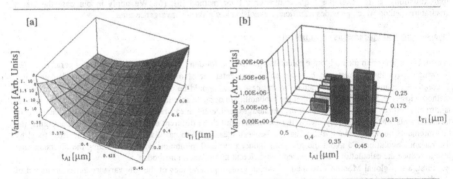

Figure 3 : Figure [a] and [b] presents respectively the evolution of the variance in regard to Al and Ti thickness and an optimization path in 8 steps for the Al and Ti thickness (light contrast represents low variances).

Results and discussion

This point will be centered on the thickness determinations and TiAl₃ detection on a bilayer system. We recall that the experiments were performed on systems with and without annealing, which produces a partial or a total transformation.

Firstly, we shall notice that when considering a multilayer sample, the P.M allows to determine in one step several parameters (thickness or resistivity) of every constitutive layer, depending from the number of experimental measurement. Contrary to the square resistance method, no additional hypothesis like geometrical coupling between the layers are needed.

The second comment arises, while considering the evolution of the apparent resistivity with the transformed fraction. When the parameter α increase (relation 11), the sensitivity of the Wenner arrangement improves (See Figure 4).

For example, a classical square resistance arrangement (with equal spaces among the four electrodes, $\alpha=2$) is more than six times less sensitive than the Wenner which allows to reduce the distance r (we can reach $\alpha=15$ with our experimental device, See Figure 4).

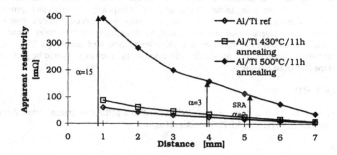

Figure 4 : Apparent resistivity evolution in regard to the thermal treatments performed on an Al/Ti bilayer. SRA indicates a distance which corresponds to a typical experimental condition for a classical square resistance method.

Resistivity values measured for the Al/Ti bilayer, correspond to :

$\rho_{Al}=2.9\mu\Omega\,cm$, $\rho_{Ti}=63\mu\Omega\,cm$, $\rho_{TiAl3}=23\mu\Omega\,cm$ (for full transformation). These values are similar to those of literature and used by other authors, performing thickness determination by the square resistance method [9].

When comparing the thickness values obtained by both methods, (square resistance and potential methods) some discrepancies arise. We reported on Table 1 the results of thickness determinations (in Å).

From Table I, we note that the square resistance measurements indicate a full transformation for the two annealing conditions, while the potential method, indicates only a partial transformation in our case.

(This point should be discussed more precisely, in particular with regard to the deposition conditions of the layers, and their annealing conditions).

	Square Resistance [8]			Potential Method		
	Al	Ti	TiAl3	Al	Ti	TiAl3
Al/Ti Ref	3950	1600		4000	1500	
Al/Ti 11h/430°C annealing	Full transformation			2600	1000	2000
Al/Ti 16h/500°C annealing	Full transformation			Full transformation		5100

Table I : Thickness determinations of a bilayer before and after reaction producing TiAl₃

129

With respect to the partially transformed sample, we shall note that the potential method is in agreement :
- with results which were obtained by XRD phase analysis [8];
- with our first results from acoustic microscopy surface wave velocity measurements [8].

Besides, for the partially transformed sample we have determined an approximate rate of transformation of about 34%. Then the corresponding thickness are close to stoechiometry : for 1400 Å of Al consumed, the stoechiometric relation gives 500 Å Ti consumed and 1800 Å $TiAl_3$ formed. The 10% of incoherence between measured and calculated $TiAl_3$ thickness may have different causes :
- it could be attributed to the interfacial roughness which appears during transformation. This transformation is detected because the Wenner arrangement is more sensitive,
- it could also be attributed to the formation of other compounds, which were detected by XRD phase analysis [8]. The compound would then act as another perturbing layer.

When considering the totally transformed system we detected a volume shrinkage of about 6.3%, which is characteristic of $TiAl_3$ transformation.

CONCLUSION

From the previous results, we can deduce, that accurate thickness determinations are possible by using the P.M, even when considering multilayer samples.Furthermore, the method permits to perform measurements at a short distance r (see Figure 4) corresponding to the non linear range of response. This gives an improved sensitivity, for resistivity/thickness determinations. The method should be convenient to detect and determine structural and physical parameters for thin layers, which may correspond, for example, to the first stages of nucleation during interfacial reactions. One of our next objectives is to characterize these nucleation steps. This will be done by improving the analytical treatment, with finite difference calculations. Therefore, it could be possible to describe the evolution of the interfacial roughness.

ACKNOWLEDGMENTS

This work is possible thanks to a donation from INTEL, and to a MRI/DGA student grant. Authors are also grateful to G. Parat and C. Puget from LETI, for technical assistance.

REFERENCES

[1] Sze S.M., Physics of semiconductor devices, Wiley&Sons, p : 31, 1981

[2] Dieter K., Paterson N.R., Grant F.S., Geophysics, vol 34, n°4, p : 615, 1969

[3] Zohdy A.A.R. , Geophysics, Vol 54 n°2, p : 245, 1989

[4] Stefanesco S., Schlumberger C., Journal de physique, n°4, p : 132, 1930

[5] Oppenheim A.V., Frisk G.V., Martinez D.R., Proc. IEEE, vol 66 n°2, p : 264, 1978

[6] Anderson W.L., Geophysics, vol. 54 n°2, p : 263, 1979

[7] Hooke R., Jeeves T.A., J. Assoc. Comput. Mach., 8, p : 212, 1961

[8] Voiron F., Ignat M., work in progress, 1997

[9] Marieb T., Fletcher E., Mach A., Kelsey J., INTEL internal technical memo, 1996

Part III

Magnetic and Polymeric Materials

GIANT MAGNETOIMPEDANCE : A RELEVANT APPLICATION OF IMPEDANCE SPECTROSCOPY

K.L. García and R. Valenzuela*
Institute for Materials Research, National University of Mexico, Mexico. *e-mail: monjaras@servidor.unam.mx

ABSTRACT

The impedance response of amorphous ferromagnetic wires is measured in the 5 Hz-13 MHz frequency range by applying a small AC current of 1 mA (RMS). A DC magnetic field (in the 0-6.7 kA/m range) is simultaneously applied. The impedance response is analyzed by means of simple series and parallel equivalent circuits. An association of circuit elements and physical parameters of the sample is proposed.

INTRODUCTION

Giant magnetoimpedance (GMI) refers to significant variations in the impedance response of ferromagnetic materials (submitted to a small AC current) when a DC magnetic field is applied. The technological applications of this phenomenon in magnetic and current sensors [1,2] has raised a strong interest in recent years. GMI has been explained in terms of classical electromagnetism [3], as a result of the interaction between the small AC magnetic field generated by the AC current and the magnetic structure of the material.

We have recently proposed an alternative approach [4] based on the modeling of GMI by means of equivalent circuits. In this paper, we complete this approach by including into the analysis the effects of the spin rotation process.

EXPERIMENT

Amorphous ferromagnetic wires of nominal composition $(Co_{0.96}Fe_{0.04})_{72.5}B_{15}Si_{12.5}$, with a diameter of 125 µm, obtained by the in-rotating water quenching technique [5], were kindly provided by Unitika Ltd, Japan. They were measured in the as-cast state, in pieces of 5 ~cm long. Electrical contacts were made with Ag paint, after cleaning the wire ends with a soft acid. Frequency measurements were carried out at room temperature by means of a system including a HP 4192A Impedance Analyzer with a 5 Hz - 13 MHz frequency range, controled by a PC computer. A small AC current of 1 mA (RMS) was used to excite the wire. A DC magnetic field in the 0 - 6.7 kA/m (0 - 83 Oe) was applied with a 200-turn, 10 cm solenoid powered by a DC power supply.

RESULTS

In the GMI literature [1-3], impedance results are often presented in the form of $\Delta Z/Z_0$ at a given frequency, where ΔZ is the difference between the zero-field value of total impedance, Z_0, and the high-field value of total impedance, Z_H. While this figure of merit is convenien to assess the magnitude of GMI and therefore its potential applications, it gives no clear clues into the physics involved. In the analysis of ferromagnetic materials, inductance formalisms give far more insight than impedance formalisms, since magnetic permeability and inductance are related simply by the corresponding geometrical factor. We therefore make the transformation,

Mat. Res. Soc. Symp. Proc. Vol. 500 © 1998 Materials Research Society

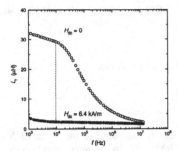
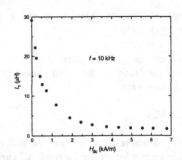

Fig. 1. Spectroscopic plot of the real inductance, L_r, at zero-field and at 6.7 kA/m (data at $f < 1$ kHz showed some dispersion and are not shown).

Fig. 2. Behavior of the real part of inductance at 10 kHz, as a function of the DC field.

$$L = (-j / \omega) Z \tag{1}$$

where L $(= L_r + j\, L_i)$ is the complex inductance, j the basis of imaginary numbers $((-1)^{1/2})$, ω the angular frequency and Z $(= Z_r + j\, Z_i)$ the complex impedance. Note that due to j, there is a crossing of terms: real inductance, L_r, depends on imaginary impedance, Z_i, and imaginary inductance, L_i, is determined by real impedance, Z_r. The analysis of experimental results is therefore carried out in inductance formalisms, and by separating their real and imaginary contributions.

Spectroscopic plots of the real part of inductance at 0 and 6.4 kA/m of DC field show the effects of the latter, Fig. 1. The zero-field experiment exhibits a larger value at low frequencies, followed by a dispersion. The high-field results are virtually independent of frequency, and show a small L_r value very close to the zero-field results at high frequency (higher than the dispersion). From these results, a characteristic plot of GMI can be obtained at a constant frequency (at 10 kHz, for instance), by considering the variations in L_r as a function of DC field, Fig. 2. This plot clearly shows the sensitivity of GMI to small fields.

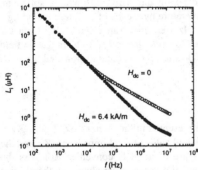
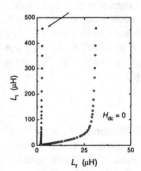

Fig. 3. Spectroscopic plot of the imaginary part of inductance, for $H_{dc} = 0$ and 6.7 kA/m.

Fig. 4. Complex plane representation of inductance data, for zero- and maximum DC field.

Imaginary inductance also shows the influence of the *DC* field, Fig. 3. Both experiments (at zero- and high-field) lead to a slope very close to -1 (in a log-log plot) up to frequencies of 10 kHz; the zero-field results then exhibit a lower slope. The largest changes therefore appear at higher frequencies.

DISCUSSION

The representation of data in the complex L_i-L_r plane can be useful in the search for an equivalent circuit to model the sample behavior. At high *DC* fields, a "spike" is observed in this representation, while the zero-field results show a more complex behavior, Fig. 4. It can be shown [4] that the presence of a spike is associated with a *RL* series circuit. Figures 1 and 2 for the high-field results point also to a *RL* series circuit. On the other hand, the dispersion observed in the real part of inductance for the zero-field results (Fig. 1) seems to represent a relaxation behavior, related with an arrangement which should include a *RL* parallel circuit.

We make therefore the assumption that the equivalent circuit is a R_sL_s series, in series with a R_pL_p parallel for the zero-field condition. Also, that the equivalent circuit becomes simply a series R_sL_s for the high-field experiment, Fig. 5. Then, if high-field data are subtracted point-by-point for each corresponding frequency from the zero-field results, a R_pL_p parallel circuit should be obtained. Such a circuit is easy to recognize, since it should give a simple relaxation behavior in L_r (*f*) plots, a maximum in L_i (*f*) plots for the frequency at which L_r value is ½ of the low frequency value, and a semicircle in the complex plane plot. These appear in Fig. 6.

The "good" equivalent circuit has a clear association of circuit elements with physical parameters of the material. The observed behavior can be interpreted on the basis of the interaction of the *AC* circular magnetic field (generated by the small *AC* current) with the magnetic structure of these amorphous wires as follows. As a result of the extremely rapid cooling to prevent crystallization in the wires, a complex state of stress appears. For compositions including a high Co content, the magnetic structure is formed by an inner-core with magnetic domains oriented close to the axial direction, and an outer-shell with alternate domains with circumferential magnetization, Fig. 7.

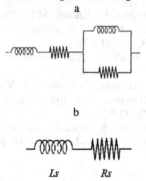

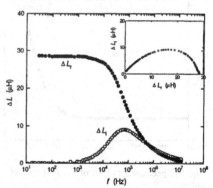

Fig. 5. Equivalent circuits to model the inductance response of wires. a) A combination of *RL* series and *RL* parallel circuit for $H_{dc} = 0$; b) The high field $H_{dc} = 6.7$ kA/m eliminates the parallel arm; only the series part remains.

Fig. 6. Differences in real and imaginary parts of inductance of experiments at zero field, and at the highest field, $\Delta L = L_0 - L_H$, represented as a spectroscopic plot; inset, complex plane.

135

For the $H_{dc} = 0$ condition, the AC circular magnetic field interacts with the magnetic structure, resulting in a periodical movement of domain walls. This inductive coupling is quite strong, and leads to a significant contribution to impedance. Domain walls are subject to a certain damping, which defines a relaxation frequency, f_x. As frequency increases above f_x, they become unable to follow the excitation field and a dispersion appears.

When a strong DC magnetic field is applied, the wire is saturated; i.e., magnetic domain walls are eliminated and all the spins become oriented in the DC field direction. A spin rotation process appears, with a smaller permeability (and therefore inductance) value than the corresponding value for the domain wall mechanism. Also, this spin process is characterized by the well-known ferromagnetic resonance dispersion, with typical resonance frequencies in the GigaHertz range. Since this frequencies are well above the upper limit of our measuring system, the equivalent circuit becomes essentially a RL series circuit.

The evidence of the spin resonance process appear on spectroscopic plots of real impedance, Z_r, Fig. 7, instead than on inductance plots. A RL series circuit should exhibit only a real impedance independent of frequency, since $Z = R + j\omega L$, and therefore $Z_r = R$. The experimental plot, however, shows an increase as frequency increases. Infortunately, our measuring system is unable to reach the GigaHertz range.

CONCLUSIONS

We have shown that Impedance Spectroscopy can lead to a straightforward analysis of the Giant magnetoimpedance phenomena, by suggesting the relevant equivalent circuits and allowing a resolution of the several magnetization processes involved.

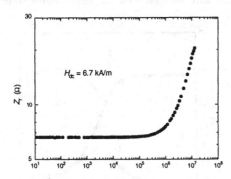

Fig. 7. Spectroscopic plot of the real part of impedance.

ACKNOWLEDGMENTS

This work has been partially supported by CONACyT-Mexico, under Grant 3101P-A9607. K.L.G. acknowledges a student scholarship from the same Grant.

REFERENCES

1. K. Mohri, and L.V. Panina. Sens. & Act. **A59**, p. 1 (1997).
2. M.Vázquez, M. Knobel, M.L : Sánchez, R. Valenzuela and A.P. Zukhov. Sens. & Act. **A59**, p. 20 (1997).
3. R.S. Beach and A.E. Berkowitz. Appl. Phys. Lett. **64**, p. 3652 (1995).
4. R. Valenzuela, M. Knobel, M. Vázquez and A. Hernando. J. of Phys. C: Appl. Phys. **28**, p. 2404 (1995).
5. Y. Waseda, S. Ueno, M. Hagiwara and K.T. Austen. Prog. Mater. Sci. **34**, p. 149 (1990).

ELECTROMAGNETIC PROPERTIES OF SOILS

J. C. SANTAMARINA, K. A. KLEIN
CEE Department, Georgia Institute of Technology, Atlanta, GA 30332-0355

ABSTRACT

Electromagnetic waves can be used to characterize geomaterials and to monitor geo-processes. Permittivity, conductivity, and magnetic permeability measurements provide complementary information. Furthermore, events at different frequencies, such as the various polarization mechanisms, suggest multiple internal scales within materials. Three laboratory studies are presented: characterization of kaolinite-water mixtures with permittivity data, monitoring soil-cement hydration with conductivity measurements, and characterization of kaolinite-iron mixtures with magnetic permeability data. Laboratory techniques face inherent limitations, in particular, low frequency permittivity measurements of highly conductive specimens are not feasible. Likewise, field techniques are restricted by the compromise between the desired resolution and the achievable skin depth.

INTRODUCTION: ELECTROMAGNETIC PROPERTIES

Soil-water mixtures consist of a continuous highly conductive fluid phase with a large volume fraction of mineral particles. Due to their dependence on the chemical makeup of geomaterials and internal fabric, electromagnetic waves can be used to monitor geo-processes such as pressure and chemical diffusion, hardening of soil-cement slurries, and corrosion. Electromagnetic techniques are also useful to identify the presence of magnetic or electrically conductive materials (e.g. rocks containing conductive sulphide minerals and buried metallic drums).

Three material properties can be obtained from electromagnetic measurements: permittivity ε, conductivity σ or resistivity ρ, and magnetic permeability μ. These properties vary as a function of frequency, thus, spectra provide valuable information about material composition and internal scales, both temporal and spatial. Table I presents typical values for different materials.

Table I Electromagnetic Properties of Materials

material	real permittivity, κ' @ 1 MHz	relative magnetic permeability, μ_r	conductivity, σ [S/m]
water	78.30 [1]	0.999910 [3]	σ=f(c) *
quartz	4.7 - 12.8 [1]	0.999985 [2]	10^{-5} - 10^{-3} [4]
pyrite (semiconductor)	------	1.0015 [2]	10^3 [1]
iron (99.91% Fe)	------	5000 [2]	10^7 [1]

* The conductivity of electrolytes is a function of the ionic concentration - see below.

<u>Permittivity</u>

The dielectric permittivity characterizes the polarizability of a material subjected to an electric field. The relative permittivity is a complex quantity:

$$\kappa^* = \kappa' - j\kappa''_{eff} \qquad (1)$$

The real relative part κ' reflects the polarizability of the material and it is the number of dipole moments per unit volume. The effective imaginary permittivity κ''_{eff} reflects conduction and polarization losses:

$$\kappa''_{eff} = \kappa''_{polarization} + \frac{\sigma}{\omega \varepsilon_0} \qquad (2)$$

where σ is the dc conductivity, ε_0 is the permittivity of free space ($\varepsilon_0 = 8.85 \cdot 10^{-12}$ F/m), and ω is the angular frequency [rad/s].

The mechanisms that cause polarization depend on the frequency of the applied electric field and the composition of the material. Single-phase, homogeneous materials only experience high frequency polarization mechanisms including electronic, ionic, and molecular polarization [5]. Electronic and ionic polarization manifest as resonance at ultraviolet and infrared frequencies, respectively.

Multi-phase, heterogeneous mixtures experience polarization mechanisms at both high and low frequencies. Low frequency polarization mechanism include: Maxwell-Wagner interphasial-spatial polarization, bound water polarization, and double layer polarization [5, 6, 7, 8, 9]. Interfacial-spatial polarization occurs due to charge accumulation at the interface between two materials with different electrical properties; electrochemical effects may develop at the interface between the host medium and a metal inclusion. Bound water polarization is the relaxation of water molecules adsorbed onto the surface of a soil particle. Double layer polarization occurs when the double layer surrounding negatively charged clay particles is displaced due to the application of a low frequency electric field; charges within the double layer have non-uniform ionic mobility ("membrane polarization"). These polarizations manifest as relaxations. Multiple relaxations may occur within one material if more than one polarization mechanism is present.

Conductivity - Resistivity

In the context of this publication, conductivity is treated as a real parameter. The micro-level interpretation of conductivity involves the availability and mobility of ions. The conductivity σ_0 of a mono-ionic electrolyte is:

$$\sigma_0 = c\,z\frac{v_{ion}}{E}F = c\,z\,u\,F \qquad [S/m] \qquad (3)$$

where v_{ion} is the velocity of the ion [m/s], E is the strength of the electric field [V/m], c is concentration [mol/m^3], $u = |v|/|E|$ is ionic mobility [m^2/V·s], F is Faraday's constant [96484.6 C/mol], and z is the valence of the ion [10]. Conduction in geomaterials is largely electrolytic, taking place within connected pore spaces [4]. The ionic concentration near surfaces confers "surface conductivity" to the mineral grains.

In water-bearing geomaterials, conductivity is often expressed as a function of porosity ϕ according to Archie's law [4]:

$$F = \rho_r / \rho_e = a\phi^{-m} \qquad (4)$$

138

where F is the formation factor, ρ_r and ρ_e are the resistivities of the rock and the electrolyte respectively, 'a' is a constant referred to as the coefficient of saturation (typically between 0.6 and 1.0), and 'm' is the cementation factor (i.e. tortuosity of porous pathways, typically between 1.4 and 2.2). Equation (4) shows that the resistivity of the geomaterial increases as electrolyte resistivity increases, degree of saturation increases, and amount of cementation decreases. Archie's law shows no dependence on frequency.

Many models have been developed to predict the conductivity of two-phase media including: parallel, series, Maxwell-Wagner, Lichtenecker's rule, percolation law, Hashin-Shtrikman, and general effective medium theories [11]. In all cases, the overall conductivity of the mixture is a function of the conductivity of the pore fluid, the conductivity of the mineral phase, and the characteristics of conduction paths (orientation, size, and tortuosity).

Magnetic Permeability

Magnetic materials consist of randomly distributed magnetic dipoles. A net magnetic-dipole moment results when magnetic materials are placed in a magnetic field. The net magnetic-dipole moment per unit volume is defined as the magnetization or magnetic polarization of the material and it is related to the internal atomic structure of the material. The magnetic permeability of a material μ is expressed in terms of the magnetic permeability of free space μ_o [$4\pi \cdot 10^{-7}$ H/m] and the magnetic susceptibility χ:

$$\mu = \mu_o(1+\chi) = \mu_o\mu_r \qquad (5)$$

where μ_r is the relative magnetic permeability (in general, this is a complex quantity). Materials can be classified by their relative magnetic permeability μ_r. It is generally assumed that non-magnetic materials have $\mu_r = 1.0$. More accurately, paramagnetic materials have μ_r slightly greater than 1.0 and diamagnetic materials have μ_r slightly less than 1.0. Ferromagnetic materials typically have $\mu_r \gg 1.0$.

LABORATORY MEASUREMENTS

Electromagnetic measurements combine the permittivity, conductivity, and permeability properties of the specimen. Yet, depending on the type of material being tested and the frequency at which the measurement is made, specific parameters dominate. For example, in geomaterials which typically have low amounts of ferromagnetic materials, permittivity values are calculated based on the assumption that $\mu=\mu_o$. Additionally, while calculated effective imaginary permittivity values include polarization losses and conductivity effects, polarization losses dominate at high frequencies (GHz) while conductivity dominates at low frequencies (Hz to MHz - Equation 2). A brief review of electromagnetic measurement techniques is presented below.

Permittivity/Conductivity Measurements

Different measurement systems are required to obtain broad-band real permittivity and effective imaginary permittivity data of soil-electrolyte mixtures (includes conductivity effects). Coaxial probe systems can be used at MHz to GHz frequencies, while two-terminal and four-terminal systems can be used at lower frequencies (Hz to MHz). Difficulties in obtaining low frequency measurements of highly conductive soil-electrolyte mixtures occur due to the manifestation of electrode polarization in two-terminal systems and the lack of equipment resolution for phase angle measurements in four-terminal systems with decreasing frequency. Klein and Santamarina [12] present equations for the frequency at which these phenomena significantly affect permittivity measurements. The inductance in peripherals affects the measurements with capacitor-type systems at frequencies $>\approx 1$ MHz. Problems with coaxial probe systems include the presence of an air gap between the probe and the specimen, the size of particles relative to the cross sectional area of the transmission line, and the formation of a water layer at the probe-soil interface for saturated specimens.

<u>Magnetic Permeability Measurements</u>

The most common technique for measuring the magnetic permeability of paramagnetic and diamagnetic materials is based on the change in weight of the specimen when it is placed in a magnetic field (e.g. magnetic susceptibility balance). The magnetic permeability of ferromagnetic materials can be measured by placing the specimen in a magnetic field produced by a primary coil or magnetizing winding and measuring the induced flux density in a secondary winding [13]. Alternatively, the inductance of a coil filled with the specimen can be used to compute permeability.

FIELD IMPLEMENTATION: RESOLUTION AND SKIN DEPTH

Various geophysical electrical techniques are used to characterize the subsurface (e.g. ground penetrating radar). The implementation of these techniques is restricted by the trade-off between the desired resolution and the achievable skin depth. Maxwell's equations for electromagnetism lead to the wave equation for electromagnetic phenomena. If a solution of the form:

$$E = E_0 e^{(j\omega t - \gamma^* x)} \tag{6}$$

is adopted and substituted into the wave equation, a complex "propagation constant" γ^* emerges:

$$\gamma^* = \frac{2\pi f}{c_0} \sqrt{-\left(\mu'_r - j \cdot \mu''_r\right)\left(\kappa' - j \cdot \kappa''_{eff}\right)} \quad [\text{m}^{-1}] \tag{7}$$

where c_0 is the velocity of electromagnetic waves in free space [$c_0 = 3 \cdot 10^8$ m/s], and μ'_r and μ''_r are the real and imaginary relative magnetic permeabilities ($\mu^*_r = \mu^*/\mu_0$). For non-ferromagnetic materials, it is assumed that $\mu'_r = 1$ and $\mu''_r = 0$:

$$\gamma^* = \frac{2\pi f}{c_0} \sqrt{-\left(\kappa'_{(assumed)} - j \cdot \kappa''_{eff(assumed)}\right)} \quad [\text{m}^{-1}] \tag{8}$$

The error associated with this assumption for a material with $\mu'_r \neq 1$ and $\mu''_r \neq 0$ is:

$$\kappa'_{(assumed)} = \mu'_r \kappa' - \kappa''_{eff} \mu''_r \tag{9a}$$

$$\kappa''_{eff(assumed)} = \mu'_r \kappa''_{eff} + \kappa' \mu''_r \tag{9b}$$

High frequency waves can provide information with high spatial resolution, but they have limited penetration depths. The order of magnitude for spatial resolution can be estimated from the wavelength λ, which is related to the frequency f and the velocity v of wave propagation in the material as $\lambda = v/f$. From Equations (6) and (7):

$$\lambda = \frac{2\pi}{Im\left[\gamma^*\right]} = \frac{c_0}{f} \cdot \frac{1}{Im\left[\sqrt{-(\mu'_r - j \cdot \mu''_r)(\kappa' - j \cdot \kappa''_{eff})}\right]} \quad [m] \tag{10}$$

where Im[ ] indicates the imaginary part of the quantity in brackets. In the case of zero polarization and Ohmic losses, $\kappa''_{eff} \approx 0$, and non-ferromagnetic materials, $\mu^* = \mu_0$, Equation (10) reduces to $\lambda = c_0 / f\sqrt{\kappa'}$.

The depth of penetration into the medium can be estimated by computing the "skin depth" S_d [m]:

$$S_d = \frac{1}{Re\left[\gamma^*\right]} = \frac{c_0}{2\pi f} \cdot \frac{1}{Re\left[\sqrt{(\mu'_r - j \cdot \mu''_r)(-\kappa' + j \cdot \kappa''_{eff})}\right]} \tag{11}$$

where Re[ ] indicates the real part of the quantity in brackets. For non-ferromagnetic materials, Equation (11) can be written in terms of the loss tangent, $\tan\delta = \kappa''_{eff}/\kappa'$:

$$S_d = \frac{c_0}{2\pi f} \cdot \frac{1}{Re\left[\sqrt{-\kappa' + j \cdot \kappa''_{eff}}\right]} = \frac{\lambda_0}{2\pi} \cdot \left[\frac{\kappa'}{2}\left(\sqrt{1 + \tan^2\delta} - 1\right)\right]^{-1/2} \tag{12}$$

EXAMPLES: MATERIAL CHARACTERIZATION AND PROCESS MONITORING

This section presents examples of permittivity, conductivity, and magnetic permeability data for various mixtures. Low frequency conductivity and permittivity data (5 Hz to 13 MHz) presented in this paper were obtained using a two-terminal parallel-plate type capacitor and an HP-4192A Low Frequency Impedance Analyzer. Permittivity data at frequencies between 20 MHz and 1.3 GHz were obtained using an HP-8752A network analyzer in conjunction with an HP-85070A dielectric coaxial termination probe. The magnetic permeability was determined by measuring the inductance of a coil with an HP-4192A Low Frequency Impedance Analyzer.

Permittivity: Characterization of Kaolinite-Water Mixtures

Permittivity and conductivity values of soil-water mixtures are dependent upon the type of soil and the water content of the specimen. The effective permittivity (polarizability and conductivity) was measured for kaolinite specimens at three different water contents: oven-dry (w/c $\approx$ 0%), air-dry (w/c = 0.2%), and saturated (w/c = 33%). Measurements were also performed on deionized water.

Permittivity and effective conductivity increase as moisture content increases (Figure 1). Electrode polarization affects all measurements (below 3 kHz for the water specimen); the frequency at which electrode effects dominate the measured impedance is a function of the conductivity and permittivity of the mixture [12]. For example, in the measurements shown in

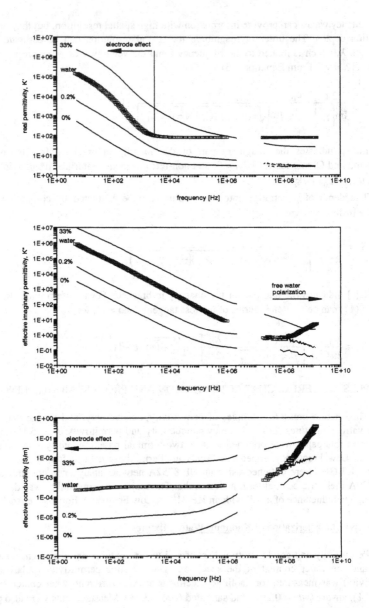

Figure 1   Permittivity and conductivity: characterization of kaolinite-water mixtures. Data for
deionized water are provided as reference. Numbers indicate the gravimetric moisture
content of the kaolinite-water mixtures.

142

Figure 1, Maxwell-Wagner and double layer polarizations coexist at low frequencies, but are overshadowed by electrode effects below 1-10 kHz. The polarization of free water molecules manifests as an increase in imaginary relative permittivity at approximately 100 MHz for the deionized water specimen; the contribution of molecular polarization losses are clearly seen in the effective conductivity data for all specimens.

Conductivity: Monitoring Soil-Cement Hydration

Bentonite-cement slurries are used to form permanent structural elements such as piles, retaining walls, or stabilized cut-off barriers [14, 15, 16, and 17]. See also the compilation of papers in [18]. Geophysical techniques offer a viable alternative for non-destructive in-situ monitoring of the continuity, integrity, and homogeneity of cut-off walls.

The hydration of cement and bentonite-cement slurries was monitored with conductivity measurements. A cement slurry at water content 40% and two bentonite-cement slurries at water contents 1150% and 2300% were tested. The cement was Type I Portland cement from the St. Lawrence Cement Company and the bentonite was a light green sodium bentonite (LL = 250 and PL = 50). Conductivity data are presented in Figure 2. Variations in conductivity correspond to the different stages of hydration. The cement slurry reached a low conductivity value after about 4 days, while the conductivities of the soil-cement slurries are still decreasing after more than 100 days.

Magnetic Permeability: Characterization of Kaolinite-Iron Mixtures

The magnetic permeability of soils varies depending on the amount of ferromagnetic material present. Magnetic permeability measurements may be useful in determining the purity of soils, such as kaolinite. Variations in the magnetic permeability of air dry kaolinite mixed with different amounts of fine iron particles Fe (1%, 10%, 25%, 50%, and 75% by weight) were determined over the frequency range 5 Hz to 13 MHz. Permeability was found to remain constant over the frequency range 0.1 to 100 kHz. Data show that the relative magnetic permeability of the mixtures is a power function of the percent iron Fe (Figure 3), so that the susceptibility of the mixtures is $(\mu_r-1) = \chi \approx 2 \cdot Fe^2$.

CONCLUSIONS

The electromagnetic parameters, conductivity σ, permittivity ε, and permeability μ provide complementary information about geomaterials. Assumptions made during the back-analysis of electromagnetic parameters affect the computed values. Conductivity and permittivity are often inverted together, obtaining "effective" parameters. If non-ferromagnetic behavior is assumed, the error in inverted permittivity values is proportional to μ'_r (assumes μ''_r is low).

The variation of electromagnetic parameters in time is a clear indicator of ongoing changes within the medium; hence electromagnetic measurements present unique opportunities to monitor processes.

For a given specimen, the variation of permittivity with frequency reflects different polarization mechanisms, which in turn suggest different spatial or temporal scales within the specimen, e.g. free water polarization versus Maxwell-Wagner interphasial polarization. Electrode polarization and limitations in phase resolution restrict low frequency permittivity measurements of wet soil specimens. The higher the conductivity of the mixture, the higher the limiting frequency.

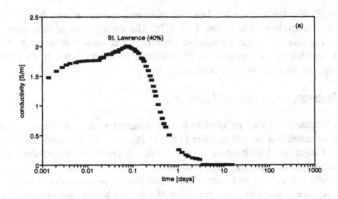

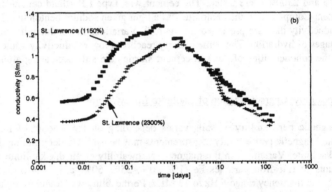

Figure 2   Monitoring soil-cement hydration with conductivity: (a) cement slurry;
(b) bentonite-cement slurry

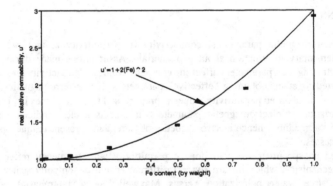

Figure 3   Permeability: Characterization of kaolinite-iron filing mixtures. Magnetic permeability
data were obtained at 10 kHz. The mixture porosities ranged from 0.64 to 0.73.

144

Powerful field tools for site characterization have been developed using electromagnetic waves. Their application is subjected to the compromise between resolution and penetration depth.

ACKNOWLEDGMENTS

Dr. Rosario Gerhardt's encouragement is appreciated. This study is part of a research program on wave-based characterization of geomaterials, which is partially supported by the National Science Foundation.

REFERENCES

1. R. Olhoeft in Physical Properties of Rocks and Minerals edited by Y. S. Touloukian, W. R. Judd, and R. F. Roy (McGraw-Hill Book Co., Toronto, 1981) pp. 257-329.
2. S. Carmichael, CRC Practical Handbook of Physical Properties of Rocks and Minerals (CRC Press, Boca Raton, Florida, 1989) p. 741.
3. A. Omar, Elementary Solid State Physics, (Addison-Wesley Publishing Company, Reading, Massachusetts, 1975) p. 669.
4. H. Ward in Geotechnical and Environmental Geophysics, Volume I: Geotechnical, Investigations in Geophysics No. 5, edited by S. Ward (Society of Exploration Geophysics, Tulsa, Oklahoma, 1990) pp. 147-189.
5. von Hippel, Dielectrics and Waves, John Wiley & Sons, Inc., New York, 1954) p.284.
6. S. Sumner, Principles of Induced Polarization for Geophysical Exploration, (Elsevier Scientific Publishing Company, New York, 1976).
7. I. Parkhomenko, Electrical Properties of Rocks, (Plenum Press, New York, 1967) p. 313.
8. P. de Loor, IEEE Trans. on Geoscience and Remote Sensing GE-21 (3), 364-369 (1983).
9. C. Santamarina and M. Fam, J. of Env. and Eng. Geophysics 2 (1), 37-51 (1997).
10. H. Reiger, Electrochemistry (Prentice-Hall, Inc., Englewood Cliffs, New Jersey, 1987).
11. J. Christensen, R. T. Coverdale, R. A. Olson, S. J. Ford, E. J. Garboczi, H. M. Jennings, and T. O. Mason, J. Amer. Ceramic Society 77 (11), 2789-2804 (1994).
12. Klein and J. C. Santamarina, Geotechnical Testing Journal, GTJODJ 20 (2), 168-178 (1997).
13. Vigoureux and C. E. Webb, Principles of Electric and Magnetic Measurements (Prentice-Hall, Inc., New York, 1936) p. 392.
14. P. Xanthakos, Slurry Walls (McGraw-Hill Inc., New York, 1979) p. 622.
15. P. Xanthakos, Slurry Walls as Structural Systems (McGraw-Hill, New York, 1994) p. 855.
16. G. H. Boyes, Structural and Cut-off Diaphragm Walls (John Wiley, New York, 1975) p. 181.
17. A. Millet, J. Y. Perez, and R. R. Davidson in ASTM STP 1129 Slurry Walls - Design, Construction, and Quality Control edited by D. B. Paul, R. R. Davidson, and N. J. Cavalli (ASTM, Philadelphia, 1992) pp. 42-66.
18. B. Paul, R. R. Davidson, and N. J. Cavalli, ASTM STP 1129 Slurry Walls - Design, Construction, and Quality Control (ASTM, Philadelphia, 1992) p. 425.

MAGNETIC TRANSITIONS STUDIED BY ELECTRICALLY BASED METHODS IN Mn-Zn FERRITE

P. Gutiérrez, A. Peláiz*, A. Huanosta and R. Valenzuela**
Institute for Materials Research, National University of Mexico, Mexico.
*On leave from University of Havana, Cuba.
**e-mail: monjaras@ servidor.unam.mx

ABSTRACT

The thermal variations in electrical conductivity of polycrystalline Mn-Zn ferrites was investigated by impedance spectroscopy. Two well defined semicircles were observed in complex impedance plots, which were associated with grain (or bulk) and grain boundary impedance response, for the high and low frequency ranges, respectively. From these results, the characteristic relaxation frequencies were obtained for each process. The grain boundary frequency exhibited a monotonous behavior as a function of temperature, while the bulk frequency showed a maximum at 132 °C. By a magnetic method, it was verified that this temperature corresponds to the Curie transition. These results show therefore that a magnetic phase transition can be studied by means of electrical methds.

INTRODUCTION

Electrical measurements, when carried out in a wide frequency range, can be extremely useful for the characterization of many materials [1]. In polycrystalline materials, which is the case of many ceramics, a resolution [2] of the impedance contributions from grains (or "bulk"), grain boundaries, and sometimes, electrodes, can be made. This allows the detailed investigation [3] of each of these parameters of polycrystals, separately.

Magnetic materials offer an additional interest and complexity, since they show magnetic order-disorder phenomena. These magnetic changes are related to a more basic change in electronic structure, which would affect as well the electrical conductivity mechanisms. In this paper, we report preliminary results showing that the relaxation frequency associated with the bulk conductivity exhibits evidence of such phase change in MnZn polycrystalline .ferrites

EXPERIMENT

Spinel ferrites of nominal composition $Zn_xMn_{1-x}Fe_2O_4$ (with $x = 0.5$) were prepared by the ceramic method in a toroidal shape. During sintering, typically at 1180°C for 12 h, an oxidant atmosphere (O_2 100%, 1 atm) was maintained. A single spinel phase was observed on x-ray diffraction patterns, with unit cell parameters in good agreement with the ones reported for this composition. A D-5000 Siemens system was used to obtain the diffraction patterns.

Frequency measurements were carried out at room temperature by means of a system including a HP 4192A Impedance Analyzer with a 5 Hz - 13 MHz frequency range, controled by a PC computer. Gold electrodes were previously formed on the surfaces of the samples by thermal treatment from a gold paste. Frequency measurements at different temperatures were carried out in an electric furnace with a temperature control leading to temperature variations ±3°C about the preset temperature. Typically, the sample was left to stabilize for 20 min at the desired temperature before obtaining the corresponding frequency run. A series of measurements were

Mat. Res. Soc. Symp. Proc. Vol. 500 © 1998 Materials Research Society

made when heating, and then another series for the cooling process. Both measurement runs were verified to be consistent.

The Curie temperature was determined by measuring the initial magnetic permeability in a system [4] that uses the sample as a transformer core. It can be shown that the thermal variations of permeability correspond to variations in the secondary voltage, which is connected to the Y-axis of a X-Y recorder. A thermocouple, connected to the X-axis, is used to monitor the thermal variations of the sample.

RESULTS AND DISCUSSION

Impedance measurements, represented in the complex Z_i-Z_r plane, showed the tendency to two slightly deformed semicircles, Fig. 1. It has been shown [2] that the high-frequency semicircle (the one close to the origin) can be associated with the impedance response of grains (or the bulk), and the low-frequency one represents the impedance due to the grain boundaries.

A simple equivalent circuit (inset in Figure 1) can be used to model these results: a parallel R_gC_g arm in series with another $R_{gb}C_{gb}$ parallel arrangement, where the subindex g stands for grains, and gb for grain boundaries, respectively. This circuit presents two relaxation dispersions, corresponding to the condition $Z_i \approx Z_r$ for each parallel arm, at frequencies f_g and f_{gb}, for the grains and grain boundaries arm, respectively. These appear clearly on Z_i (f) plots, Fig. 2.

The circuit element values can be extracted as follows: the resistors's values, R_g and R_{gb}, are simply the diameter for the corresponding semicircle. The relaxation frequencies, f_g and f_{gb}, are determined for each contribution from Z_i (f) plots. The capacitors values are calculated from $Z_i = Z_r$, as $C_g = 1/(2\pi f_g R_g)$ for the grain contribution, and the corresponding equation for the grain boundary contribution.

Frequency measurements were carried out as a function of temperature, from room temperature to 180 °C. On all this temperature range, two semicircles were observed on complex plane plots, with decreasing diameters as temperature increased, as expected for a temperature-activated conduction mechanism. An Arrhenius-type plot can be obtained.

The grain-boundary relaxation frequency, f_{xgb}, as a function of temperature is shown in Fig. 3. A monotonous behavior close to a straight line is observed when plotted in a semi-log scale. The relaxation frequency associated with grains, f_{xg}, however, showed a maximum at 132 °C, Fig. 4.

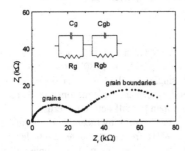

Fig. 1. Complex impedance plane Z_i-Z_r at room temperature.The equivalent circuit used to interpret the data is shown in inset.

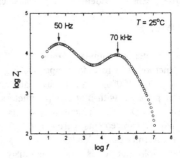

Fig. 2. Behavior of the imaginary part of impedance, Z_i, at room temperature.

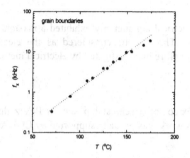

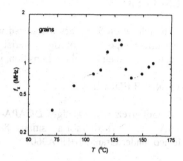

Fig. 3. Grain boundary relaxation as a function of temperature.

Fig. 4. Relaxation frequency associated with grains, as a function of temperature.

In order to elucidate the origin of this maximum, the initial magnetic permeability, μ, of the samples was measured from room temperature to 180 °C. The thermal behavior this property has shown [5] to be the best method to determine the Curie point, i.e., the transition from the ordered, ferrimagnetic phase, to the disordered paramagnetic state, Fig. 5. The Curie temperature, T_c, is 133 °C, which corresponds quite well to the temperature at which a maximum in f_{xg} appeared. This maximum is therefore closely related with the magnetic transition. Similar results have been obtained also for another type of magnetic ceramics, the Ni-Zn ferrites.

Magnetic transitions are not an isolated phenomenon; they are associated with other changes such as a discontinuity in the heat capacity coefficient, for instance, Fig. 6 [6]. From a thermodynamic point of view, magnetic transitions are considered as 2nd order transitions. Since they involve a change in the electronic structure as a whole, it is easy to understand that such a transition should lead to a change in the conductivity process. Electrical conductivity is the most sensitive property in nature; a ratio of about 10^{30} can be found between the best conductor and the best insulator. It is therefore expected that the magnetic phase affects this property in some way. It is important to note that the grain-boundary relaxation frequency showed no change at T_c; since this is a zone of the sample with a considerable disorder, its properties are significantly different from the grains.

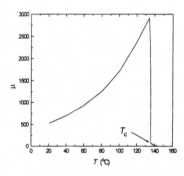

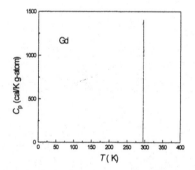

Fig. 5. Thermal behavior of the initial magnetic permeability.

Fig. 6. Heat capacity for Gd [6].

149

CONCLUSIONS

The relaxation frequency associated with grain electrical conductivity exhibited a maximum at the Curie temperature of the material investigated. This can be considered as the electrical evidence of a magnetic phase transition, which can therefore be investigated by electrical methods.

ACKNOWLEDGMENTS

P. Gutiérrez acknowledges DGAPA-UNAM México for a scholarship and A. Peláiz thanks DGIA-UNAM México for a visiting fellowship. This work was partially supported by CONACyT México, under Grant 3101P-A9607.

REFERENCES

1. J.R. McDonald. *Impedance Spectroscopy*. (J. Wiley and Sons, NewYork, 1987).
2. H.F. Cheng. J. Appl. Phys. **56**, 1831-1837 (1984).
3. J.T.S. Irvine et al. J. Am. Cer. Soc. **73**, 729-732 (1990)
4. E. Cedillo, V. Rivera, J. Ocampo and R. Valenzuela. J. Phys. E: Sci. Instr. **13**, 383 (1980).
5. R. Valenzuela. J. Mater. Sci. **15**, 3173-3174 (1980).
6. H.E. Nigh, S. Legvold and F.H. Spedding. Phys. Rev. **132**, 1092-1097 (1963).

APPLICATION OF BROAD-BAND DIELECTRIC SPECTROSCOPY FOR INVESTIGATIONS OF LIQUID CRYSTAL - POROUS MEDIA MICROCOMPOSITES

G.P. SINHA, B. BATALLA and F.M. ALIEV
Department of Physics and Materials Research Center, PO BOX 23343, University of Puerto Rico, San Juan, PR 00931-3343, USA

ABSTRACT

We applied ultra broad-band dielectric spectroscopy in the frequency range from 10^{-3} Hz to 10^9 Hz to investigate the effect of size, shape and volume fraction of the pores in the porous matrices on the dielectric properties of liquid crystals (LC) dispersed in these matrices. Measurements in such a broad frequency range make it possible to obtain detailed information on the important aspects of the electrical behavior of heterogeneous materials such as: conductivity, surface polarization, and influence of confinement on dynamics of molecular motion of polar molecules forming LC. We investigated alkylcyanobiphenyls in the isotropic, nematic and smectic phases dispersed in porous glasses (average pore sizes - 100 Å and 1000 Å) which have randomly oriented, interconnected pores, and anopore membranes (pore diameters - 200 Å and 2000 Å) with parallel cylindrical pores. Dispersion of LC resulted in qualitative changes of their dielectric properties. Analysis of broad-band dielectric spectra shows that in organic (LC) - inorganic (porous matrix) heterogeneous composites conductivity plays an important role at $f < 1$ Hz. We observe the appearance of new dielectric modes: a very slow process with characteristic frequency $\simeq (1 - 10)$ Hz and a second process in frequency range about $(10^3 - 10^6)$ Hz. The slow process arises due to the relaxation of interfacial polarization at pore wall - LC interface. The origin of this could be due to absorption of ions at the interface. Another possibility is the preferential orientation of the permanent dipoles at pore surface. The second new mode is due to the hindered rotation of the molecules near the interface. Additionally we observed two bulk like modes due to the rotation of the molecules around their short and long axii which are modified.

INTRODUCTION

Dielectric spectroscopy method contribute significantly to overall characterization of porous materials [1] in general, and investigations of condensed matter confined to porous matrices [2-11] in particular. Applications of dielectric spectroscopy to confined liquid crystals [2,8-11] and glass-forming liquids [3-7] revealed new information on the changes in the molecular mobility, broadening of the distribution of relaxation times as well as changes in phase and glass transition temperatures. Investigations of dielectric properties of condensed matter dispersed in porous media requires very broad frequency range, much broader than the range in which the pure material has frequency dependence of it's dielectric properties. This is because in two component heterogeneous systems new dispersion regions might arise due to existence of interface, possible surface induced polar ordering and surface polarization. Broad band dielectric spectroscopy in the frequency range $(10^{-3} - 10^9)$ Hz is a powerful method for investigations of all possible frequency dependent dielectric processes in such systems.

In this paper we present the results of investigations of dielectric properties and behavior of pentylcyanobiphenyl (5CB) and octylcyanobiphenyl (8CB) liquid crystals confined to porous glass matrices with randomly oriented, interconnected pores and to Anopore membranes with parallel oriented cylindrical pores. Broad band dielectric spectroscopy in the frequency range 10^{-3}Hz - $1.5 \cdot 10^9$Hz was applied to investigate the influence of the confinement on dynamical behavior of liquid crystals in isotropic, nematic and smectic phases. We found that confinement has a strong influence on the dielectric properties of LC which resulted in the appearance of a low frequency relaxational process ($f \leq 10$ KHz) absent in bulk. We also observed a strong modification of the bulk-like modes due to the molecular rotation and molecular tumbling.

EXPERIMENT

Measurements of the real (ϵ') and the imaginary (ϵ'') parts of the complex dielectric permittivity in the frequency range 10^{-3} Hz to 1.5 GHz were carried using two sets of devices. In the range from 10^{-3} Hz to 3 MHz we used Novocontrol Broad Band Dielectric Converter (BDC-S) with an active sample cell in combination to the Schlumberger Technologies 1260 Impedance/Gain-Phase Analyzer. The BDC-S and the active sample cell containing the sample holder, the sample capacitor, a high precision reference capacitors and active electronics optimize the overall performance and reduces the typical noise in measurement, particularly at low frequencies. Computer controlled measurements customarily were performed at one hundred different frequencies in the frequency range 10^{-3} Hz to 3 MHz. For measurements in the frequency range 1 MHz - 1.5 GHz we used Hewlett-Packard 4291A RF Impedance Analyzer in combination with a high temperature test head and a calibrated Hewlett-Packard 16453A Dielectric Material Test Fixture. In this frequency range measurements were performed at two hundred different frequencies with averaging done over three hundred times at each frequency. Both equipment allow to apply D.C. bias voltage. The accuracy of the temperature stabilization was better than $\pm 0.1^\circ C$.

We used matrices with randomly oriented, interconnected pores (porous glasses with average pore sizes of 100 Å and 1000 Å) and Anopore membranes with parallel oriented cylindrical pores 200 Å in diameter. The samples were porous glass plates, of dimension $2cm \times 2cm \times 0.1cm$ and Anopore membranes 60 μm thick, impregnated with 5CB and 8CB. The temperatures of phase transitions of 5CB in the bulk are $T_{CN}=295$ K and $T_{NI}=308.27$ K and for 8CB the bulk phase transition temperatures are: $T_{CSm}=294.2$ K, $T_{SmN}=306.6$ K and $T_{NI}=313.9$ K. Both matrices have practically negligible electrical conductivities, and their dielectric permittivities are independent of temperature and frequency over a wide range of frequencies.

RESULTS AND DISCUSSION

The dielectric properties of the bulk 4-n-alkyl-4'-cyanobiphenyls have been studied extensively [12-16] and have been quite clearly understood. In the nematic phases of pentylcyanobiphenyl (5CB) [12] and octylcyanobiphenyl (8CB), in a geometry in which the electric field E is parallel to the director n i.e. E∥n, [12,13] the real (ϵ') and imaginary (ϵ'') parts of the dielectric permittivity have a dispersion region with characteristic frequency at about 5 MHz. This dispersion is due to the restricted rotation of the molecules about their short axii and is of Debye type, i.e. it has a single relaxation time. The temperature dependence of the corresponding relaxation times (τ) obey empirical Arrhenius equation. For the geometry in which the electric field E is perpendicular to the director n, E⊥n the most prominent relaxational process with characteristic frequency about 70 MHz was observed, which has been attributed to the tumbling of the molecules about their molecular short axis [9].

The dielectric behavior of confined LC that we observed is different from it's bulk behavior. It is illustrated in Fig.1. This figure is a typical example of broad-band dielectric spectra we observed for all confined liquid crystals under investigation.

It is clear from Fig.1 that for 8CB confined in 100 Å five dispersion regions could be obviously identified as: a very slow process (10^{-3} Hz - 1 Hz), a second wide process (1 Hz - 10^4 Hz), a very clear process in MHz frequency range and the last one in the frequency range $f > 30$ MHz which is also visible without detailed analysis. Immediately after the first two slow processes, in 100 Å pores an additional relaxation preceding the MHz process is observed.

In 1000 Å the highest frequency relaxational process is more obvious than in 100 Å pores however the relaxation preceding the MHz process was barely seen.

For the quantitative analysis of the dielectric spectra the Havriliak-Negami function [17] has been used. For the case of more than one relaxational process and taking into account the contribution of the D.C. conductivity to the imaginary part the Havriliak-Negami function is given as:

$$\epsilon^* = \epsilon_\infty + \sum_j \frac{\Delta\epsilon_j}{[1 + (i2\pi f\tau_j)^{1-\alpha_j}]^{\beta_j}} - i\frac{\sigma}{2\pi\epsilon_0 f^n}, \tag{1}$$

where ϵ_∞ is the high-frequency limit of the permittivity, $\Delta\epsilon_j$ the dielectric strength, τ_j the mean relaxation time, and j the number of the relaxational process. The exponents α_j and β_j describe the symmetric and asymmetric distribution of relaxation times. The term $i\sigma/2\pi\epsilon_0 f^n$ accounts the contribution of conductivity σ, with n as fitting parameter. In the case of pure ohmic conductivity $n = 1$. (Presence of electrode polarization results in the decrease of n i.e. $n < 1$.) The decrease of n i.e. $n < 1$ could be observed, as rule, if additionally to the contribution to ϵ'' from conductivity, there is an influence of electrode polarization.

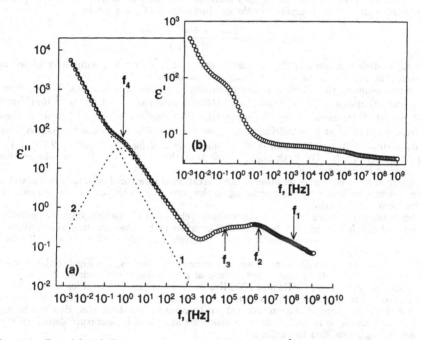

Figure 1: Broad band dielectric spectrum of 8CB in 100 Å random pores, log-log scale: $T = 295K$. (a) - imaginary part, (b) - real part of dielectric permittivity. The solid line in (a) is result of fitting the low frequency part of the spectrum using Havriliak-Negami function in combination with conductivity: (1)-contribution from conductivity, (2)-Havriliak-Negami process. The arrows identify characteristic frequencies of relaxational processes.

First of all for simplicity we analyze data at $f < 1$ kHz separately. Application of formula (1) clearly shows that the strong frequency dependence of ϵ'' at $f < 0.1$ Hz is due to ohmic conductivity and electrode polarization is negligibly small. The contribution from conductivity is perfectly described by the dotted line (1) in Fig.1 with parameters: $\sigma = 6.6 \times 10^{-10} Ohm^{-1} \cdot m^{-1}$ and $n = 0.97$. After removing the contribution described by the second term in formula 1 the frequency dependence of ϵ'' is represented by the dotted lines (2) in Fig.1. This dependence has relaxational origin and quantitatively is described by Havriliak-Negami formula with the following parameters: for 8CB in 100 Å $\alpha = 0; \beta = 0.8; \Delta\epsilon = 76.6 and \tau = 4.6 \times 10^{-1}$s. The main features of these low relaxation processes observed in 1000 Å and 100 Å, absent in bulk, are: (a) these processes are not described by Debye relaxation function, (b) the dielectric strength is very high and (c) these relaxation times are temperature dependent (for details of these temperature dependence for 5CB in 1000 Å and 100 Å pores see [8].) The existence of this low frequency relaxation process accompanied by huge increase in the dielectric strength suggests that both these facts are results of interfacial polarization arising at pore wall-liquid crystal interface. Generally two main different reasons may cause appearance of the interfacial polarization: (a) - absorption of ions or (b) - dipole polarization due to polar interactions of molecular dipoles with solid surface.

Ion absorption in two component heterogeneous materials gives rise to dispersion of dielectric permittivity known as the Maxwell-Wagner (M-W) effect [18]. The relaxation time due to the M-W effect, taking into account that the conductivity of empty porous matrix is negligibly small compared to the conductivity of LC, is given by

$$\tau_{MW} = 2\pi\epsilon_0 \frac{2\epsilon_m + \epsilon + \omega(\epsilon_m - \epsilon)}{\sigma(1 - \omega)}, \tag{2}$$

where ϵ_m is dielectric permittivity of matrix material, ϵ dielectric permittivity of second component, in our case LC and ω is volume fraction of pores.

One can estimate the relaxation time according to formula (2) using $\epsilon_m = 3.95$, determined conductivities $\sigma = 2 \times 10^{-9} Ohm^{-1} \cdot m^{-1}$ (1000 Å pores) and $\sigma = 6.6 \times 10^{-10} Ohm^{-1} \cdot m^{-1}$ (100 Å pores). If we use as dielectric permittivity of second component LC, either of these three values obtained at $T = 295K$:$\epsilon_\parallel \approx 13, \epsilon_\perp \approx 5.5, \langle\epsilon\rangle \approx 8.0$, we obtain for 1000 Å pores relaxation times 0.70 s, 0.52 s and 0.58 s respectively and for 100 Å pores: 2.1 s, 1.49 s and 1.71 s respectively. For both pores we see that these times are 10 times slower than determined in experiment.

Possibly at low frequencies we observe the relaxation of interfacial polarization due to formation of a surface layer with polar ordering on the pore wall. In this case a new cooperative and slow process may arise.

Even though the above analysis tends towards relaxation mechanism due to orientation of permanent dipoles at pore wall-liquid crystal interface rather than due to Maxwell-Wagner mechanism, it should be noted that the above analysis was very qualitative and all values are just estimations.

We assign the process with characteristic frequency identified as f_3 in Fig.1 to the rotation of molecules, located in the surface layer formed at pore walls, about their short axii.

The last two processes in MHz and 100 MHz frequency range observed for both matrices are "bulk like". The first one is due to molecular rotation about its short axis and the second one is due to the tumbling of the molecules about their molecular short axis. However having the same mechanism as in bulk these two processes and their temperature dependence in pores are strongly modified by confinement.

These two modes were quantitatively analyzed by using equation (1). The analysis of bulk like modes shows that the parameters α and β in Havriliak-Negami function are not equal to 0 and 1 respectively and thereby differing from Debye type behavior. The temperature dependence of relaxation times obtained as result of this analysis is presented in Fig. 2 for 5CB in 200 Å cylindrical pores and for 8CB in 1000 Å random pores. There are marked changes in the temperature dependence of relaxation times compared to the bulk behavior.

Rigorous consideration of the temperature dependencies of relaxation times due to molecular rotation in different pores shows that there are several differences: (a) - LC in pores, does not crystallize even at temperatures 20 degrees below the bulk crystallization temperature; (b) - in the temperature range corresponding to nematic (or nematic-like) extended phase $ln\tau$ is not a linear function of $1/T$; (c) - the change in relaxation time at nematic-isotropic phase transition in pores is not as sharp as in the bulk.

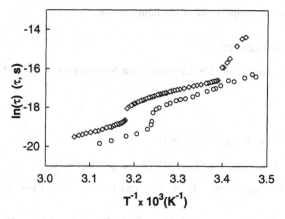

Figure 2: Temperature dependence of relaxation times corresponding to the molecular rotation around short axis for 5CB in 200 Å cylindrical pores (open circles) and 8CB in 1000 Å random pores (open diamonds).

Relatively smooth and small changes in τ at phase transition in pores suggest that the "isotropic" phase of LC in pores is not bulk like with complete disorder in molecular orientations, and some orientational order still persists.

The relaxational process at the frequencies $f > 50$ MHz was observed in 5CB and 8CB for an orientation of the electric field perpendicular to the director. In random pores, there are always molecules oriented perpendicular to the electric field regardless of the molecular alignment inside pores. For the Anopore membrane with cylindrical pores, the applied electric field was parallel to the pore axis. Since at frequencies greater than 1 MHz the process corresponding to the rotation of the molecules around short axis dominates, we conclude that majority of the molecules are oriented along the pore axis. However due to the fact that the high frequency process (at $f > 30$ MHz) is also present, we assume that small fraction of molecules presumably at the pore walls are tilted with respect to the pore axis direction.

So we suppose that the mode observed in the frequency range $f > 30$ MHz has the same origin as was found earlier for bulk alkylcyanobiphenyls. However this mode in pores like other modes observed at lower frequencies is not Debye like.

CONCLUSION

Application of broad band dielectric spectroscopy is powerful method for investigations of dielectric properties of liquid crystal - porous media microcomposites. This method makes it possible to obtain complete information about different aspects of dielectric behavior of two component heterogeneous systems. The separation of modes of different physical origin such as modes due to molecular rotation, surface polarization modes as well as dispersion due to conductivity can be performed. Detailed qualitative and quantitative information characterizing these modes can be obtained. The spatial confinement has a strong influence on the dielectric properties of LC. Slow relaxational process which does not exist in the bulk phase was observed. All observed relaxational modes, even bulk like modes are characterized by spectrum of relaxation times and are not described by Debye equation. The dielectrically active modes were not completely frozen even at temperatures at least 20 °C below the bulk crystallization temperature.

These investigations should be extended to the temperature range which is much below the bulk crystallization temperature of the liquid crystal.

ACKNOWLEDGEMENTS

This work was supported by US Air Force grant F49620-95-1-0520 and NSF grant OSR-9452893.

REFERENCES

1. R. Hilfer, Phys. Rev. B **44**, p. 60 (1991).

2. F.M.Aliev, M.N.Breganov, Sov. Phys. JETP **68**, p. 70 (1989).

3. J. Schuller, Yu.B. Mel'nichenko, R. Richert, and E.W. Fischer, Phys. Rev. Lett. **73**, p. 2224 (1994).

4. M. Arndt and F. Kremer in: Dynamics in Small Confining Systems II, edited by J.M. Drake, J. Klafter, R. Kopelman, S.M. Troian, (Mater. Res. Soc. Proc. **363**, Pittsburgh, PA 1995), pp. 259-263.

5. Yu. Mel'nichenko, J. Schuller, R. Richert, B. Ewen and C.-K. Loong, J. Chem. Phys. **103**, p. 2016 (1995).

6. J. Schuller, R. Richert, E.W. Fischer, Phys. Rev. B **52**, 15232 (1995).

7. M. Arndt, R. Stannarius, W. Gorbatschow, and F. Kremer, Phys. Rev. E **54**, 5377 (1996).

8. F.M. Aliev and G.P. Sinha in: Electrically based Microstructural Characterization, edited by R.A. Gerhardt, S.R. Taylor, and E.J. Garboczi (Mater. Res. Soc. Proc. **411**, Pittsburgh, PA 1996), pp. 413-418.

9. S.R. Rozanski, R. Stanarius, H. Groothues, and F. Kremer, Liquid Crystals **20** p. 59 (1996).

10. G.P. Sinha and F.M. Aliev, MCLC **304**, p. 309 (1997).

11. G.P. Sinha and F.M. Aliev in: Dynamics in Small Confining Systems III, edited by J.M. Drake, J. Klafter and R. Kopelman (Mater. Res. Soc. Proc. **464**, Pittsburgh, PA 1997), pp. 195-200.

12. P.G. Cummins, D.A. Danmur, and D.A. Laidler, MCLC **30**, p. 109 (1975).

13. D.Lippens, J.P.Parneix, and A. Chapoton, J. de Phys. **38**, p. 1465 (1977).

14. J.M. Wacrenier, C. Druon, and D. Lippens, Molec. Phys. **43**, p. 97 (1981).

15. T.K. Bose, R. Chahine, M. Merabet, and J. Thoen, J. de Phys. **45**, p. 11329 (1984).

16. T. K. Bose, B. Campbell, S. Yagihara, and J. Thoen, Phys. Rew. A **36**, p. 5767 (1987).

17. S. Havriliak and S. Negami, Polymer **8**, p. 101 (1967).

18. B.K.P. Scaife, Principles of Dielectrics, (Clarendon Press, Oxford, 1989).

STUDY OF THE MOLECULAR MOBILITY OF POLYSACCHARIDE SOLID THIN LAYERS BY DIELECTRIC RELAXATION SPECTROSCOPY

K. LIEDERMANN[*], Ľ. LAPČÍK, JR.[**], S. DESMEDT[***]

[*] Department of Physics, Faculty of Electrical Engineering and Computer Science, Technical University of Brno, Technicka 8, CZ -616 00 Brno, Czech Republic, liederm@dphys.fee.vutbr.cz

[**] Department of Physics and Material Science, Faculty of Technology in Zlín, Technical University of Brno, nám. TGM 275, CZ-762 72 Zlín, Czech Republic, lapcik@zlin.vutbr.cz

[***] Laboratory of General Biochemistry and Physical Pharmacy, Faculty of Pharmaceutical Sciences, University of Gent, Harelbekestraat 72, B-9000 Gent, Belgium, Stefaan.Desmedt@rug.ac.be.

ABSTRACT

Temperature dependence of measured dielectric relaxation spectra (DRS) in the frequency range 20 Hz – 1 MHz of hydroxyethylcellulose (HEC) are in the temperature range 100 – 350 K of Arrhenius character with one relaxation process at 150 – 250 K. This process reflects most probably β-relaxation of the side chain groups. Calculated activation energy of this process was 5730 kJ/mole. Four types of polysaccharides were studied at 293 K temperature: hyaluronic acid (HA), chondroitin sulfate (CHS), HEC and carboxymethylcellulose (CMC), in the low-frequency range 10^{-5} - 10^{0} Hz. Measured dielectric spectra were interpreted as sum of one A.C. conductivity process and of up to two relaxation processes. The relaxation processes were described by means of the Havriliak--Negami formula and their parameters were related to the molecular structure of the polymers. The low value of α in CHS is related to its strong coupling due to the presence of two polar groups in its monomeric unit, whereas low values of $\alpha \times \beta$ are interpreted as being due to the strong steric hindrances caused by long pendants present in HEC.

INTRODUCTION

From the physicochemical point of view it is desirable to know, to understand and to quantify the molecular mobility of the polymer main chain, its side–chain groups and its different skeletal segments. For the application as coatings of the affinity sensors, it is desirable to cover the surface of the measuring device with the membrane having both controllable pores size distribution as well as the specific affinity to various diffusants. Suitable materials, which find increasing application for these purposes seems to be polysaccharides, particularly hyaluronic acid. Molecular mobility properties of polysaccharides may be studied by means of the Dielectric Relaxation Spectroscopy (DRS), as both the main chain and the side chain groups are of a polar character. DRS method is particularly suitable for the analysis of polymers with a high degree of structure similarity, as the dielectric relaxation spectra reveal even minor differences in the dipolar constitution.

EXPERIMENT

Dielectric spectra were obtained from the time–domain measurements. Measuring apparatus consisted of a digital electrometer Keithley 617, stabilized D.C.–power supply, relay control circuitry, and personal computer interconnected *via* IEEE–488 bus. For the majority of the measurements, the charging period was set to be 24 h. For temperature dependent measurements in the frequency range of 20 Hz – 1 MHz a digital electrometer HP4284 was used.

Prior to charging, samples were kept short-circuited for at least 12 h in order to remove any stray charges. Measured samples were of the circular or rectangular shape with graphite electrodes of the size about 12 mm × 20 mm pasted onto their surface. Samples were fixed in the sample holder and placed into the thermostatic chamber. The measuring chamber was provided with a simple installation based on the oversaturated solution of K_2CO_3 to keep constant humidity during the course of the measurement.

Four different types of polysaccharides were studied, hyaluronic acid extracted from rooster combs in the form of sodium salt (Sigma), chondroitin sulfate C (Sigma), hydroxyethylcellulose Natrosol 250 H (Hercules), and carboxymethylcellulose Blanose D in the form of sodium salt (Hercules).

THEORY

Results of the time–domain measurements were obtained in the form of the time dependence of the discharge current. The Fourier Transformation (FT) relates time-domain and frequency–domain results between each other. The time-domain data were transformed into the frequency domain according to the formula [1]

$$\hat{\varepsilon}(\omega) = \varepsilon_\infty + (\varepsilon_s - \varepsilon_\infty) \int_0^\infty \varphi(t) \exp(-j\omega t) dt \tag{1}$$

or after the separation into the real (ε') and imaginary part (ε'') :

$$\varepsilon'(\omega) = \varepsilon_\infty + (\varepsilon_s - \varepsilon_\infty) \int_0^\infty \varphi(t) \cos(\omega t) dt \tag{2}$$

$$\varepsilon''(\omega) = (\varepsilon_s - \varepsilon_\infty) \int_0^\infty \varphi(t) \sin(\omega t) dt \tag{3}$$

where $\varphi(t)$ is the decay function of the dielectric related to the discharge current $i(t)$ by the formula

$$i(t) = -\varepsilon_0 (\varepsilon_s - \varepsilon_\infty) \frac{U A}{d} \varphi(t) \tag{4}$$

In the above formulas the symbols have their usual meaning; $\hat{\varepsilon}(\omega)$ is the frequency – dependent complex permittivity, ε_∞ optical permittivity of the sample (corresponding to the electronic polarizations), ε_s static permittivity of the sample (corresponding to the total polarization including dipole and interfacial contributions), $\varphi(t)$ decay function, U applied voltage, A sample surface area, and d sample thickness.

In view of the slowly decreasing discharge current with only small variations of the slope of the $i(t)$ curve in the logarithmic coordinates, it is possible to use a simplified form of the Fourier Transformation, the so–called Hamon approximation [1]

$$\varepsilon''(f) = \frac{i\left(\dfrac{0.1}{f}\right)}{2\pi C_0 U f} \tag{5}$$

where $f = \omega/2\pi$, C_0 is the geometrical capacitance of the sample and i discharge current. Dielectric spectra were analyzed in terms of conductivity and relaxation processes. The contribution of the conductivity component was modeled using the formula

$$\varepsilon''_c(\omega) = A\omega^{-n} \tag{6}$$

where the value of n equal to one corresponds to the purely D.C. conductivity mechanism. The contribution of the relaxation process to the complex permittivity was modeled with *Havriliak–Negami* formula [2, 3]

$$\hat{\varepsilon}_r(\omega) = \varepsilon_\infty + \frac{\varepsilon_s - \varepsilon_\infty}{\left[1 + (j\omega\tau_0)^\alpha\right]^\beta} \tag{7}$$

where τ_0 indicates the position of the relaxation on the time scale, α describes the slope of the low–frequency side of the relaxation, and the product $\alpha \times \beta$ describes the slope of the high–frequency side of the relaxation.

Dielectric spectra obtained by means of eqn (5) were fitted by nonlinear least–squares method with a sum of up to two conductivities and two relaxation processes described by formulas (6) and (7).

RESULTS

Time dependencies of the discharge current of all samples exhibit a steady decrease of the discharge current with time. This decrease is characteristic with the changing slope suggesting thus the presence of the relaxation process. The results of the fits of the measured relaxation spectra are summarized in Table 1. In view of eqn (5) frequency interval of the dielectric spectrum after Fourier Transformation was $10^{-5} - 10^0$ Hz. The values of A and of the relaxation strength $(\varepsilon_s - \varepsilon_\infty)$ are only relative values, since they were not normalized to different thickness of the measured samples. In most cases one relaxation process and one conductivity component is sufficient for a reasonable agreement between the fit and the measured results. Only in the case of the carboxymethylcellulose the fit with one relaxation process was apparently not satisfactory enough to reach the optimum minimization conditions. Therefore the fit with two relaxation mechanisms was necessary to perform (Table 1).

Value of the exponent n in eqn (6) was for all samples studied lower than 1, which indicates the presence of mechanism of A.C. conductivity [2]. The origin of the A.C. conductivity is usually interpreted as being due to the motion of loosely bound electric charges that form temporary neutral pairs during the conduction process. These neutral pairs have no electric charge and therefore during the period of their temporary existence they do not contribute to the conductivity [5, 6].

Table 1. Calculated Parameters of Measured Dielectric Spectra of Studied Polysaccharides at 293 K.

Parameter	Measured sample			
	Hyaluronic acid	Chondroitin sulfate C	Hydroxyethyl -cellulose	Carboxymetyl -cellulose
$10^3 (\varepsilon_s - \varepsilon_\infty)$	2.06	4.8	2.3	37.6 2.2
α	0.73	0.29	0.68	0.60 0.97
$\alpha \times \beta$	0.84	1.17	0.68	0.60 0.97
$10^{-3} \tau_0 /s$	1.0	3.4	5.1	68.1 3.8
A	3×10^{-18}	6×10^{-22}	1×10^{-18}	3×10^{-19}
n	0.47	0.92	0.54	0.70

Scaling parameters are related to the molecular mobility. They reflect the existence of some ordering in the molecular system. Parameter α is known to decrease with the increase of the intermolecular interactions and long-range correlation of segmental motion [4]. Hence, the intermolecular interaction and long-range correlation are the strongest in chondroitin sulfate ($\alpha = 0.29$) and the weakest in hyaluronic acid ($\alpha = 0.73$). The product $\alpha \times \beta$ decreases with the increase of hindrances for conformational transitions in the polymer chain. Therefore, the effect of steric hindrances seems to be most pronounced in the case of hydroxyethylcellulose ($\alpha \times \beta = 0.68$). This is most probably due to the presence of the long side chain groups in the monomeric unit of studied hydroxyethylcellulose.

Measured relaxation times τ_0 for all samples at 293 K are all of the order of 1000 s, although the side groups of individual polysaccharides are different. For this reason the origin of the measured relaxation was ascribed to the large scale rearrangements of the main macromolecular chain, so called α-relaxation process. It can be proposed that the relaxation times of such a large scale rearrangements would not be much different because of the high similarity of the macromolecular main chain structural components of the studied biopolymers [5,6].

Temperature dependent measurements of DRS in the range of 100 – 350 K of HEC show in the temperature range of 150 – 250 K a broad relaxation process in the frequency range of 20 Hz – 1 MHz. Calculated activation energy of this process was 5730 kJ/mole. Measured temperature dependence of DRS was of the Arrhenius character, suggesting that the observed relaxation process is due to the β-relaxation of the side chain groups.

CONCLUSIONS

Four types of polysaccharides were studied: hyaluronic acid (HA), chondroitin sulfate (CHS), hydroxyethylcellulose (HEC) and carboxymethylcellulose (CMC), in the low-frequency range 10^{-5} - 10^{0} Hz. Observed dielectric spectra were interpreted as sum of one A.C. conductivity process and of up to two relaxation processes. The relaxation processes were described by means of the Havriliak--Negami formula and their parameters were related to the molecular structure of the polymers. The low value of α in CHS is related to its strong coupling due to the presence of two polar groups in its monomeric unit, whereas low values of α × β are interpreted as being due to the strong steric hindrances caused by long pendants present in HEC. Temperature dependent measurements of DRS in the range of 100 – 350 K of HEC show in the temperature range of 150 – 250 K a broad relaxation process in the frequency range of 20 Hz – 1 MHz. Calculated activation energy of this process was 5730 kJ/mole. Measured temperature dependence of DRS was of the Arrhenius character.

ACKNOWLEDGMENTS

Authors would like to express their gratitude for financing of this research by Grant Agency of Czech Republic (grant No. 101/97/0976), Ministry of Education, Youth and Sports of Czech Republic (grant No. VS 96 108) and to Technical University of Brno (grant No. FU470004/97). One of the authors K.L. would like to express his special gratitude to DAAD (Germany) for providing support for temperature dependent measurements

REFERENCES

1. Hamon, B. V., Proceedings IEE **99**, Part IV, 151 (1952).

2. Havriliak, S. and Negami, S., J. Polym. Sci., C, Polym. Symp. **14**, 99 (1966).

3. Havriliak, S. and Negami, S., Polymer **8**, 161 (1967).

4. Feldman, Y. and Kozlovich, N., TRIP **3**, 53 (1995).

5. Liedermann K., Lapčík, Ľ., Jr., Demeester J., Proceedings of the 5-th International Conference on Properties and Applications of Dielectric Materials May 25 – 30, 1997, Seoul, Vol. 1, pp. 541 – 544.

6. Liedermann K., Lapčík Ľ., Jr., IEEE 1996 Annual Report, Conference on Electrical Insulation and Dielectric Phenomena, Vol. I, October 20 – 23, 1996, San Francisco, pp. 213 – 215.

Part IV
Dielectrics and Ferroelectrics

Universality of Dielectric Response as an Aid to Diagnostics

Andrew K Jonscher

Royal Holloway, University of London
Egham, Surrey, TW20 0EX

Abstract

The advent of frequency response analysers and the formulation of the universality of dielectric responses have between them greatly enhanced the usefulness of dielectric measurements as a diagnostic tool. The available frequency range was extended to between eight and twelve decades, making it possible to determine spectra with some precision. In addition, we now have a much more complete understanding than was available previously of the significance of various spectral shapes. For charge carrier systems the universal approach is based on the fractional power law of frequency dependence of the real and imaginary components of the susceptibility $\tilde{\chi}(\omega) = \chi'(\omega) - i\chi''(\omega) \propto (i\omega)^{n-1}$, where the exponent falls in the range $0 < n < 1$. Values of n close to unity correspond to low-loss behaviour, values close to zero to very lossy processes dominated by low-frequency dispersion (LFD). Examples will be presented of both extremes and an indication will be given of the theoretical significance of these results. A brief discussion will be given of the physical principles of low-loss dielectrics showing "flat" or frequency-independent χ and of the opposite limit of LFD. It will be shown how the presence of universality simplifies the analysis of data and their interpretation.

Introduction

Measurements of dielectric parameters have often been used to characterise a wide range of materials and a considerable volume of literature has appeared on this subject while much work remains unpublished because its presentation does not measure up to the standards required in modern literature. The available range of measurements was expanded in the past twenty years to sub-Hertz frequencies with the advent of Frequency Response Analysers, opening an entirely new territory.

Dielectric measurements characterise the capacitance and the loss of a sample and because of the added dimension of frequency in the chosen range, which may encompass between six and ten decades, they provide a very large source of information compared with the conventional static information. Because of this wealth of information, it is essential to use appropriate forms of representation, without which it may be difficult to realise fully the potential of the data. The main purpose of the present paper is to discuss this matter of presentation and to draw attention to considerable advantages which may be gained from its proper use.

Mat. Res. Soc. Symp. Proc. Vol. 500 © 1998 Materials Research Society

Dielectric measurements characterise primarily *rate processes* governing the response of a material to alternating electric fields. The fundamental difference between direct current (dc) and alternating current (ac) or dielectric measurements consists in the fact that the former relate to the *steady state* condition under a static electric field, while the latter probe the *transient* response at various frequencies, as it were in the process of reaching the steady state but never quite getting there. Thus the dc conductivity, σ_0 is determined by movements of charge carriers all the way from one electrode to the other, while the ac conductivity $\sigma'(\omega)$ at a radian frequency ω is determined by *partial* movements within each cycle, without requiring the complete traversal of the sample. In particular, ac measurements reveal clearly the movements of *bound* charges in the form of *dipoles*, which cannot give rise to dc conduction but easily give *charge displacement* leading to the concept of *capacitance, ε*. The dynamics of these partial displacements constitutes the essence of dielectric measurements.

Dielectric measurements may be carried out in the time domain (TD) as functions of time under step-function field E, in which the time-dependent current $i(t)$ determines the relaxation function $f(t) \equiv i(t)/E$, or in the frequency domain (FD) where the charge displacement $Q(\omega)$ within a cycle of the alternating field $E(\omega)$ at a frequency ω defines the dielectric permittivity $\varepsilon(\omega) = Q(\omega)/E(\omega)$.

Basic Relations

While TD responses are always real numbers as functions of real time, FD response has to take into account not only the *amplitude* but also the *phase* of the resulting signal and this means that there are now two parameters and they are expressed as the real and imaginary components of the complex response - the following presentation is an abridged rendering of a much more comprehensive discussion which may be found in references [1,2]

$$\tilde{\varepsilon}(\omega) = \varepsilon'(\omega) - i\varepsilon''(\omega) \tag{1}$$

The imaginary component $\varepsilon''(\omega)$ represents the loss of energy per radian and is therefore related to ac conductivity through

$$\varepsilon''(\omega) = \sigma'(\omega)/\omega \tag{2}$$

Defining the dielectric susceptibility

$$\tilde{\chi}(\omega) = \left[\tilde{\varepsilon}(\omega) - \varepsilon_\infty\right] / \varepsilon_0 \equiv \chi'(\omega) - i\chi''(\omega) \tag{3}$$

where ε_∞ is the high frequency limit of ε and ε_0 is the permittivity of free space, we note that the TD and FD quantities are Fourier transforms of one another

$$\chi'(\omega) = \int_0^\infty f(t)\cos(\omega t)dt$$

$$\chi''(\omega) = \int_0^\infty f(t)\sin(\omega t)dt \tag{4}$$

Since these equations define $\chi'(\omega)$ and $\chi''(\omega)$ in terms of the same physical variable $f(t)$, it is possible to express them directly in terms of one another, which leads to the following equations known as Kramers-Kronig (KK) relations:

$$\chi'(\omega) = \tfrac{2}{\pi} \int_0^\infty \frac{x\chi''(x)}{x^2 - \omega^2}\, dx$$

$$\chi''(\omega) = -\tfrac{2}{\pi} \int_0^\infty \frac{x\chi'(x)}{x^2 - \omega^2}\, dx \tag{5}$$

These define either of the components in terms of the entire frequency spectrum of the other. They are of fundamental significance and are valid completely generally, regardless of any particular physical models. We shall be making extensive use of KK relations in our treatment of the universal response.

A common form of representation is the complex plane plot of ε'' vs ε' with frequency as the implicit variable. This type of representation, known as the Cole-Cole plot, has superficial advantages of simplicity in dipolar situations - see below - but its drawback is the emphasis it places on the loss peak region with the neglect of the "wings".

Special cases of dielectric response

There are two basic types of dielectric response - dipolar in which there is a clear loss peak in the spectrum at a frequency ω_p and carrier-dominated which do not show any loss peaks.

The classic example of dielectric behaviour is represented by the Debye response, which corresponds to the analytical expression

$$\bar{\varepsilon}(\omega) = \varepsilon_\infty + \frac{\Delta\varepsilon}{1 + i\omega / \omega_p} \tag{6}$$

This gives a symmetric peak in FD and a semicircle in the Cole-Cole plot, as shown in Figure 1. The TD equivalent of this behaviour is the exponential relation

$$f(t) \propto \exp(-\omega_p t) \tag{7}$$

Real materials have more or less distorted Cole-Cole plots and we are not going into this aspect in further detail, beyond stating that they are well represented by the empirical formula due to Havriliak and Negami [3] (HN)

$$\tilde{\chi} \propto \left[1 + \left(i\omega / \omega_p \right)^m \right]^{\frac{n-1}{m}} \tag{8}$$

which contains two parameters m and n falling between 0 and 1, leading to a loss peak that is asymmetric and broader than for the case of Debye which may be regarded as corresponding to $m = 1$ and $n = 0$. This has the limiting power-law frequency dependence

$$\chi'(\omega) = \tan(n\pi/2)\, \chi''(\omega) \propto \omega^{n-1} \tag{9}$$

so that both χ' and χ'' are the same functions of frequency, at least at high frequencies $\omega \gg \omega_p$, while the low-frequency behaviour is

$$\chi''(\omega) = \tan(m\pi/2) \left[\chi(0) - \chi'(\omega) \right] \propto \omega^m \qquad (10)$$

This corresponds to an asymmetric broadened loss peak as shown in Figure 2a) with the corresponding Cole-Cole plot in diagram b).

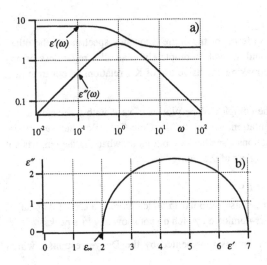

Figure 1 The response of an ideal Debye system with $\varepsilon_\infty = 2$, a) in the FD and b) in the complex plane plot. Note the logarithmic slopes of ±1 in loss.

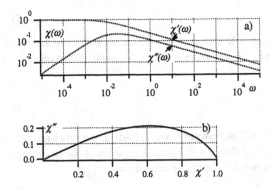

Figure 2 a) The frequency spectrum of a broadened dipolar system obeying the HN function with $1 - n = 0.3$ and $m = 0.7$, showing the constant ratio beyond the loss peak. b) shows the corresponding polar plot with its characteristic straight line portion on the high-frequency side.

A general form of TD response which has been widely invoked particularly in carrier-dominated systems is the *stretched exponential* or Kohlrausch-Williams-Watts (KWW) function containing one extra parameter $1 - n$

$$f(t) \propto \frac{d}{dt} \exp\left[-(\omega_p t)^{1-n}\right] \approx t^{-n} \tag{11}$$

the second approximate relation being valid for short times, $t \ll 1/\omega_p$. The limit $n \to 0$ corresponds to the Debye process.

The common property of these "non-Debye" response functions is the power-law TD behaviour. While the Fourier transform of the complete KWW function into the FD is not simple, its high-frequency branch is the same power law as in eqn (9). An overview of dielectric responses found in various simple models is given in Table I.

TABLE I
Types of polarising transitions and their frequency dependence

Charge carrier systems	Dipolar systems	Resulting law
Non-interacting carriers hopping between pairs of sites	Non-interacting permanent dipoles	Debye in FD or exponential in TD
Interacting electronic or ionic charges hopping between pairs of sites	Interacting permanent dipoles	Two-exponent HN in FD, deviation from Debye increasing with strength of interaction.
Highly dilute interacting hopping systems	Highly dilute interacting dipolar systems	Limiting "flat" loss with $\chi'(\omega)$ and $\chi''(\omega)$ independent of frequency
Ionic conductors with ions hopping between nearest-neighbour sites interstitially or substitutionally		KWW stretched exponential in TD
Hopping "high" density ionic or electronic systems		Low-Frequency Dispersion

Systems in which hopping charge carriers, as distinct from dipoles, make the dominant contribution to dielectric response have no loss peak in their spectra which is replaced by a strongly dispersive low-frequency power law with a much higher value of the exponent $1 - n_2$ close to unity, while the high frequency exponent $1 - n_1$ is close to zero. A typical response is shown in Figure 3, with a dispersive low-frequency branch known as the Low-Frequency Dispersion (LFD)[2] and a less dispersive high-frequency branch. It is important to appreciate the difference between LFD and the conventional dc conduction - the latter has a loss component equal to σ_0/ω while the real part is strictly independent of frequency, the former has both components proportional to ω^{n-1}, however small n may be. The physical reason for this is that, by definition, dc conduction cannot give rise to any charge storage, hence $\varepsilon' = $ const, while the steeply rising ε' in LFD points to a finite amount of charge storage. The precise mechanism of this storage is not always evident [2] but it is very likely related to some inhomogeneities in the system.

The TD equivalent of the fractional power law (9) is another fractional power law in time

$$f(t) \propto t^{-n} \tag{12}$$

the TD equivalent of LFD is an almost - but not quite - constant current, stressing its close relationship to dc conduction, which should be completely independent of time.

One final type of very typical dielectric response which is found in many low-loss materials is the frequency-independent or "flat" loss, with a correspondingly flat $\chi'(\omega)$. This corresponds to the limit of $n \to 1$ with a very large ratio $\chi'(\omega)/\chi''(\omega)$. Our empirical definition of flat loss accepts variation of $\chi''(\omega)$ by a factor of not more than 2 or 3 in, say, four or five decades of frequency.

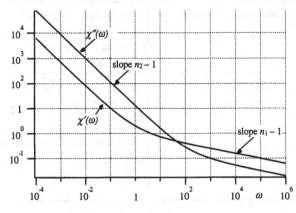

Figure 3 The spectrum of a carrier-dominated material with Low-Frequency Dispersion (LFD) followed at higher frequencies by a conventional low-loss universal process. Here n_1 is close to unity, 0.7 in the present instance, n_2 is close to zero, 0.2 in this instance. The constant ratios $\chi'(\omega)/\chi''(\omega)$ are given by the respective $\tan(n\pi/2)$ according to eqn (9).

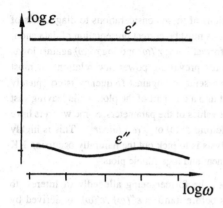

Figure 4 A schematic representation on logarithmic scales of a "real-life" flat loss material, showing a slight but finite variation of $\varepsilon''(\omega)$ and a virtually constant $\varepsilon'(\omega)$. The ε plots shown here instead of the χ plots emphasise the ratio $\varepsilon'(\omega)/\varepsilon''(\omega)$ which is larger than the corresponding ratio $\chi'(\omega)/\chi''(\omega)$ because of the contribution of ε_∞. The ticks on the axes are intended to signify orders of magnitude.

A schematic representation of flat spectrum is shown in Figure 4, where a finite "waviness" is shown on $\varepsilon''(\omega)$ which is always present in real life situations but which does not alter the basic flatness. On a logarithmic scale, the real part $\varepsilon'(\omega)$ is virtually constant because of its much higher value. One way of picturing how this might arise in practice is to consider a dipolar loss process which is being progressively reduced by purification of the material, giving rise to a lowering of the loss peak, as shown in Figure 5. The appearance of a flat low loss proves that there is an underlying constant loss mechanism which does not depend on the presence or absence of any dipolar processes - in fact there has been hardly any discussion of the phenomenon of flat loss in the literature.

The precise physical cause of this widely observed "flat" loss is not understood, although there is a suggestion[4] that this may be due to dipolar screening in materials with low, but not necessarily very low, dipole densities.

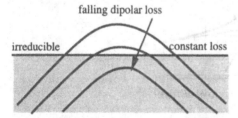

Figure 5 A schematic representation of the formation of a "flat" low loss in a system in which the density of impurity dipoles is being progressively reduced by suitable purification. As the dipolar contribution decreases, the loss does not fall to zero but instead is limited to an irreducible constant level.

Diagnostic implications

We are now able to look at the practical implications of these considerations to diagnostics of dielectric materials. The starting point of any analysis must be correct presentation of data and it is evident that the overwhelmingly most suitable format is $\log\chi'(\omega)$ and $\log\chi''(\omega)$ against $\log\omega$, since only this format corresponds naturally to the prevailing power-law relations of most actually observed dielectric functions. Linear representation against frequency is completely useless, since it compresses the low-frequency end into a tiny part of the plot, while leaving vast expanses of the frequency range to almost constant values of the parameters. Some workers have the tendency to plot loss *logarithmically* while plotting $\varepsilon'(\omega)$ or $\chi'(\omega)$ *linearly* . This is highly confusing, since the essence of any dielectric analysis is to pick out the naturally occurring KK relations and this cannot easily be done between linear and logarithmic plots.

Another common mistake is to plot $\tan\delta$ and ε', the former being allegedly of interest to electrical engineers. The same problem arises as before: $\tan\delta = \varepsilon''(\omega) /\varepsilon'(\omega)$ is derived by forming the ratio of two independent parameters and their ratio is less informative than the total information content of both together.

The principal advantage of the universal approach is the recognition of the correct general pattern of behaviour - most materials have an underlying universal trend and it is helpful to discover what this trend is, by subtracting from the general results of measurements the perturbing features - the better the "guess" at the nature of the residues, the easier it will be to obtain a satisfactory analysis

An example of a simple analysis of experimental data clearly suggesting the presence of a loss peak which is skewed by a sloping trend is shown in Figure 6. Here the guess that the slope is in fact a universal trend proves correct and the residue turns out to be a pure Debye process.

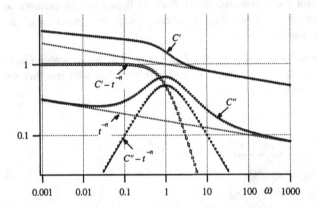

Figure 6 The analysis of data marked as C' and C'' which show a loss peak superimposed on a univer-sal trend t^{-n}. Subtracting this trend and its KK counterpart from the original data we are able to recover the residual $C' - t^{-n}$ and $C'' - t^{-n}$ which here is a pure Debye. Note the lack of parallelism of C' below and after the loss peak.

As an example, take the loss spectrum of a ferroelectric ceramic covering a range of temperatures[5] and shown in Figure 7. There is a visible rise of $C''(\omega)$ at the highest frequencies and this may be due to a series resistance or a high-frequency dipolar process. Not enough is known about this from the data, so the simplest course is to eliminate it by subtracting a series resistance. This process will inevitably influence $C'(\omega)$ to a slight extent - these data are not shown here, but they are essential for the analysis. The next step is to eliminate the steep rise of $C''(\omega)$ at low frequencies which may be due to two causes: dc if there is no corresponding rise in $C'(\omega)$ or LFD if there is. Analysis shows in this case LFD with $1 - n = 0.90$. If there is a peak in loss, then this should be eliminated by subtracting a suitable dipolar process from both $C'(\omega)$ and $C''(\omega)$, and this involves some trial and error, since we do not know the shape in advance. In many situations it is sufficient to subtract a Debye process at some suitable frequency, but if the agreement is not sufficient, it may be necessary to use the full HN formula. By this time the residues of the real and imaginary components may just differ by a constant C_∞ which has to be subtracted giving a pair of KK-consistent residues with the right slopes and ratio given by eqn (9), as shown for two temperatures in Figure 8. The residue represents therefore the underlying trend of data and shows that this has the universal fractional power law nature with a finite value of the exponent, suggesting that the lattice itself is fairly lossy.

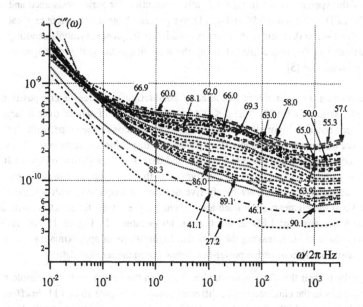

Figure 7 The complete loss spectrum for a range of temperatures of a ferroelectric ceramic, showing a complex behaviour with strong rise towards low frequencies, a loss peak in the middle region and flattening towards high frequencies. Tags denote temperature in °C. From reference [5]

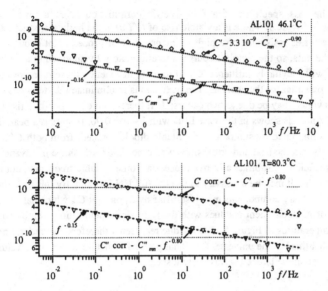

Figure 8 Two of the spectra shown in figure 7, after correction for series resistance and subtraction of C_∞, LFD with $1 - n = 0.90$ and of a Debye process. Note the excellent residual universal trend with $1 - n = 0.15$ and 0.16, over six decades of frequency, symbols denoting the calculated values and the dotted lines showing the KK compatible real and imaginary components. From reference [5].

Another example of data analysis refers to ionically conducting ceramics of composition $Na_{x-\delta} Fe_x Ti_{2-x} O_4$, with $x = 0.875$, $0 \le \delta \le 0.44$,[6] the loss data of which over a large temperature range are shown in Figure 9. There a continuous variation of the slope, with steep LFD rise at the higher temperatures and a less steep power-law line at higher frequencies. These data were analysed on similar lines to those of Figure 7, and the results are shown in Figures 10 and 11 for lower and higher temperatures, respectively. At the lower temperatures we note excellent KK-compatible power laws over nearly five decades of frequency, with a slope of approximately –0.3 and the elements subtracted are C_∞ only at the lowest temperature, with a dipolar element and the onset of LFD at the highest temperature. In Figure 11 we have eliminated the lower slope of –0.3, leaving the dominant higher slope of approximately –0.91. Again there is an excellent KK-compatible power-law behaviour in the residual data.

It is necessary to make it clear that there is no sudden change in the behaviour of the sample at 197°C, the difference lies in the elimination of a different process in Figure 10 and 11. In effect, our analysis highlights separately the low-loss universal law at lower temperatures and the high-loss power law at higher temperatures, while in fact Figure 9 shows clearly that these two power laws go gradually one into the other, as indicated schematically in Figure 3. The strength of our analysis lies in the fact that it is capable of sorting them out and showing that each of these is applicable over the entire available frequency range.

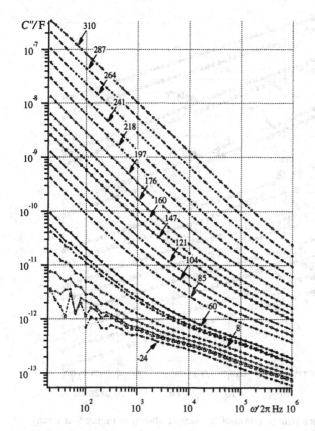

Figure 9 The complete loss spectra of an ionically conducting ceramic of for a range of temperatures in °C indicated by tags. The low-frequency response is dominated by a highly lossy LFD – the slope is distinctly not –1, as would be the case for dc conduction - at high frequencies a less lossy behaviour dominates. There is a smooth transition from the low-loss to the high-loss process. From Reference [6].

From the presence of LFD or dc conduction we may infer the presence of significant hopping transport, while the presence of Debye-like or similar dipolar processes shows the existence of dipoles in the system. In this way it is possible to obtain the temperature dependence of the various parameters such as dc, LFD or the loss peak frequency of any dipolar processes - these data are obtained with much higher precision than would be the case with reading them off "by eye" from the raw data.

177

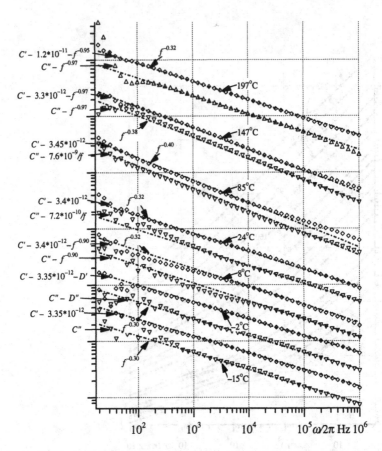

$C' - 1.2*10^{-11} - f^{-0.95}$
$C'' - f^{-0.97}$

$C' - 3.3*10^{-12} - f^{-0.97}$
$C'' - f^{-0.97}$

$C' - 3.45*10^{-12}$
$C'' - 7.6*10^{-9}/f$

$C' - 3.4*10^{-12}$
$C'' - 7.2*10^{-10}/f$

$C' - 3.4*10^{-12} - f^{-0.90}$
$C'' - f^{-0.90}$

$C' - 3.35*10^{-12} - D'$

$C'' - D''$

$C' - 3.35*10^{-12}$

C''

197°C
147°C
85°C
24°C
8°C
2°C
−15°C

$10^2 \quad 10^3 \quad 10^4 \quad 10^5 \, \omega/2\pi \, \text{Hz} \, 10^6$

Figure 10 The residual dielectric response of the sample shown in Figure 9 at a range of lower temperatures. The secondary processes subtracted are indicated in the tags on the margin, the figure following C' being C_∞, that following C'' being either in the form G_0/f for the dc contribution or f^{n-1} for LFD elements. The Debye components are denoted by D' and D''. Points correspond to the reduced data, lines are the fitted power laws which are drawn in KK compatible positions, the high-frequency exponents being indicated for every pair of data. In the interests of clarity each consecutive set of data is displaced with respect to the previous one by $\sqrt{10}$.

The question arises regarding the mechanism of transport giving rise to these clear power laws with different exponents. The materials in question are ionic conductors with channel-like paths of easy flow and with much less probability of transport normal to the channels. Relatively low losses are incurred when ions move in the channels, while higher losses arise in movement across the channels and this accounts for the shape of the response shown schematically in Figure 3 which may be regarded as symbolic combination of Figures 10 and 11. A more detailed analysis is needed to account for the universal nature of the response, but the general argument is as explained here.

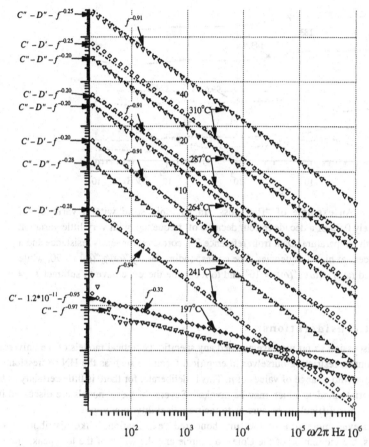

Figure 11 The corresponding higher temperature data, comparable to those shown in Figure 10 but with the 197°C set reproduced to provide a reference. The multiplying factors for data from 264°C onwards are indicated by numbers.

We conclude by returning to the discussion of the other extreme form of dielectric relaxation - the "flat loss" which is found very widely in many low-loss materials and which was already mentioned earlier. Figure 12 shows an example of low-loss ionic material[7] in which the variation of loss is on average 1/80th of a decade per decade of frequency, and not necessarily a power law. This is an extreme example, but others are shown in reference [2]. The fact that this behaviour is found very widely in low-loss materials suggests the existence of some form of general principle leading to this result. We have already mentioned the possibility that there is a mechanical process coupling to the dielectric process and giving rise to the flat loss.

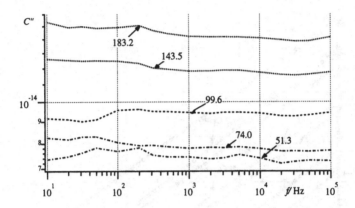

Figure 12 An example of "flat" loss in an ionic solid with a variation of approximately 0.05 of a decade in four decades of frequency, with very little variation of slope with temperature. Data from reference [7], corrected for series resistance and a dipolar process at high frequencies. The corresponding ratio $\chi'(\omega)/\chi''(\omega)$ is 90, while the measured ratio $\varepsilon'(\omega)/\varepsilon''(\omega) \approx 10^3$ and to reconcile these we have to subtract $\varepsilon_\infty = 0.92\varepsilon'$.

Theoretical considerations

In our whole discussion so far we have avoided any specific theoretical models of the universal or dipolar responses, restricting ourselves to empirical formulae such as the HN expression or the universal laws with a range of values of n. This is deliberate, for there is little certainty as to the most appropriate models which may be invoked. Several of these models are discussed in reference [2] but it is difficult to be sure as to their applicability in any given situation. In our opinion, this is even more true of the time-honoured "explanations" like distributions of relaxation times, the inclination of the Cole-Cole circle and the width of the loss peak, and so on, which give a purely qualitative impression of the shape of the response, but mean absolutely nothing in terms of any theory.

One interpretation which we find particularly appealing by its simplicity is the so-called "energy criterion" [2] which states that the inevitable consequence of universality is the constancy of the ratio (9) which states that for any universal model with exponent $1 - n$:

$$\chi''(\omega)/\chi'(\omega) = \cot(n\pi/2) = \text{constant independently of frequency.} \tag{13}$$

and this may be interpreted as the independence of frequency of the ratio

$$\frac{\text{energy lost per radian}}{\text{energy stored at peak}} = \cot(n\pi/2) \tag{14}$$

If it is possible to show that a given process does satisfy this requirement, then the only possible frequency dependence is the universal one. In our particular examples, it would be necessary to specify what the process of energy loss might be for ions moving in the channels and across them and it is intuitively evident that the former are less energy consuming than the latter. This energy loss is most likely related to the *mechanical* deformation involved in the movement of an ion. At this point in time we are not able to discuss this in more detail, suffice it to say that there does not appear to be any other evident theory which could be invoked to explain the universality of the response of these, or any other for that matter, materials.

We would point out that there are many theories of the frequency dependence of ionic hopping[8,9] and the results predicted generally differ from the universal law which we believe to be applicable in most cases. It is an interesting conjecture whether there is any significance in the apparent similarity, if not identity, of theoretical predictions from very different starting points.

Rather than search for such theories, it is much better to give a reliable analysis of the constituent elements of the observed behaviour, and this is most conveniently done on the basis of our universal approach, leaving the more fundamental analysis until a time when a more reliable theory will have been be established.

Conclusions

This review is aimed at showing the importance of the "universal" approach in the analysis of dielectric data, since it facilitates the extraction of significant trends from a variety of overlaying secondary processes. It makes use of the empirically established fact that most dielectrics or, for that matter other materials showing frequency-dependent behaviour, follow "universal" fractional power laws of frequency and time dependence. It is pointed out that, despite a very large volume of theoretical and experimental work on these subjects, there is as yet no general understanding of their physical significance.

References

[1] A K Jonscher, *Dielectric Relaxation in Solids*, Chelsea Dielectrics Press, London, 1996

[2] A K Jonscher, *Universal Relaxation Law* , Chelsea Dielectrics Press, London, 1996

[3] S Havriliak Jr and J S Havriliak, *Dielectric and Mechanical Relaxation in Materials - Analysis, Interpretation and Application*, Hanser Verlag 1996

[4] A K Jonscher, submitted to J Materials Science

[5] A Isnin and A K Jonscher, Ferroelectrics 210, 47-65 (1998)

[6] A Kuhn, C León, F Garcia-Alvarado, J Santamaria, E Moran and M A Alario-Franco, IEEE Trans on Dielectrics and Insulation, to be published

[7] B S Lim, A V Vaysleyb and A S Nowick, Applied Physics A **56**, 8-14 (1993)

[8] J C Dyre, J Non-Crystalline Solids, **135**, 219-226 (1991)

[9] K Funke, Zeitschrift Fur Physikalische Chemie-International Journal Of Research In Physical   Chemistry & Chemical Physics, **188**, No.Pt1-2, Pp.243-257 (1995)

Acknowledgement

I wish to thank Professor Jacobo Santamaria and his colleagues for their permission to use Figures 9, 10 and 11 which are based on their experimental results.

THE DIELECTRIC LOSS OF SINGLE CRYSTAL AND POLYCRYSTALLINE TIO$_2$

XIAORU WANG, ALAN TEMPLETON, STUART J. PENN AND NEIL MCN. ALFORD
SEEIE, South Bank University, 103 Borough Road, London SE1 0AA

ABSTRACT

The dielectric loss of single crystal and polycrystalline TiO$_2$ has been studied. In polycrystalline TiO$_2$ the dielectric loss is determined by both the microstructure and by the oxygen stoichiometry. Experiments have been carried out to determine the influence of both the microstructure (particularly porosity) and the oxygen stoichiometry. The TiO$_2$ powder has been doped with partially stabilised zirconia, an oxygen ion conductor, in order to modify the oxygen stoichiometry. Sintered discs have been examined for loss as a function zirconia doping, pore volume and as a function of temperature. The behaviour of the doped and undoped titania powders is significantly different. Since many microwave dielectric materials contain Ti eg Ba-Ti-O, Ba-Nd-Ti-O, (Ba-RE-Ti-O, RE=Rare Earth), Zr-Sn-Ti-O etc it is essential to understand the role of the titanium, particularly as it can exist in mixed valence states, and the role of oxygen and its influence on the dielectric loss.

INTRODUCTION

For dielectric resonator applications there are three main requirements. The first is that the dielectric loss should be very low. The second is that the dielectric constant should be high so as to aid miniaturisation and the third is that the Tcf should be controlled (usually to be as near zero as possible). The achievement of all three in one material is a formidable problem. In this paper, the base material of many dielectric resonator ceramics, titanium oxide, is chosen as a model material and studied in order to shed light on the dependence of the dielectric loss on factors such as pore volume and oxygen stoichiometry.

Ti - containing ceramic materials have found uses as dielectric resonator materials[1]. In order to sinter the material, temperatures between 1000°C-1500°C are required. This can cause the TiO$_2$ to lose oxygen producing a high concentration of oxygen vacancies. One of the main issues affecting the loss tangent is the non-stoichiometry believed to be the result of oxygen vacancies.

TiO$_2$ itself can have a high Q at a high dielectric constant but as a dielectric resonator material it is unsatisfactory because of the magnitude of the temperature coefficient of the resonant frequency (Tcf) which is influenced by changes in dimension and changes in the dielectric constant due to thermal effects. The pioneering work of O'Bryan[2,3,4] demonstrated that Ti bearing ceramics, notably barium titanate could be prepared with compositions in which the Tcf could be controlled. O'Bryan also found that the phase composition in Ba$_2$Ti$_9$O$_{20}$ was strongly influenced by the oxygen stoichiometry.

EXPERIMENT

High purity TiO$_2$ powder was used. (Pikem, UK). The size of the powder was determined by laser diffraction using a Coulter LS230 and was found to be 2-3μm.
The powder was formed into discs by uniaxial pressing at 100MPa in a 13mm diameter stainless steel die. The discs were sintered at temperatures ranging from 1400°C-1050°C in a

Mat. Res. Soc. Symp. Proc. Vol. 500 ©1998 Materials Research Society muffle furnace in air. The density of the discs was determined by mass and dimensions and compared with that of dense TiO_2 at $4.23Mgm^{-3}$ ($4.28Mgm^{-3}$ calculated from x-ray data[5]). The tan δ was measured at approximately 3GHz by a resonant cavity method using the TE_{018} mode. The sample was placed in a copper cavity on a 4 mm high low loss quartz spacer. The surface resistance of the copper was calculated from the Q of the TE_{011} resonance of the empty cavity to allow the results to be corrected for the loss due to the cavity walls[6]. The sintered TiO_2 samples were approximately 10 mm diameter, 4 mm thick discs. The TE_{018} mode was examined using a vector network analyser (Hewlett Packard HP8720C) with 1 Hz resolution. The dielectric constant, ε was measured using the same equipment but without the quartz spacer and with the cavity height close to the size of the sample. The discs were measured in the "as-fired" condition, i.e. without surface preparation such as polishing. For comparison a single crystal of rutile supplied by ESCETE (The Netherlands) was also measured and this was found to have a tan δ of 5.5×10^{-5} at 3.4 GHz and at room temperature. Low temperature measurements were performed by placing the cavity on a closed cycle cryo-cooler.

RESULTS AND DISCUSSION

Dielectric Constant

Fig 1 shows the results for the dielectric constant of the TiO_2 sintered in air. The dielectric constant behaves as expected in a dielectric containing porosity[7,8]. It was found to fit the equation well:

$$\varepsilon = \varepsilon_1 [1 - \frac{3(\varepsilon_1 - 1)}{2\varepsilon_1 + 1} P] \qquad (1)$$

where ε_1 is the dielectric constant of pure TiO_2, P is the volume of pores within the sample and ε is the dielectric constant of the sintered sample.

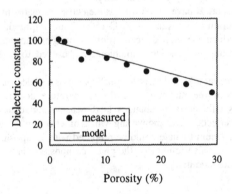

Fig.1 Dielectric constant as a function of porosity at 3GHz

Dielectric Loss

In fig 2 the unloaded-Q (=1/tan δ) of the sintered TiO_2 is plotted as a function of sample porosity. An interesting observation is made here. The Q rises and then falls as densification proceeds. One possible explanation for this is oxygen deficiency. At near full density the TiO_2 is tan coloured suggesting a slight deficiency in oxygen. At a temperature of 1400°C the sample is near fully dense and on cooling there is very little free surface capable of absorbing oxygen. At 95% of theoretical density there is still no interconnected porosity and as a consequence the Q is reduced further due to the presence of the porosity[7,8]. At 94% of theoretical density the Q increases sharply and this can be explained by the fact that there is now a interconnected network of pores reaching more of the surface of the TiO_2. The material is better oxygenated and the Q increases. Below a density of 94% the Q then drops due to the presence of increasing amounts of porosity. The colour of the more porous samples is white suggesting that there is more oxygen present. The colour of the dense TiO_2 was tan coloured but on polishing the sample it was found that a well defined dark region existed, strongly suggestive of a lower oxygen content and also suggesting that oxygen had diffused only a short distance into the surface of the sample. Such effects have been observed in barium titanate[9]. Wavelength dispersive electron probe microanalysis however, did not reveal any significant differences between the darker region and elsewhere. The accuracy of this technique is ±0.2% suggesting that the degree of reduction is below this level. The results suggest that there is only very limited reduction of the TiO_2 but that this is sufficient to cause a severe deterioration in the dielectric loss. Negas et al[9] found that in $BaTi_4O_9$ and $Ba_2Ti_9O_{20}$ based ceramics low concentrations of lattice defects caused by reduction degraded the Q factor by almost 100%.

The Addition of Yttria Stabilised Zirconia

On the assumption that the drop in Q at high density was indeed an oxygenation problem, 3 mole % yttria stabilised zirconia (YSZ) powder (Mandoval HSY3) was added to the TiO_2 powder. YSZ is well known for its oxygen ion conduction properties. If YSZ were added to the TiO_2 powder at the 5-10wt% level the reasoning was that such a level is approximately the percolation threshold and thus the YSZ would effectively act as a conduit for oxygen during cooling in the furnace. The diffusion of oxygen[10] in dense TiO_2 at 1400°C is 2.5 x10^{-11}cm^{2}s^{-1}

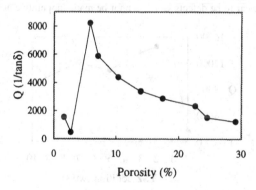

Fig.2. Q (=1/tanδ) as a function of the porosity at 3GHz

whereas in calcia stabilised zirconia (which we will assume is approximately similar to YSZ) it is $2 \times 10^{-6} \text{cm}^2 \text{s}^{-1}$. Thus in a 10mm diameter sample of thickness 4mm it would be impossible to oxygenate the sample in a reasonable time with a diffusion length of 2mm (approximately 28 years) but with a diffusion length on the order of the grain size eg 5 microns, a short time would be required on the order of 0.4 hours for TiO_2 and only 0.035 seconds for partially stabilised zirconia.

Fig 3 shows that adding YSZ does indeed have a significant effect and the Q is enhanced even at very low porosity. However, the original premise that the percolation threshold of approximately 5% porosity would be the appropriate volume fraction of zirconia is clearly incorrect. The enhancement in Q is observed at very low dopant concentrations of YSZ implying that grain boundary segregation has taken place and that this is a very effective diffusion path for oxygen.

In addition, TiO_2 doped with unstabilised ZrO_2 were made for comparison. It was found that the Q was not improved. It can be believed because of the fact of that the content of oxygen is poor in comparison with YSZ.

Low Temperature Measurements

The dielectric loss and the dielectric constant were also measured as a function of temperature. Here it is seen that this type of measurement is very sensitive to the nature of the original powder used to manufacture the discs. It gives us important information on how to tailor materials in order to obtain low dielectric loss. In much the same way as impurities in Al_2O_3 were found to influence the tan δ significantly[7], it now appears that the same is true for TiO_2. Fig 4 shows the tan δ vs. temperature plot for dense discs of the TiO_2 and it is clear that there are differences in the behaviour. The magnitude of the dielectric loss is different at room temperature and in the undoped material the loss is fairly constant until a temperature of approximately 100K where there is a well defined change of slope in the tan δ vs. T curve. At this temperature the tan δ drops sharply. In the doped material the loss is lower throughout most of the temperature range but again falls sharply at a temperature of around 20K. It is noted that the YSZ doped TiO_2 possesses a higher loss at 20K than the undoped material which achieves very low tan δ at 20K and below.

This is rather unusual behaviour. In contrast, when single crystal Al_2O_3 is measured at low temperatures, there appear to be defects which cannot be made inoperative at low temperature[11]

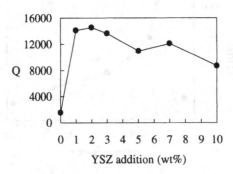

Fig.3 Q as a function of addition of YSZ at 3GHz

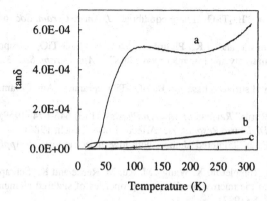

Fig.4 tanδ as a function of temperature for TiO2 at
3GHz (a): undoped (b):doped with YSZ 10wt%
(c):single crystal (001)

and lead to a substantial residual loss at low temperatures. For comparison the behaviour the single crystal rutile is also shown. Here it is seen that there is no such inflection at low temperature and the loss drops steadily.

CONCLUSIONS

The microwave dielectric properties of TiO_2 single crystal and polycrystalline discs have been examined. The dielectric constant of the sintered materials in the form of 10mm diameter discs reduces as expected as the pore volume increases. The dielectric loss was measured as a function of pore volume was found to increase as the pore volume reduced, as expected. However, the Q vs pore volume displayed unusual behaviour reaching a peak at approximately 5-7% porosity, coincident with the onset of interconnected porosity. It was found that the reduction in the Q of TiO_2 at high density was due to oxygen deficiency. Addition of partially stabilised zirconia to this sample caused an increase in the Q at high density and this was believed to be due to the high oxygen diffusion coefficient in partially stabilised zirconia. When unstabilised zirconia was added, no increase in Q at high density was observed, consistent with the fact that oxygen in unstabilised zirconia is poor in comparison with YSZ.

ACKNOWLEDGEMENTS

This work is partially supported by the EPSRC Grant No. GR/K70649 and by the European Community Brite-Euram project DiHiMiCo. Thanks are given to Dr Stephen Reed, Department of Earth Sciences, Cambridge University for the electron microprobe data. Thanks are also given to Mr Karthigeyan for assistance with experimental work.

REFERENCES

[1] Y. Konishi "Novel dielectric waveguide components - microwave applications of new ceramic materials" *Proc. IEEE* **79** (6)726-740 (1991)

[2] H. O'Bryan and J. Thomson. "$Ba_2Ti_9O_{20}$ Phase equilibria" *J. Amer. Ceram. Soc* **66**(1) 66-68 (1983)

[3] H.M. .O'Bryan, J Thomson and J.K. Plourde. "A new $BaO-TiO_2$ compound with temperature-stable high permittivity and low microwave loss" *J. Am. Ceram. Soc* **57** (10) 450-453 (1974)

[4] H. O'Bryan "Identification of surface phases on $BaTiO_3-TiO_2$ ceramics" Am. Ceram. Soc. Bull **66**(4) 677-80 (1987)

[5] F.A. Grant "Properties of Rutile" *Reviews of Modern Physics* 31 (3) 646-674 (1959)

[6] D. Kajfez and P. Guillon, *Dielectric Resonators*, (Artech House, Zurich, 1986)

[7] N.McN. Alford and S.J. Penn "Sintered alumina with low dielectric loss" *J. Appl. Phys.* **80** 5895-5898 (1996)

[8] S.Penn, N.McN Alford, A. Templeton, X. Wang , M. Xu, M. Reece and K. Schrapel. "Effect of porosity and grain size on the microwave dielectric properties of sintered alumina" *J. Am. Ceram. Soc.* **80** (7) 1885-1888 (1997)

[9] T. Negas, G Yeager, S Bell, N. Coats and I. Minis. "$BaTi_4O_9/Ba_2Ti_9O_{20}$ - based ceramics resurrected for modern microwave applications. *Am. Ceram. Soc. Bull.* **72**(1) 80-89 (1993)

[10] W.D. Kingery, H.K. Bowen and D.R. Uhlmann. Introduction to Ceramics 2nd edition (Wiley, New York 1976)

[11] V. B. Braginsky. "Experimental observation of fundamental microwave absorption in high quality dielectric crystals" V. S. Ilchenko and K. S. Bagdassarov. Physics Letters A **120** p300-305 (1987)

VERY LOW LOSS CERAMIC DIELECTRIC RESONATOR MATERIALS

N MCN ALFORD[1], S J PENN[1], A TEMPLETON[1], X WANG[1], P FILHOL[2], N KLEIN[3], C. ZUCCARO[3] AND J C GALLOP[4]

[1] South Bank University, 103 Borough Road, London SE1 0AA, UK, alfordn@sbu.ac.uk
[2] Tekelec Components, Pessac, France
[3] Forschungszentrum Jülich GmbH, postfach 1913, D-5170 Germany
[4] National Physical Laboratory, Teddington, UK

ABSTRACT

The huge increase in the use of microwave communications places severe constraints on the operating performance of the microwave filters. Dielectric resonators are a very attractive option as they can be prepared with very low dielectric loss, moderately high dielectric constant and a low temperature coefficient of the resonant frequency. In filters they possess very low intermodulation products and are capable of handling high power. Certain dielectric oxide single crystals display very low loss at microwave frequencies and on cooling the loss is generally observed to drop. The Q of sapphire at 10 GHz exceeds 10^7 at low temperatures of around 10K. However, single crystals are expensive and the purpose of this research is to explore inexpensive, sintered polycrystalline alternatives. By very careful attention to purity, processing and microstructure Q values approaching those of single crystals have been achieved. The loss of polycrystalline ceramics of Al_2O_3, $Ba(Mg_{1/3}Ta_{2/3})O_3$ (BMT) and $Zr_{0.875}Sn_{0.25}Ti_{0.875}O_4$ (ZTS) has been studied. Alumina, Al_2O_3 has been studied as a model material for dielectric loss. Theory predicts that the loss in single crystal sapphire should follow a T^5 dependence. However at low temperatures the loss is dominated by extrinsic losses due to crystal imperfection, residual dopant atoms, dislocations and other lattice defects and the T^5 dependence does not hold. In polycrystalline alumina the intrinsic loss is immediately masked by these extrinsic losses, even at room temperature, and a simple T dependence is observed. Results on polycrystalline alumina show that a Q of $> 5\times10^4$ at 10 GHz and at room temperature are possible and Q's well in excess of 10^5 at 10 GHz and 77 K can be achieved.

INTRODUCTION

The requirement for miniature low loss microwave filters has led to the dielectric loading of cavity resonators to form dielectric resonators. The miniaturisation factor can be expressed as

$$D \approx \frac{\lambda_0}{\sqrt{\varepsilon_r}}$$

where D is the diameter of the dielectric resonator, λ_o is the free space wavelength at the resonant frequency and ε_r is the dielectric constant of the dielectric.

There are three key parameters for the dielectric materials in filter applications: the dielectric constant (ε_r), the quality factor ($Q = 1/\tan\delta$) and the temperature coefficient of resonant frequency (τ_{cf}). The larger the dielectric constant, the smaller the filter. The higher the Q the lower the insertion loss and steeper the cut-off. A τ_{cf} close to zero ($< \pm 10$ ppm/K) is required for stability against ambient temperature change.

Even though single crystals have a higher Q than the equivalent polycrystalline material they are not suitable for most applications due to their high cost. Some of the higher dielectric constant materials cannot be made as single crystals. The focus of this work has been to improve the properties of existing dielectrics through careful processing. Three materials have been studied here, Alumina ($\varepsilon_r = 10$), $Zr_{0.875}Sn_{0.25}Ti_{0.875}O_4$ (ZTS) ($\varepsilon_r = 37$) and $Ba(Mg_{1/3}Ta_{2/3})O$ (BMT)($\varepsilon_r = 24$) .

Cryocoolers have provided another opportunity to improve filter performance as the dielectric loss of dielectrics generally decreases with temperature. The use of cooling also provides temperature stability which can be used to reduce the requirements for a very small τ_{cf}. The relaxing of this constraint broadens the materials options considerably and has made real the possibility of utilising cold dielectrics in filter assemblies. With sufficient cooling, high temperature superconductors can be used as the cavity housing and thus reduce the size still further.

EXPERIMENTAL

Samples of the dielectrics were produced by pressing appropriate powder in 13 mm diameter stainless steel dies at a pressure of 100 MPa. The resulting disks were sintered in air at between 1000 °C and 1600 °C for between 5 and 1800 minutes.

Alumina discs were produced from commercial high purity alumina powders. BMT was obtained from the Murata Manufacturing Company and was also prepared in our laboratories by mixed oxide routes. ZTS was prepared in our laboratories by mixed oxide routes.

The Q and was measured by a resonant cavity method using the TE_{018} mode. The sample is placed in a copper cavity on a 4 mm high low loss quartz spacer. The surface resistance of the copper has been calculated from the Q of the TE_{011} resonance of the empty cavity to allow the results to be corrected for the loss due to the cavity walls[1]. The sintered samples were approximately 10 mm diameter, 4 mm thick discs. The TE_{018} mode was examined using a HP8720C vector network analyser with 1 Hz resolution. The dielectric constant, ε_r was measured using the same equipment but without the quartz spacer and with the cavity height close to the size of the sample. The measurement cavity is attached to a closed cycle cooler to allow measurements to be made from 10 K to 300 K. The discs were measured in the "as-fired" condition, i.e. without surface preparation such as polishing.

ALUMINA

Alumina (Al_2O_3) is the polycrystalline equivalent of sapphire with a dielectric constant of about 10. It has been studied here for several reasons. First, it is a useful dielectric in it own right and is capable of having a very high Q. However, with its low dielectric constant its use is limited to high frequency applications (ideally satellite communications > 10GHz) due to size considerations. Another reason for studying alumina is to investigate the causes of dielectric loss. Commercial materials, such as ZTS and BMT, are complex multiphase materials which are very sensitive to the processing conditions. As a result it is difficult to separate the effect of different sources of loss. Alumina, however, is a single phase material which is available in a very pure form. It has been determined [2] that purity is a very important factor in dielectric loss. Also single crystals, i.e. sapphire, are readily available for comparison and study of intrinsic loss. Fig 1 shows the behaviour of single crystals [4] and TiO_2 doped alumina [2]. The T^5 dependence predicted for single crystal sapphire is completely absent in the polycrystalline alumina and fails even in single crystals at low temperatures.

Samples of sintered alumina were produced from commercially available high purity alumina powder. Optimising the processing conditions [2] produced material with a Q of 37,000

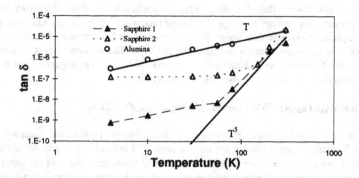

Figure 1 The temperature dependence of the tan δ (=$1/Q$) of alumina and two different sapphire single crystals. The lines marked T and T^5 show a linear and T^5 temperature dependence respectively. Single crystal data from [4].

at 9 GHz and room temperature. The porosity, purity and grain size of the sintered sample all influence the Q [3]. Small amounts of porosity (< 3%) significantly reduce the Q.

It was found [2] that doping alumina with TiO_2 increased the Q by over 30%, a most unexpected result. The room temperature Q at 9 GHz reaches a maximum of 55,000 (tan δ = 1.82 x 10^{-5}) at a dopant level of 0.1% -0.3% when sintered in flowing oxygen as shown in Fig 2. The precise reason for this is unclear. The TiO_2 used has a far higher dielectric loss than alumina (Q ≈2000 at 300K, 10GHz) but does aid sintering, reducing the temperature required and producing a fine grained dense microstructure. At present the modification of microstructure is the preferred explanation for the enhancement in Q. Also seen in Fig 2 is the significant effect of sintering in oxygen in reducing the dielectric loss. The temperature dependence of the resonant frequency of this material shows that the τ_{cf} at room temperature is around -60 ppm/K which is too large for some applications. However this improves with cooling so that by about 77 K it has fallen to -8 ppm/K. Coupled with the fact that in a cryogenic system the temperature stability can easily be maintained to better than 0.5 K, the stability of alumina is sufficient for most cryogenic devices.

The tan δ (=$1/Q$) of alumina decreases approximately linearly with temperature as shown in

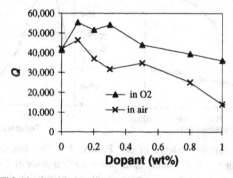

Figure 2 Effect of TiO_2 doping in alumina and influence of sintering atmosphere.

191

Fig. 1 which also shows data for single crystal sapphire taken from Braginsky et al [4]. Theory[4] predicts that the tan δ of sapphire should follow a T^5 dependence. This appears to be true for temperatures greater than about 100 K. However, at low temperature the temperature dependence is much weaker, similar to that of the polycrystalline alumina. This is because at low temperature the dielectric loss is dominated by extrinsic effects such as impurities and defects. It is these factors that limit the Q of polycrystalline materials.

Tan δ of Ba(Mg$_{1/3}$Ta$_{2/3}$)O$_3$ (BMT) and Zr$_{0.875}$Sn$_{0.25}$Ti$_{0.875}$O$_4$ (ZTS).

ZTS is a commonly used microwave dielectric with a dielectric constant of 37. Commercially available materials are not usually pure ZTS but are doped with other dielectrics in order to obtain a τ_{cf} close to zero. These dopants tend to decrease the Q of the material. This makes optimising the material with respect to Q very difficult especially since it is not clear what the dominant causes of dielectric loss are. Discs of ZTS with the composition Zr$_{0.875}$Sn$_{0.25}$Ti$_{0.875}$O$_4$ were produced from powders developed in-house using a mixed oxide route. No dopants were used. As with alumina the tan δ of the ZTS decreases approximately linearly with temperature as shown in fig 3. Results for the temperature dependence of the tan δ of BMT are shown in Fig. 3. The temperature dependence of the TCF of ZTS is shown in Fig. 4. The results have been normalised to 10 GHz by assuming f /tan δ = constant (i.e. $Q \times f$ = constant). The BMT results are for a commercially available disc (Murata Manufacturing Company, Ltd.), diameter 7.69 mm and height 3.04 mm, measured at 8.34 GHz. The ZTS results are for a disc produced from our own powder, diameter 6.57 mm and height 3.44 mm, measured at 7.5 GHz. The temperature dependence of the tan δ of the materials have been fitted to a power law. The tan δ of the BMT was found vary as $T^{0.6}$ and the ZTS as $T^{0.9}$. This compares with a linear T dependence for alumina.

USE OF SUPERCONDUCTORS TO MINIATURISE FILTERS

The experiments described above indicate that the dielectric loss can be reduced significantly on cooling the dielectric. Perhaps even more important, the τ_{cf} becomes less of a constraint if the temperature is controlled. Reducing the losses in the housing cavity by using

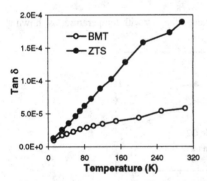

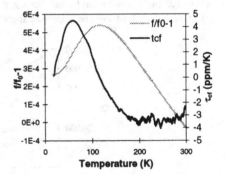

Figure 3. The temperature dependence of the tan δ of polycrystalline BMT and ZTS normalised to 10 GHz.

Figure 4. The temperature dependence of resonant frequency of ZTS. The value of the τ_{cf} is at an acceptably low level over the full temperature range being within ± 4 ppm/K

Table I Summary of properties at 20 K, 60 K, 77 K and 300 K normalised to 10 GHz [6]

Material	ε_r	tanδ @ 20 K	tanδ @ 60 K	tanδ @ 77 K	tanδ @ 300 K
Alumina, Tan δ	9.5	1.42×10^{-6}	2.94×10^{-6}	3.71×10^{-6}	2.50×10^{-5}
Ba(Mg,Ta)O, Tan δ	24	1.14×10^{-5}	2.32×10^{-5}	2.71×10^{-5}	4.81×10^{-5}
ZrSnTiO, Tan δ	38	1.79×10^{-5}	4.54×10^{-5}	5.88×10^{-5}	1.89×10^{-4}
YBaCuO Thick film, R_s (mΩ)		0.544	0.945	1.45	-

high temperature superconductors is now a possibility. A dielectric resonator was constructed from superconducting $YBa_2Cu_3O_x$ thick film [5] plates and a 10.7 mm diameter, 5.4 mm thick disc of alumina. The gap between the alumina and the top plate was 1.5 mm. At 77 K the Q of the resonator at 11 GHz was 101,000. Fig 5 shows the Q of the resonator versus temperature. It also show the contributions to the Q from dielectric and the films. The effect of the films on the Q of the resonator can be reduced by placing a spacer between the plates and the dielectric. Thus the Q can be increased at the expense of size. With a 5 mm spacer top and bottom the effect of the HTS plates on the Q is negligible and the Q of the resonator equals the Q of the dielectric. For dielectrics with a higher dielectric constant and lower Q a much smaller gap is required to achieve the same effect.

CONCLUSION

Polycrystalline dielectrics are capable of making small high Q resonators which can be combined to form filters. Smaller size can be achieved at the expense of lower Q's. Alumina can be produced with very high Q's at room temperature approaching to within 50% that obtained from a single crystal. The tan δ of alumina and ZTS decrease linearly with temperature. Alumina has a high τ_{cf} at room temperature (-60 ppm/K). However, it falls with temperature bringing it to acceptable levels below about 80 K. The τ_{cf} of ZTS and BMT is within acceptable levels over the entire temperature range. At 60K, well within the range of small closed cycle coolers, the Q of Al_2O_3 $Q = 340,000$, of BMT $Q = 43,000$ and of ZTS $Q = 22,000$.

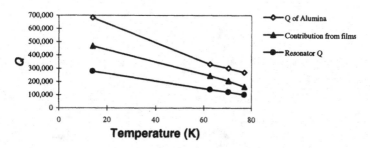

Figure 5 Q at 11 GHz of a dielectric resonator made from HTS thick films and a polycrystalline alumina dielectric. The graph show the contributions from the alumina and the superconducting films.

The use of high temperature superconductors in combination with sintered dielectric resonators will result in a reduction in the size of the filter assembly.

ACKNOWLEDGEMENTS

This work is partially supported by the EPSRC Grant No. GR/K70649 and by the European Community Brite-Euram project DiHiMiCo.

REFERENCES

1. D. Kajfez and P. Guillon, Dielectric Resonators, (Artech House, Zurich, 1986)
2. Alford N McN and Penn S J, "Sintered alumina with low dielectric loss", J. Appl. Phys., **80** p. 5895-5898 (1996)
3. S. Penn, N. McN. Alford, A. Templeton, M. Xu, X. Wang, M. Reece and K. Schrapel, "Effect of Porosity and Grain Size on the Dielectric Properties of Sintered Alumina", J. Am. Ceram. Soc. **80** p. 1885-1888 (1997)
4. V. B. Braginsky, V. S. Ilchenko and K. S. Bagdassarov. Physics Letters A **120** p300-305 (1987)
5. N. McN. Alford, T. W. Button, M. J. Adams, S. Hedges, B. Nicholson and W. A. Philips. "Low Surface Resistance $YBa_2Cu_3O_x$ Melt-Processed Thick Films" Nature **349** p. 680-683, (1991)
6. N.McN. Alford, SJ Penn, A Templeton and X Wang "Low loss sintered dielectric resonators with HTS thick films" J Supercond **10** (5) 467-472 (1997)

COMPLEX DIELECTRIC SPECTROSCOPY CHARACTERIZATION OF A Li$_{0.982}$Ta$_{1.004}$O$_3$ FERROELECTRIC SINGLE CRYSTAL

Ming DONG and Rosario A. GERHARDT
School of Materials Science and Engineering, The Georgia Institute of Technology,
Atlanta, GA 30332-0245, USA

ABSTRACT

The dielectric properties of a c-oriented ferroelectric Li$_{0.982}$Ta$_{1.004}$O$_3$ single crystal have been investigated. The frequency and the temperature dependence of the dielectric properties have been measured from 500 to 650°C at frequencies ranging from 5 to 10^6 Hz. Both blocking and non-blocking electrodes were used for separating the electrode effect from the crystal bulk dielectric response. A low-frequency dispersion was identified to be due to the contribution of Li$^+$ ionic carriers. Based on the electrical measurement data and complex nonlinear least squares fitting, an equivalent circuit is proposed to represent the dielectric properties of the single crystal.

INTRODUCTION

Lithium tantalate exists as a solid solution of Li$_{1-5x}$Ta$_{1+x}$O$_3$ (with 49.6 ~ 56% Ta$_2$O$_5$ content, depending on the growth temperature) [1-3]. The Curie temperature varies markedly with composition and provides an accurate measure of composition [3-6]. The presence of the lithium vacancies within the crystal is favorable to the simultaneous occurrence of ferroelectric and ionic conduction properties. Previous measurements showed a strong low-frequency dielectric dispersion in the neighborhood of T$_c$ [6-8]. At temperatures close to T$_c$, the Li$^+$ ionic conduction is significant. The electrode effect influences the low frequency dielectric response. It may occur due to the blocking of Li$^+$ ions at the electrode-crystal interfacial region. In the present work, we analyze the dielectric properties of a Li$_{0.982}$Ta$_{1.004}$O$_3$ crystal along the polar c-axis using both blocking and non-blocking electrodes.

EXPERIMENT

The crystal used was a parallelepiped fragment ($10 \times 10 \times 1.54$ mm^3), the large face being perpendicular to the polar c-axis. Graphite paste electrodes were deposited on the crystal for the first measurement. The second measurement was carried out using silver paste electrodes. A preliminary heating and cooling cycle was done to ensure measurement reproducibility. The measurements were carried out in vacuum during the cooling process. Each temperature was maintained by a Eurotherm-808 Controller with an accuracy of ± 0.2°C for 20 minutes before collecting the data. A frequency sweep from 5 Hz to 10^6 Hz with a nominal applied voltage of 0.5 V was achieved using a HP-4192A Impedance Analyzer. Electrical data measured as impedance Z* were converted into complex resistivity ρ*, permittivity ε* and electrical modulus M* data using the relations:

$$\rho^* = Z^*(S/d) = \rho' - j\rho'' \quad \text{and} \quad \varepsilon^* = 1/M^* = (j \omega C_0 Z^*)^{-1} \tag{1}$$

where S is the electrode surface area, d is the thickness of the sample, ω is the angular frequency (ω = 2πf) and C$_0$ is the vacuum capacitance. When both measurements (using silver or graphite electrodes) are very similar, we show the data without indicating the electrode type.

RESULTS AND DISCUSSION

The temperature dependence of the permittivity ε' is shown in Fig. 1. A strong dielectric anomaly appears at 600°C, the value of the maximum of ε' is about four orders of magnitude higher than the room-temperature value. It indicates a ferroelectric Curie temperature T_c equal to 600°C. It is consistent with the composition $Li_{0.982}Ta_{1.004}O_3$ [3-5]. The permittivity ε' present s a strong low-frequency dispersion, suggesting the existence of significant ionic conduction in this crystal.

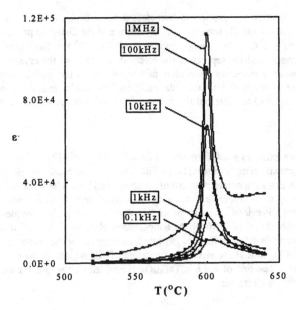

Fig. 1 Variation of the permittivity ε' with the temperature at different frequencies.

The resistivity plane plots and complex modulus plane plots are shown in Fig. 2 for three temperatures around the T_c. In the resistivity plane plot only one semicircle is apparent for each temperature. But the complex modulus plots show two semicircles clearly at each temperature. These results are also confirmed in the frequency explicit plots (Fig. 3). At each temperature, the resistivity vs. frequency plots show only one peak, the position of the peak changes with the temperature. In contrast to this, two peaks can be seen in the modulus vs. frequency plots. The position and the magnitude of modulus change significantly with the ferroelectric phase transition. The modulus plots clearly suggest that two polarization processes exist in this single crystal sample.

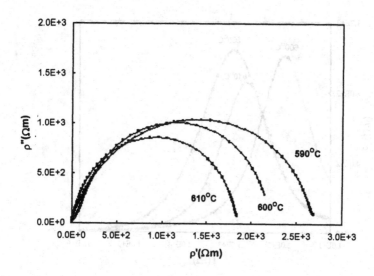

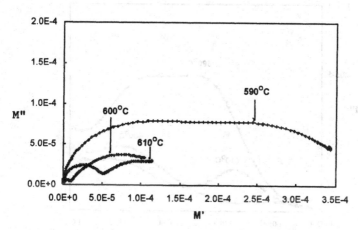

Fig. 2 Resistivity plane plots (up) and Complex modulus plots (down) for three temperatures.

Since long-range Li$^+$ ion diffusion occurs in these ferroelectric crystals, two polarization mechanisms are possible, both polarizations are related to Li$^+$ ion motions through an octahedron face:

1) the dielectric relaxation (or local response) due to the short-range lithium displacements along the polar axis across an oxygen triangle common to two octahedra. These backward and forward motions are partially responsible for the inversion of the ferroelectric spontaneous polarization.

2) the conductivity relaxation (or charge carrier response) due to Li$^+$ ionic conductivity. Such a long-range effect is intensified by the lithium vacancies in the LiTaO$_3$-type crystal network.

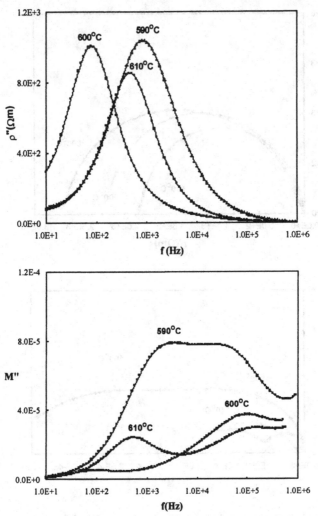

Fig.3 Frequency explicit plots of ρ" (up) and M" (down) for three temperatures.

The behavior of most dielectric materials (without ionic conduction) is often described by the following expression [10]:

$$\varepsilon^*(\omega) = \varepsilon_\infty + \frac{\varepsilon_s - \varepsilon_\infty}{1 + (i\omega\tau)^{1-\alpha}} \tag{2}$$

where graphically, the parameter α represents the tilting angle $(\alpha\pi/2)$ of the circular arc from the real axis in the complex permittivity plane.

On the other hand, for ionic conducting materials, the conductivity frequency dependence can be represented by the Almond-West expression [11]:

$$\sigma^*(\omega) = \sigma_0[1 + (i\omega\tau_c)^n] \tag{3}$$

where σ_0 is the dc conductivity, τ_c is the ionic hopping time and n is a dimensionless exponent $(0 < n < 1)$. So the contribution of ionic conduction to the dielectric response can be presented as:

$$\varepsilon^*(\omega) = \frac{\sigma_0}{i\varepsilon_0\omega}[1 + (i\omega\tau_c)^n] \qquad (4)$$

Consequently, $\varepsilon^*(\omega)$ may be described as follows:

$$\varepsilon^*(\omega) = \varepsilon_\infty + \frac{\varepsilon_s - \varepsilon_\infty}{1 + (i\omega\tau_d)^m} + \frac{\sigma_0}{i\varepsilon_0\omega}[1 + (i\omega\tau_c)^n] \qquad (5)$$

Simulation of the experimental data has been carried out based on the above expression using LEVM complex nonlinear least squares program and good agreement has been obtained [12-14]. The above expression could be considered as a general dielectric response expression. It has also been applied successfully to a high oxygen pressure annealed $Ba_2Li_2Ti_2Nb_8O_{30}$ ferroelectric ceramic [13-14].

The dielectric response of the sample with graphite and silver electrodes are shown in Fig. 4.

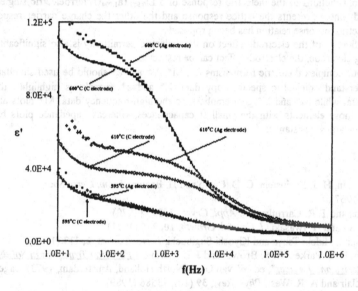

Fig. 4 The permittivity ε' vs. log f plots at three temperatures for two measurements.

Here the silver electrode is a blocking electrode for Li^+ ions. The graphite is non-blocking for Li^+ ions. The low-frequency dispersion is larger in the sample with silver electrode. We propose that the total dielectric response for the whole sample can be represented by the following equivalent circuit:

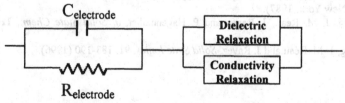

The total permittivity $\varepsilon^*(\omega)$ can then be written as:

$$\frac{1}{\varepsilon^*(\omega)} = \frac{1}{\varepsilon^* \text{electrode}} + \frac{1}{\varepsilon^* \text{crystal}} \tag{6}$$

Using the LEVM program, we have carried out a simulation of the measured data for the sample with silver electrode. The simulation results show that if we subtract the electrode contribution from the Ag electrode experimental data, the result curve matches that of the graphite electrode. The calculated ε' of the crystal part for the sample with silver electrode agrees well with the total measured data for the sample with the graphite electrode.

CONCLUSIONS

Two contributions to the dielectric response of a $Li_{0.982}Ta_{1.004}O_3$ ferroelectric single crystal are identified: one represents the lattice response and the other the charge carrier response. A general dielectric response relation has been proposed.

The influence of the electrode effect on the dielectric permittivity is also significant. Using non-blocking electrode, the electrode effect can be restricted.

The four complex dielectric formalisms Z^*, M^*, Y^* and ε^* should be used simultaneously to best understand dielectric spectroscopy data. Z^* and ε^* formalisms highlight the low-frequency data, while M^* and Y^* give emphasis to the high-frequency data. M^* plots also give emphasis to those elements with the smallest capacitances, whereas impedance plots highlight those with the largest resistances.

REFERENCES

1. A. A. Ballman, H. J. Levinstein, C. D. Capio and H. Brown, *J. Am. Cera. Soc.*, **50**, 657 (1967).
2. R. L. Barns and T. R. Carruthers, *J. Appl. Cryst.*, **3**, 395 (1970).
3. S. Miyazawa and H. Iwasaki, *J. Crystal Growth*, **10**, 276 (1971).
4. Yoshio Fujino, Hideki Tsuya and Kiyoshi Sugibuchi, *Ferroelectrics*, **2**, 113 (1971).
5. R. S. Roth, H. S. Parker, W. S. Brower and J. L. Waring, *"Fast Iion Transport in Solids, Solid State Batteries and Devices"*, ed. W. van Gool (North Holland, Amsterdam, 1973). page 217.
6. D. C. Sinclair and A. R. West, *Phys. Rev.*, **39** (18), 13586 (1989).
7. A. Huanosta and A. R. West, *J. Appl. Phys.*, **61**, 5386 (1987).
8. J. Ravez, G. T. Joo, M. Dong and J.-M. Réau, *Phys. Stat. Sol.* (a), **146**, K71 (1994).
9. L. T. S. Irvine, D. C. Sinclair and A. R. West, *Adv. Mater.*, **2**,132 (1990).
10. K. S. Cole and R. H. Cole, *J. Chem. Phys.*, **9** (1951) 351.
11. D. P. Almond and A. R. West, *Solid State Ionics*, **23** (1987) 27.
12. J. R. Macdonald, *"Impedance Spectroscopy - Emphasizing Solid Materials and Systems"*, (Wiley, New York, 1987).
13. M. Dong, J. M. Réau, J. Ravez and P. Hagenmuller, *J. Solid State Chem.*, **116** 185-192 (1995).
14. M. Dong, J.-M. Réau and J. Ravez, *Solid State Ionics*, **91**, 183-190 (1996).

Part V

Varistors

TIME DOMAIN RESPONSE OF ELECTRICAL CERAMICS
MICRO TO MEGASECONDS

F. A. Modine
Solid State Division, Oak Ridge National laboratory, Oak Ridge, TN 37831-6030

ABSTRACT

The electrical properties of ceramics can be measured in either the time domain or in the frequency domain. But for electrically nonlinear ceramics such as varistors, time-domain measurements provide insights that are different and more relevant to material performance as well as being more physically incisive. This article focuses specifically on the electrical properties of ZnO varistors, but much of it is of relevance for other materials, in particular those materials with grain-boundary barriers and disordered ceramics or glasses. The interpretation of electrical measurements in the time domain is profoundly influenced by such practical matters as source impedance and waveform characteristics. Experimental results are presented for both high and low source impedance relative to that of a test varistor, and the difference in experimental difficulty and ease of interpretation is described. Time-domain measurements of capacitance and of the inductive response of varistors to large, fast electrical pulses are presented and their implications for varistor theory are given. Experimental evidence is given of short- and long-term memory in varistors. These memory phenomena are ascribed respectively to the life time of holes that become trapped in barriers and to polarization currents originating from deep electron traps. Polarization current measurements are presented for a wide range of time and temperature. The power-law time dependence and "universal" behavior of these currents is discussed. The exponent that describes the power law behavior is seen to change with temperature, and the change is interpreted as a double transition from diffusive to dispersive transport that originates with current from two different electron traps.

INTRODUCTION

Electrically based characterizations of ceramic materials can be done in either the time domain or in the frequency domain. When a material's electrical response is linear, the same information is obtained, at least in principal. Frequency-domain measurements are convenient and most often employed. But for electrically nonlinear ceramics such as varistors, time-domain measurements provide insights that are different and more directly relevant to material performance as well as more physically incisive. This article describes the time-domain measurements done on ZnO varistors. Varistor measurements are profoundly influenced by the test apparatus that is used, and this linkage between test apparatus and device response is discussed. Some measurement problems of the time domain are described. The understanding of varistors obtained from time domain measurements is presented. Comparisons of time- and frequency-domain results are given to illustrate similarities and differences.

LUMPED PARAMETER CIRCUIT DESCRIPTION

Figure 1 is a simplified circuit description of a time-domain test apparatus that includes a lumped parameter portrayal of a varistor. An approximately square electrical pulse with a width varying from about 1 μs to 1 Ms is applied to the varistor from a supply that has a resistive output impedance denoted as R_s. The voltage applied to the varistor and the resulting current flow are measured and interpreted to obtain an electrical characterization of the varistor. Though almost any wave form can be used, interpretation is easiest with square pulses. For fast pulses, the current and voltage measurements require fast digitizers and parasitic reactance becomes a big

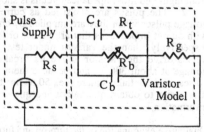

Figure 1. A simplified description of a varistor in a test circuit.

problem. For slower pulses, simple electrical meters suffice, but stability is a concern. The output impedance of the pulser (i.e., R_s) is ideally much higher or much lower than that of the varistor that is measured. But this condition is difficult to achieve because the impedance of a varistor changes with voltage.

Varistors are ceramic grain-boundary-barrier devices, and they have an electrical response that can be only roughly described by lumped parameters. The nonlinear resistance that is the essence of varistors is contributed by Schottky-like grain boundary barriers. Since this resistance changes with voltage by 12 orders of magnitude or more, it is denoted in Fig 1 as the variable resistor R_b. However, the change in R_b with voltage is not instantaneous; R_b evolves in time and introduces phase shifts into the electrical response. As is discussed below, these phase shifts can cause R_b to appear to be either a capacitor or an inductor. The Schottky barriers also exhibit an easily measured capacitance C_b that is contributed by barrier depletion layers of ionized donors. Although the varistor grains are highly conductive compared to the grain boundaries they contribute a small resistance R_g (e.g., 1 ohm cm) that can profoundly influence the electrical response in some measurements. In addition, there are electronic traps in varistors that contribute a capacitance C_t which is much larger than C_b. One conduction path represents the contributions of all electron traps in Fig. 1. But, in actuality, there are several different traps, and each trap should be represented by a parallel conduction path. The response of deep electronic traps can be very slow; so in a lumped parameter model C_t is associated with a large resistance R_t. However, the traps do not exhibit the exponential time response suggested by the lumped parameter description. Instead, they have a distinctly nonlinear – often termed a non-Debye – response to the applied voltage. In the strict sense, neither the trapping phenomena nor the nonlinear resistance of varistors can be described in terms of lumped parameter circuits or even within linear response theory. The lumped parameter description of Fig. 1 is nevertheless the jargon of electrical measurements and it is useful in the discussions of varistor measurements that follow.

FAST PULSE RESPONSE

The response of varistors to fast electrical pulses is a test of their performance as surge arresters that also reveals a lot about the fundamental mechanisms that control the electrical nonlinearity. (The focus here is on fundamental understanding rather than performance testing.) The pulse response of varistors gives clear insights into varistor capacitance, and it illustrates more directly than frequency domain measurements an inductive response of varistors that reveals the influence of electron holes on varistor barriers.

The response of a varistor to a fast electrical pulses from high- and low-impedance circuits is shown in Fig 2. The pulses are classified as current or voltage pulses when they are supplied by circuits with output impedances that are higher or lower, respectively, than the varistor under test. The pulses drive the varistor above its breakdown voltage at about 300 V, and because the applied pulses are essentially square (i.e., 50 ns rise time), it is easier to obtain qualitative and quantitative information.[1]

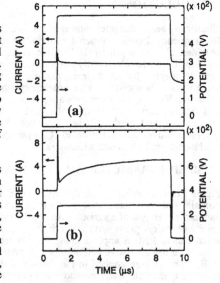

Figure 2. Varistor response to fast pulses from high- and low-impedance circuits.

Qualitative examination of the turn-on and turn-off responses as seen in the leading and trailing edges of the pulses show voltage lagging current below breakdown, but leading current above

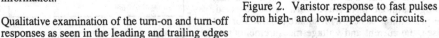

breakdown. Hence, the electrical behavior is described as capacitive below breakdown and inductive above. A quantitative interpretation follows from linear circuit theory, and the values of capacitance obtained for voltages below breakdown are given in Table I.

TABLE I. Varistor capacitance from pulse response and ac admittance.

	Voltage Rise	Voltage Fall	Current Spike	1 MHz Admittance
C (nF)	1.5	1.5	1.8	1.8

The capacitance from voltage rise in Fig. 2(a) is obtained from $dV/dt = I/C$. The same value can be obtained from the RC time constant of the decay since it is known that the output impedance of the pulse generator 1.7 k Ω with the voltage off. The capacitance obtained from the current spike of Fig. 2(b) is found by integrating over the initial spike to obtain the stored charge and using $C = Q/V$. The value may be a little high because the integration includes a little over barrier current. However, the value is the same as that found from 1 MHz admittance measurements, which also might be a little high because traps may contribute even at high frequencies. The varistor capacitance varies greatly because the contributions of traps are frequency dependent, but the capacitance found from admittance measurements is a minimum value at about 1 MHz.

It is possible to obtain values for the varistor inductance that appears when the varistor is above its breakdown voltage (i.e., ~300V) by assuming linear circuit theory and analyzing the rate of approach to steady state of either the voltage overshoot or the current undershoot in Fig. 2. But such different values as 1 μH and 50 μH are found from Figs. 2(a) and 2(b), respectively. There are two reasons for this difference: First, these values do not represent real inductance. They only reveal voltage lead relative to current that is interpreted as inductance in the linear-circuit lumped-parameter model of reality. Second, linear response theory is not applicable because the electrical conduction in the barrier breakdown region is well described by the distinctly nonlinear empirical equation:

$$I = (V/V_o)^{\alpha}, \tag{1}$$

where α is a nonlinearity coefficient that is typically 20 to 60 and V_o is a constant. A small change in voltage corresponds to a large change in current, and this difference explains the small voltage overshoot relative to current undershoot that is seen in Fig. 2. To a first approximation, Eq. 1 implies that current undershoot is greater than voltage overshoot by the factor α:

$$(I_s - I_i)/I_s = -\alpha (V_s - V_i)/V_s, \tag{2}$$

where the subscripts s and i refer to steady state and initial values.[1] Inductance values found from current undershoot and voltage overshoot measurements differ by this same factor of α.

Varistor performance testing relies on time-domain measurements, while varistor theory has relied largely upon ac immittance measurements. However, the time-domain results have many clear theoretical implications. The most accepted of the varistor theories attribute electrical breakdown to barrier collapse that occurs when holes are trapped in the grain boundaries.[2-8] And, strong evidence of holes is given by the inductive response of varistors, which is clearly seen in the time response of Fig. 2. (That minority carriers in semiconductor junctions can contribute an inductive response has been recognized for a long time,[9] and a clear explanation of the inductive response in ZnO varistors is given by Pike.[3]) Thus, while much elegant experimental and theoretical work has been done to verify the existence and explain the role of holes in varistor breakdown,[3-5,8] strong evidence is easily found in time response measurements. Moreover, the time response measurements distinguish between different interpretations of the observed change in varistor capacitance with voltage. On one hand, a huge increase in capacitance just below breakdown has been interpreted as real capacitance (i.e., real energy storage) stemming from hole accumulation in the barriers.[2] On the other hand, ac impedance data showing a capacitance that increases below breakdown before decreasing to negative values above barrier breakdown has been attributed to phase shifts that result from a modulation of the barrier height by electron and hole trapping.[3,6,8] Below the barrier breakdown voltage, Figs. 2(a) and 2(b) show only a well behaved capacitance without the charge storage that a large increase in capacitance would produce, but an inductive response – some times called negative capacitance – is seen above breakdown. Thus, the time domain measurements confirm the explanation[3] of the voltage dependence of varistor capacitance in terms of phase shifts produced by electrons and holes that respectively raise and lower the barriers when they are trapped.

INFLUENCE OF TEMPERATURE AND RESISTANCE

The interpretation of electrical measurements in the time domain is profoundly influenced by such practical matters as waveform characteristics and source impedance. The measurements are easiest to interpret when the source impedance is either high or low relative to the sample being tested. But such conditions can be difficult to achieve in measurements of electrically nonlinear materials because the sample impedance can change dramatically. Consequently, pulse response date has often been presented that is a mixture of such results as are shown in Fig. 2. Neither current nor voltage remain constant, and interpretations are less straight forward.

The source impedance has a strong influence on the decay time of voltage overshoot and on the rise time associated with undershoot. The recognition of this influence proved the key to understanding the inductive response of varistors and its temperature dependence. The current rise time is found to have a value of a few microseconds at room temperature and to change with temperature in a manner suggesting a curious negative activation energy.[10,11] As seen in Fig. 3, the voltage overshoot exhibits a similar decay time (i.e., 2.5 μs) at room temperature, but it has no significant temperature dependence. This difference is surprising and difficult to explain if the voltage overshoot and current undershoot are two aspects of the same hole trapping process. The explanation lies in the source resistance of the pulser, which is greatly different in voltage overshoot and current undershoot measurements. The influence of even small amounts of source resistance is shown in Fig. 4(a). And Fig. 4(b) is a mathematical simulation of experimental results that confirms source impedance as the origin of the difference.

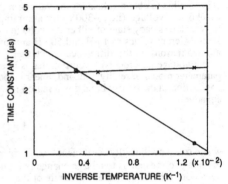

Figure 3. The temperature dependence of the time constants describing voltage overshoot (denoted x) and current under-shoot (denoted o). Thermal activation energies of -7.2 and +0.4 meV are deduced.

The simulation assumes the empirical description of the nonlinear current-voltage characteristic,

$$I(V) = I_c(V/V_c)^\alpha, \qquad (3)$$

where I_c and V_c are scaling constants and α is the nonlinearity coefficient. The varistor voltage V is reduced from the source voltage V_s by the drop across the series resistance R_s,

$$V = V_s - IR_s. \qquad (4)$$

For simplicity, it is assumed that the varistor has an exponential time response described by

$$dI(t)/dt = [I(V) - I(t)]/\tau, \qquad (5)$$

where τ is a time constant associated with the generation and trapping of holes. (It has been shown that a time lag in the response to voltage change of either the barrier resistance or the barrier capacitance can explain the inductive characteristics of varistors.[12]) For a reasonable choice of parameters (e.g., $I_c = 5$ A, $V_c = 200$ V, $\alpha = 25$, and $\tau = 7$ μs), the equations can be numerically integrated for values of R_s that correspond to different values of pulser output impedance in the case in which a rectangular voltage pulse of 8 μs duration is applied, and the results are shown in Fig. 4(b). The experimental and simulated results are in excellent agreement. Of course, the current spikes associated with the charging and discharging of the varistor capacitance were not simulated. Thus, even small amounts of source impedance have a large effect, and measurements of rise time are only accurate in the limit of zero source impedance.[11]

The explanation of the temperature dependence of the current rise time follows from the understanding of the influence of the source resistance. It originates in the temperature dependence of the grain resistivity. The grain resistivity is about 0.5 Ω cm at room temperature, but it increases by an order of magnitude at liquid-nitrogen temperature due to the decrease in the carrier density. In the varistor used for Fig. 4(a), the grain resistivity contributes about 0.1 Ω at room temperature and about 2.4 Ω at 77 K. The grain resistivity reduces the barrier voltage in proportion to the current just as an external resistance does. (See Fig. 1.) Thus, a small change in the grain resistivity with temperature completely explains the temperature dependence of the current rise time. On the other hand, the grain resistivity has little effect on the voltage overshoot because it is negligible compared to the high source resistance.

PULSE MEMORY PHENOMENA

The inductive response of a varistor is either reduced or enhanced, depending upon the competing influences of a long- and a short-term varistor memory of preceding pulses. These phenomena are complex, transitory by nature, and most easily studied in the time domain. Figure 5 illustrates the complexity of these memory effects for voltage overshoot. The first pulse of Fig. 5 exhibits a voltage overshoot of 10%, which is less than the 13% overshoot seen in the second pulse. [Overshoot is roughly calculated as 100% $(V_{max} - V_s)/V_s$ with V_s a steady-state voltage measured at 1.2 μs.] Although the second pulse is delayed by only a few microseconds, its is an

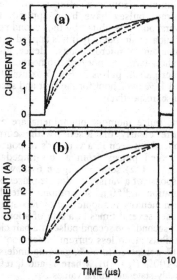

Figure 4. The influence of circuit resistance on varistor current response to rectangular voltage pulses. Parts (a) and (b) respectively show actual and simulated results. The solid, long-, and short-dashed curves are for resistances of 3, 1, and 0.1 Ω, respectively.

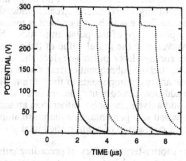

Figure 5. The voltage overshoot of a varistor measured for current pulses with a 1.2-μs duration and a 4-μs separation. The solid curve is the initial response. The dotted curve was measured after a number of other pulses had been applied, and it is arbitrarily placed on the same time scale.

example of long-term memory because this enhancement of the second pulse can be seen even when it is delayed for several minutes. However, after several large current pulses have been applied, less overshoot is exhibited by the second pulse of such a pair, but only if it is applied within some tens of microseconds. Hence, this is a short-term memory phenomenon. The latter pair of pulses in Fig. 5 exhibit 17% and 12% overshoot for the first and second pulse, respectively.

The pulse memory phenomena are even more clearly disclosed by the current undershoot seen in a varistor's response to repeated pulses from a low-impedance (e.g., ~ 1 Ω) source. Figure 6 shows the response of a varistor to double pulses of 6 μs duration and 50 μs separation. After long-term memory is suppressed by repeating the pulses several times at a rate of about once per second, the second pulse of a pair clearly exhibits much less current undershoot (i.e., 30% compared to 66%, gauging undershoot as 100% $(I_s - I_i)/I_s$ where I_s and I_i refer to steady-state and initial values).

Long-term memory is most clearly seen when the first pulse response is compared to the response of a subsequent pulse delayed by more than 10^{-4} s. (Short-term memory decays in about 10^{-4} s and long-term memory decays in about 10^3 s.) In Fig. 7, the first and tenth pulse of a sequence of pulses with 8 μs duration and 0.5 s separation are superimposed. In contrast to Fig. 6, the later pulse of Fig. 7 exhibits a greater undershoot than the first when the undershoot is measured as a percentage of the pulse amplitude. Note, however, that the initial value of the current is less for the later pulse (neglect the current

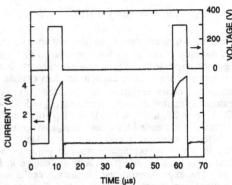

Figure 6. Current undershoot and short-term memory seen in the response of a varistor to double voltage pulses of 6-μs duration and 50-μs separation.

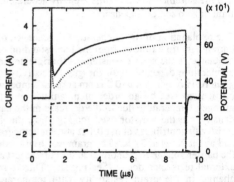

Figure 7. The response of a varistor to a sequence of voltage pulses with an 8-μs duration and a 0.5-s repetition rate. The first response (solid curve) and the tenth response (dashed curve) are superimposed.

spike) and the subsequent current rise is nearly identical for the two pulses. This result reveals that there are two components of the current with only one of these components behaving inductively. (In addition, the current spike reveals a capacitive component.) The later pulse exhibits a reduced noninductive component rather than an increased amount of inductive current. The undershoot is increased as a percentage of the pulse amplitude only because the inductive current represents an increased fraction of a diminished total.

The shorter-term memory of preceding pulses is attributed to the life time of the electron holes that become trapped in varistor barriers. Varistor breakdown occurs when trapped holes reduce the barrier height. But the drop in the barrier lags the applied voltage by the time required to generate and trap these holes. This time lag introduces a phase shift that is inductive in character. But once the hole traps are filled, the inductive character of any subsequent pulse response is diminished. (Actually, holes might immediately recombine with trapped electrons, and the extents to which they remain in trapped states is still unclear.) Thus, the shorter-term memory of preceding pulses is related to the lifetime of the holes trapped in the barriers. The longer-term memory of preceding pulses is attributed to a slow polarization-current discharge. As will be discussed below, these polarization currents can be large at short times, despite a slow-discharge classification.

POLARIZATION CURRENTS AND NON-DEBYE RESPONSE

Varistors display substantial polarization currents that are transient in nature and governed by distinctly non-Debye processes that decay slowly. Time domain measurements have provided the most definitive information on the slow decay of varistor polarization currents, and they are preferable to frequency domain measurements because of the higher information band pass they provide. As discussed above, polarization currents are significant at short times, and under breakdown conditions. But the focus here is on varistor response at low voltages, where the over-barrier currents can be neglected, and at times between 10^{-1} and 10^5 s, where depletion layer capacitance gives a negligible response. Under these conditions the polarization currents are predominant and time-domain measurements excel in revealing the nature of slow transients.[13,14]

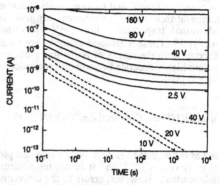

Figure 8. Transient varistor currents induced by applying voltages that differ by factors of 2. Solid line: 295 K; dashed line 77 K.

Figures 8 and 9 show charging and discharging currents for varistors at different voltages and temperatures. Figure 8 clearly shows that at room temperature about 10^3 s must elapse before polarization currents become negligible in comparison to steady-state conduction. At lower temperatures much longer times are required. Figure 9 shows the discharge currents that were measured after voltage was removed. Because there is no steady-current in this case, a current that decreases almost inversely with time is readily discerned. (A similar time dependence is seen if the steady-state current is subtracted in the results of Fig. 8.)

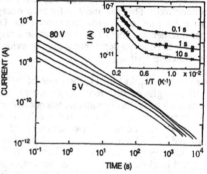

Figure 9. Transient discharge currents induced in a varistor at 295 K by removing voltages that differ by factors of 2. The inset is an Arrhenius plot of the current measured at 0.1, 1, and 10 s after voltage removal.

Because the polarization currents induced in varistors are large and decay slowly, it is tedious if not difficult to measure the steady-state current-voltage characteristics of varistors. As can be seen in Fig. 8, the approach to equilibrium is extremely slow at low voltages, and waiting a thousand seconds per data point is none too long at room temperature. At low temperatures, it can be virtually impossible to reach steady-state conditions. It also is worth noting that this slow approach to electrical equilibrium influences electrical properties measurements even at relatively high frequencies. Figure 10 shows hysteresis loops which illustrate the slow changes in capacitance and dissipation that are induced by the application of a voltage. As with measurements of I-V characteristics many hours are required to measure a steady-state capacitance as a function of voltage. This alone makes a Mott-Schottky analysis of the varistor capacitance difficult. But, there are other difficulties as well.[15]

The polarization currents are described reasonably well by the power law

$$I = I_0 e^{-E/kT}/t^m, \qquad (6)$$

where E is a thermal activation energy and the exponent m is near unity. The inset of Fig. 9 suggests two traps with activation energies of 10 meV and 150 meV, and measurements at higher temperatures than are shown suggest an additional trap with a 660 meV activation energy.[14] These activation energies are taken directly from the slopes of Arrhenius plots and are not highly accurate because differentiation of Eq. (6) gives

$$\partial \ln(I)/\partial(1/T) = -E/k - \ln(t)\,\partial m(T)/\partial(1/T), \qquad (7)$$

and as is discussed below m is the slope of a plot of ln(I) versus ln(t), and it is not temperature independent. However, errors in the activation energies contributed by the temperature dependence of m are no more than 30%. The 660 meV energy can be attributed to the deep interface traps that give rise to the barrier because it essentially identical to the measured activation energy of 680 meV energy for over-barrier conduction. The 150 meV activation energy is close to that of a bulk trap seen in DLTS measurements. The 10 mev activation energy is too low to be a trap depth, but it may be evidence of tunneling.[14,16]

As can be seen in Fig. 11, the slope of ln(I) vs. ln(t) at eighteen different temperatures varies between a value that is equal to, but not in excess of, unity as an upper bound and a value of one half as a lower bound. At two temperature (i.e., 110 and 330 K) the slope is close to one, and at two other temperatures (i.e., 230 and 450 K) m is equal or close to one half. As the temperature varies the value of m cycles back and forth between these bounds.

Jonscher[17] has pointed out that the power law dependence of Eq. (6) is nearly universal in dielectrics and its validity for ZnO varistors has been confirmed for times varying over 12 orders of magnitude.[13] Temperature variations in m are known in other materials, and changes in m between 0.5 and 1 have been reported, though not the cycling back and forth seen in varistors. Scher et al.[18] have explained the transition from m = 0.5 to 1 in terms of the convergence of moments of the distribution of transit times. In their explanation, such behavior results as the limit of a statistical distribution. Hence, no particular process is assumed, and Jonscher's universality is implicitly contained. The transition from m = 0.5 to 1.0 as temperature drops is a change from diffusive to dispersive electrical transport. In varistors the nonmonotonic behavior of m(T) can be interpreted as due to the resolution of polarization currents from two different traps (i.e., 660 and

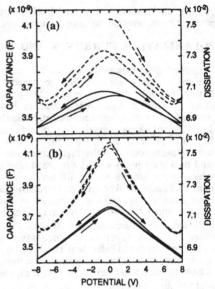

Figure 10. Capacitance (solid curves) and dissipation (dashed curves) measured at 10 kHz versus applied voltage. (a) The voltage changed at 10^{-2} V/s. (b) The voltage changed at 10^{-4} V/s.

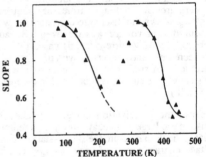

Figure 11. The slope of ln(I) plotted against ln(t). The curves are only guides to the eye.

150 meV), each of which make a current contribution that evolves in turn from diffusive to dispersive transport as the temperature is lowered.[14]

CONCLUSIONS

The time-domain characterization of electronic properties of materials offers unique and physically incisive insights, particularly for electrically nonlinear ceramics. Time-domain measurements on ZnO varistors disclose electronic transport of an exceptionally complex nature. Electrical currents in varistors are distinctly nonlinear in both their resistive and reactive components. And different types of nonlinearity appear at low and high applied voltages. At low voltages, polarization currents originating from deep electronic traps contribute to transient conduction that persists for long times, even though it is large at short times. A power-law rather than an exponential time dependence is observed. The power-law behavior is interpreted as "universal" behavior in the sense of Jonscher. The power-law exponent changes with temperature as seen in some other materials. But a unique double transition from diffusive to dispersive transport that originates in two different electron traps is observed in varistors. At high applied voltages, the electrical nonlinearity is contributed by Schottky barriers that form at grain boundaries. The time response measurements reveal an inductive behavior that is strong evidence of holes; they thereby support theoretical models that attribute the collapse of the varistor barriers at high applied voltage to hole trapping in the barriers. Also, the time response measurements distinguish between different theoretical interpretations of the observed change in varistor capacitance with voltage. Evidence is found of both a short- and a long-term memory of previously applied electrical pulses. These memory phenomena are ascribed respectively to the life time of the holes that get trapped in barriers and to the persistence of the polarization currents that originate from deep electron traps.

ACKNOWLEDGMENTS

This research was sponsored by the Division of Material Science, U. S. Department of Energy under contract No. DE-AC05-96OR22464 with Lockheed Martin Energy Research Corp.

REFERENCES

1. F. A. Modine and R. B. Wheeler, *J. Appl. Phys.* **61**, 3093 (1987).

2. G. D. Mahan, L. M. Levinson, and H. R. Philipp, *J. Appl. Phys.* **50**, 2799 (1979).

3. G. E. Pike, *Mater. Res. Soc. Proc.* **5** , 369 (1982).

4. G. E. Pike, S. R. Kurtz, P. L. Gourley, H. R. Philipp, and L. M. Levinson, *J. Appl. Phys.* **57**, 5521 (1985).

5. M. Rossinelli, G. Blatter, and F. Greuter, *British Ceram. Proc.* **36**, 1 (1985).

6. G. E. Pike, *Phys. Rev. B* **30**, 795 (1984).

7. G. Blatter and F. Greuter, *Phys. Rev. B* **33**, 3952 (1986).

8. G. Blatter and F. Greuter, *Phys. Rev. B* **34**, 8555 (1986).

9. M. A. Green and J. Shewchun, *Solid State Electron.* **16**, 1141 (1973).

10. K. Eda, *J. Appl. Phys.* **50** , 4436 (1979).

11. F. A. Modine and R. B. Wheeler, *J. Appl. Phys.* **67**, 6560 (1990).

12. F. A. Modine, R. B. Wheeler, Y. Shim, and J. F. Cordaro, *J. Appl. Phys.* **66**, 5608 (1989).

13. F. A. Modine, R. W. Major, S. I. Choi, L. B. Bergman, and M. N. Silver, *J. Appl. Phys.* **68**, 339 (1990).

14. R. W. Major, A. E. Werner, C. B. Wilson, and F. A. Modine, *J. Appl. Phys.* **76**, 7367 (199

15. M. A. Alim, *J. Appl. Phys.* **78**, 4776 (1995).

16. C. H. Seager and G. E. Pike, *Appl. Phys. Lett.* **40**, 471 (1982).

17. A.K. Jonscher, *Dielectric Relaxation in Solids*, (Chelsea Dielectrics, London, 1983).

18. H. Scher, M. F. Shlesinger, and J. T. Bender, *Phys. Today* 26 (Jan. 1991).

QUANTITATIVE ICTS MEASUREMENT OF INTERFACE STATES AT GRAIN BOUNDARIES IN ZnO VARISTORS

K. MUKAE, A. TANAKA
Fuji Electric Corporate Research and Development, Ltd., Yokosuka, 240-01 JAPAN,
mukae-kazuo@fujielectric.co.jp

ABSTRACT

Isothermal capacitance transient spectroscopy(ICTS) measurement is applied to ZnO:Pr varistors. A simple peak corresponding to the interface states at grain boundaries was obtained and the energy level of the interface states revealed to be monoenergetic and located around 0.9eV below the conduction band. The cross section was calculated as around 10^{-14} cm^2. Quantitative treatment of the ICTS intensity in relation to the density of interface states at grain boundaries was established. The density of interface states was obtained from the linear relation between ICTS intensity and reciprocal carrier density(N_D). According to experiments on series of rare-earth doped ZnO varistors, the interface states of Pr, Tb or Nd doped ZnO varistors had higher density of states than La or Er doped varistors. Moreover, application of ICTS measurement to single grain boundary using microelectrodes revealed that higher density of interface states gave higher nonlinearity in I-V characteristics.

INTRODUCTION

Electrical properties of ceramic semiconductors are strongly dependent on the double Schottky barriers(DSB) formed at grain boundaries in the ceramic semiconductors[1]. Since DSB is built by electrons trapped by the interface states at the grain boundary as shown in **Fig.1**, it is important to characterize the electronic states to investigate their electrical properties. Since isothermal capacitance transient spectroscopy(ICTS) measurement is an intensive method to characterize deep electronic states in semiconductors[2], it has been applied to characterize electronic interface states at grain boundaries in ceramic semiconductors such as zinc oxide varistors[3, 4]. However, these works have been limited to qualitative analysis of ICTS intensity. Since ICTS intensity corresponds to the density of interface states, quantitative relation between the density of interface states and I-V characteristics can be understood by quantitative analysis of ICTS intensity. Present work will discuss the characteristics of electronic interface states of DSB and establish quantitative analysis of ICTS intensity and discuss the relation between I-V nonlinearity and electronic interface states. Moreover direct characterization of single grain boundary using microelectrodes will be reported.

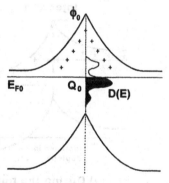

Fig. 1 Double Schottky barrier at grain boundary

THEORY

ICTS measurement is one of capacitance transient methods by which electronic states in the forbidden band of semiconductors are characterized like DLTS measurement. When we apply this method to ceramic semiconductors, we can estimate the characteristics of electronic interface states at grain boundaries. When a certain voltage is applied across a grain boundary, capacitance will decrease because electrons are injected to the interface states as shown in **Fig. 2(a)**. Then if the applied voltage is eliminated, capacitance will increase up to the initial value because captured electrons are emitted from the interface states as shown in **Fig. 2(b)**. ICTS method measures this transient recovery of capacitance and obtains the emission rate and other parameters of the interface states by mathematical processing described bellow.

ICTS signal S(t) is defined as following equation[2],

$$S(t) = t \frac{df(t)}{dt} \qquad (1)$$

where $f(t)$ is given by[4]

$$f(t) = \left(\frac{1}{c(t)} - \frac{1}{C(\infty)} \right) \qquad (2)$$

where $C(t)$ is the capacitance of the varistor at time t after the elimination of the applied voltage, $C(\infty)$ is capacitance at the steady state without the applied voltage. Since capacitance transient occurs by emission of electrons at the interface states, $f(t)$ is described as following relation.

$$f(t) \infty \exp(-e_n t) \qquad (3)$$

where e_n is thermal emission rate of electrons from the interface states which are given by

$$e_n = N_C \sigma_n v_{th} g^{-1} \exp(-E_{IS} / kT) \qquad (4)$$

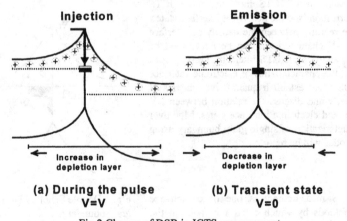

(a) During the pulse
V=V

(b) Transient state
V=0

Fig.2 Change of DSB in ICTS measurement

where N_C is the effective density of states in the conduction band, σ_n is the capture cross section, v_{th} is the thermal velosity, g is the degeneracy, E_{IS} is level of the interface states below the conduction band edge, k is the Boltzmann constant, and T is absolute temperature, respectively. From eq. (1) and (3), one can find that $S(t)$ has a peak value S_{max} at $e_n t = 1$.

$$S_{max} = -\frac{N_{IS}}{\varepsilon_s N_D Ae} \tag{5}$$

where N_{IS} is the density of interface states, ε_s is dielectric constant of semiconductor grain, N_D is shallow donor concentration, and A is the junction area. The interface states level E_{IS} or the capture cross section σ_n is obtained from the values of e_n measured at several temperatures. Since $\sigma_n N_C$ is proportional to T^2, the activation energy E_{IS} of the interface states is obtained from the slope of Arrhenius plot of $\ln(e_n / T^2)$ and $1000 / T$. Meanwhile the interface states density is obtained by peak value of ICTS spectrum using eq. (5).

EXPERIMENT

Chemical composition of ZnO-0.5 at.%Ln_2O_3-2.0 at.%Co_3O_4 were prepared where Ln indicates rare-earth metal. Powders were mixed in the plastic pot with water and ZrO_2 balls for 20 hours. In case of Ce, Pr and Tb, powders of CeO_2, Pr_6O_{11} and Tb_4O_7 were used respectively. Slurry was dried and pressed into pellets. Then pellets were fired in air at the temperature from 1300°C to 1450°C. The obtained samples were shape of 14mm in diameter and 1.2mm in thickness. Silver electrodes of 12mm in diameter was printed on both sides of samples and fired at about 600°C. Then I-V, capacitance-voltage and ICTS measurement were carried out for these samples.

Figure 3 shows the block diagram of ICTS measurement. Transient capacitance was measured by capacitance meter(MI-401 DLTS test system, Sanwa Musen Co., Ltd.) and stored in digital memory(DM-2350 Iwatsu Electric Co., Ltd.). Applied voltage was supplied by the pulse generator. Stored capacitance change was proceeded mathematically by the desk top computer to obtain ICTS spectrum.

After bulk measurements single grain boundary measurements were performed. For these measurements, microelectrodes were provided using photolithography technique. Silver electrodes were removed and the surface of samples were polished. Polished surface was thermally etched at 1050°C for 1 hour to make grain boundary visible and relax the damage to the surface by polishing. Aluminum thin film was evaporated on the surface and patterned

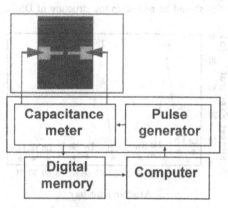

Fig 3 Block diagram of ICTS apparatus

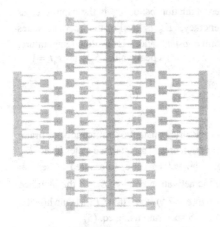

Fig. 4 Pattern of microelectrodes

by photolithography to form microelectrodes. **Figure. 4** shows the pattern of microelectrodes. There are 72 pairs of electrodes on this pattern. The tips of the electrodes are 5μm wide. The distance between a pair of electrodes is 10μm. The grain boundaries which ran between a pair of microelectrodes were confirmed by an optical microscope. SEM photograph of a single grain junction and microelectrodes is shown in **Fig. 5**.

For single grain boundary measurement, samples were placed on the heater inside the sample box. Temperature was controlled by a low noise solid state regulator. Tungsten microprobes were used for electrical contacts. I-V characteristics were measured by using curve tracer(TT-505, Iwatsu Electric Co., Ltd.). Measurements of single grain boundary ICTS spectra were performed using the same apparatus described above connecting tungsten microprobes. The applied voltage was kept constant as 1.5V.

RESULTS AND DISCUSSIONS

BULK ICTS of ZnO:Ln VARISTORS

Although rare-earth metal elements are considered to create nonlinear characteristics in ZnO:Ln varistors[5, 6], precise role of these elements are still not understood. **Figure 6** shows the nonlinear exponent α of these varistors sintered at 1350°C. From this figure, one can find nonlinearity is excellent only for varistors containing Pr or Tb. The reason for the high nonlinearity has been attributed to the high content of oxygen of these oxide materials[6]. Since the difference of nonlinear characteristics should be related to the structure of DSB at

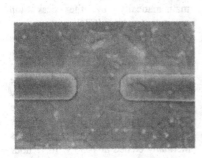

Fig. 5 SEM of single grain boundary between microelectrodes

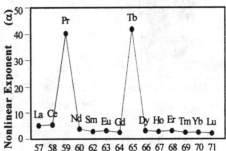

Fig. 6 Change in nonlinear exponent α with rare-earth elements

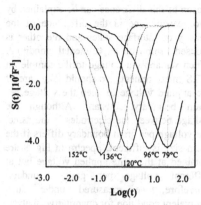

Fig 7 Measured ICTS spectra of ZnO:Pr varistor

Table I Level and capture cross section of the interface states for ZnO:Ln varistors

Element	$E_{IS}(eV)$	$\sigma_n(cm^2)$
La	0.95	1.5×10^{-14}
Pr	0.94	1.5×10^{-14}
Nd	0.91	3.3×10^{-15}
Tb	0.96	3.1×10^{-14}
Er	0.68	3.8×10^{-18}

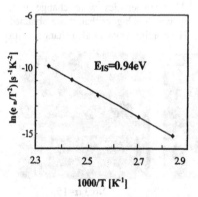

Fig. 8 Arrhenius plot ICTS peak of ZnO:Pr varistor

the grain boundary, ICTS spectra of these varistors will correspond this phenomena. A typical ICTS spectra of ZnO:Pr varistor are shown in **Fig. 7** in which ICTS signal are plotted as a function of $\log t$ in seconds. In this figure spectra measured at 79°C to 152°C are indicated. A simple peak can be seen in each spectrum. The position of peak shifts to shorter time, i.e., the emission rate increases, as the temperature. Arrhenius plot of this peak position is shown in **Fig. 8**. All data of emission rate formed a straight line. Therefore the level of interface states below conduction band edge was calculated from the slope as to be 0.94eV. The capture cross section σ_n, is also obtained as $1.5 \times 10^{-14} cm^2$. **Figure 9** shows the ICTS spectra of ZnO:La varistor. The shape of ICTS spectra are almost the same in spite of large noise and the Arrhenius plots were similarly obtained. **Table I** shows the calculated values of level of the interface states and the capture cross section for each rare earth element. The levels of the interface states were nearly same around 0.9eV which agree with literature[4] and the capture cross sections were some what different. The significance of the difference of cross section is not clear at this moment.

Although the shape of ICTS spectra were almost the same, the peak heights were different for each rare earth element. The

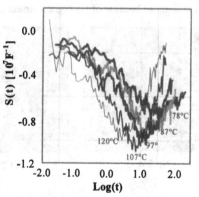

Fig.9 ICTS spectra of ZnO:La varistor

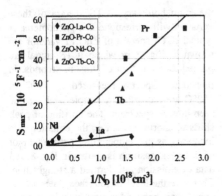

Fig. 10 Linear relation between ICTS intensity and $1/N_D$

reason of this difference can be explained by two reason. One is the difference of the density of interface states and the other is aroused from the experimental condition. When we apply a voltage to the sample on ICTS measurement, we should take care of the applied voltage to keep the voltage per grain boundary constant. Although the voltage between the electrodes is the same, the voltage per grain boundary differs if the grain size is different. Spectra in Fig. 9 are obtained at the same applied voltage but at different voltage per grain boundary. Therefore, we measured under such equivalent condition for quantitative analysis of ICTS peak intensity. In **Fig. 10**, peak intensities obtained under such equivalent conditions are plotted as a function of $1/N_D$ for varistors with different rare-earth elements and sintered at different temperatures. Since eq. (5) shows the linear relation between ICTS peak intensity and $1/N_D$, we can obtain a straight line for the points in Fig. 10 and the density of interface states N_{IS}. can be calculated from the slope by eq.(5). From Fig. 10, two lines are found and $2.4 \times 10^{11} \text{cm}^{-2}$ are obtained as the density of states for ZnO:Pr, Nd or Tb varistor which has high nonlinear characteristics. On the other hand low value of $9.7 \times 10^{10} \text{cm}^2$ for ZnO:La or Er varistor whose nonlinearity was lower. Since quantitative analysis of ICTS peak intensity was found to give the density of the interface states at grain boundaries, this analysis were continued to further investigation.

SINGLE GRAIN BOUNDARY MEASUREMENT

After bulk ICTS measurement, direct measurement of single grain boundary were carried out using microelectrodes. **Figure 11** shows I-V characteristics for several single grain boundaries on the same ceramic varistor. Their I-V characteristics were change with boundaries. The threshold voltage varied from 2V to 4V which is higher than those obtained by bulk samples because of the higher current density. The nonlinearity was also changed with

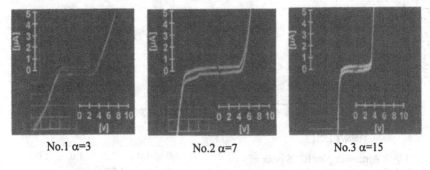

No.1 α=3   No.2 α=7   No.3 α=15

Fig. 11 I-V curves of single grain boundaries on the same ZnO:Pr varistor

Table II Calculated parameters of interface states of single grain baoundary

$E_{IS}(eV)$	$\sigma_n(cm^2)$	$N_{IS}(cm^{-2})$
0.91	3.78×10^{-14}	3.10×10^9

boundaries and their nonlinear exponent varied from 3 to 15. Characteristics of sintered material, $\alpha = 8.6$ seems to be an averaged character of these grain boundaries. **Figure 12** shows an example of obtained ICTS spectra for a single grain boundary in ZnO:Pr varistor. Only single peaks were observed and those peaks were shifted to shorter time with increasing temperature. These peaks can be regarded as the direct evidence of the interface states at the grain boundary. **Figure 13** shows the Arrhenius plot of peak temperatures and $\ln(e_n / T^2)$. The interface state level and the capture cross section were obtained from the slope and the intercept, respectively. ICTS spectra of these single grain boundaries are shown in **Fig. 14**. The density of interface states were calculated by eq. (5). **Table II** shows the interface state parameters obtained by this method. The values are average over 11 grain boundaries. The interface state level were located at 0.9eV below conduction band edge which are almost the same as those obtained by bulk measurement. These values were stable for every grain boundary. On the other hand, capture cross section and the density of interface states varied with the grain boundary. Moreover the density of interface states were quite different from those by bulk measurement. This discrepancy might be attributed to the effect of surrounding grain boundaries since their capacitance were added to the aimed single grain boundary.

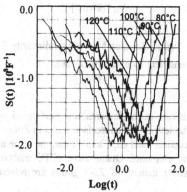

Fig 12 ICTS spectra of single grain boundary in ZnO:Pr varistor

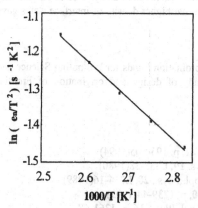

Fig. 13 Arrhenius plot of single grain boundary ICTS

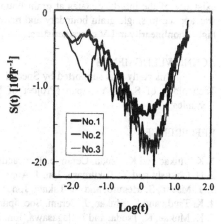

Fig. 14 Single grain boundary ICTS spectra on the same ceramic disk

Difference of I-V characteristics between single grain boundaries could be explained by the change in the properties of single grain boundaries. These change can be generated by donor concentration N_D of ZnO grain, the interface state level or the density of states. However, N_D is same for each grain boundary because of the same ceramic disk and the interface states level is quite stable as indicated above. Therefore, the density of interface states seems to be responsible for the difference of the I-V characteristics. **Figure 15** shows the relation between nonlinearity and density of the interface states for single grain boundaries. In this figure, grain boundaries with higher density exhibited higher nonlinearity.

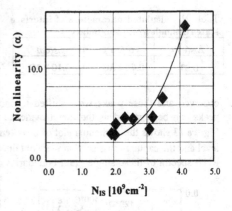

Fig. 15 Relation between nonlinearity and density of interface states

CONCLUSIONS

ICTS analysis of ZnO varistors doped with rare-earth elements was performed on both bulk ceramic disk and single grain boundary. The levels of the interface states in ZnO:Pr varistors were obtained as around 0.9eV below conduction band edge and almost the same as those in other varistors doped with other rare-earth elements. Quantitative ICTS analysis revealed that ICTS intensity and the reciprocal carrier density of ZnO grains are related linearly in accordance with the theory.

Direct measurement of ICTS and I-V characteristics of single grain boundary were carried out using microelectrodes. Single grain boundaries gave similar spectra to bulk ceramic varistors. Observed ICTS peaks can be regarded as the direct evidence of the existence of the interface states at grain boundaries. Quantitative analysis of ICTS intensity was applied to single grain boundary and revealed that higher density of interface states gave higher nonlinearity in I-V characteristics.

ACKNOWLEDGMENTS
This study was supported by Special Coordination Funds for Promoting Science and Technology of S. T. A. Japan (Project of "Study of design and origination of Frontier Ceramics")

REFFERENCES

1. K. Mukae and K. Tsuda, Ceramic. Transaction, **41**, p. 195-205(1994)
2. H. Ohkushi and Y. Tokumaru, Jpn. J. Appl. Phys. **19**, L335-38(1980)
3. T. Maeda, S. Meguro, and M. Takata, Jpn. J. Appl. Phys., **28**, L714-16(1989)
4. K. Tsuda and K. Mukae, J. Ceram. Soc. Jpn, **100**, p. 1239-44(1992)
5. K. Mukae, K. Tsuda, and I. Nagasawa, Jpn. J. Appl. Phys., 16, p. 1361-68
6. K.Mukae, Am. Ceram. Soc. Bul., 66, p. 1329-1331(1987)

CURRENT LOCALIZATION, NON-UNIFORM HEATING, AND FAILURES OF ZnO VARISTORS

M. BARTKOWIAK
Department of Physics and Astronomy, University of Tennessee, Knoxville, TN 37996-1200, and Solid State Division, Oak Ridge National Laboratory, Oak Ridge, TN 37831-6030, bartkowiakm@ornl.gov

ABSTRACT

Non-uniform heating of ZnO varistors by electrical pulses occurs on three different spatial scales: (1) microscopic (sub-micron), (2) intermediate (sub-millimiter), and (3) macroscopic (of order of millimeters or centimeters). Heating on these scales has different origins and different consequences for device failure in large and small varistors. On the microscopic scale, the heating localizes in strings of tiny hot spots. They occur at the grain boundaries in a conducting path where the potential is dropped across Schottky-type barriers. These observations are interpreted by applying transport theory and using computer simulations. It is shown that the heat transfer on a scale of the grain size is too fast to permit temperature differences that could cause a varistor failure. On an intermediate size scale, the heating is most intense along localized electrical paths. The high electrical conductivity of these paths has microstructural origin, i.e., it derives from the statistical fluctuations of grain sizes and grain boundary properties. Current localization on the intermediate size scale appears to be significant only in small varistors. On the macroscopic scale, current localization in large blocks can be attributed to inhomogeneities in the electrical properties which originate during ceramic processing. The resulting non-uniform heating is shown to cause destructive failures of large varistor blocks.

INTRODUCTION

Metal oxide varistors have highly nonlinear electrical characteristics and are widely used as devices for over-voltage protection [1, 2]. Varistor applications range from the use of small varistors to protect delicate electronic components to the use of much larger varistors for the protection of electrical-power-distribution systems. Specifically, varistors provide protection against voltage surges due to lightning strikes, switching transients, and similar disturbances. While varistors have a large capacity to absorb energy (e.g., 500 J/cm^3), they are, in fact, subject to occasional failure. The significant varistor-failure mechanisms include: electrical puncture, thermal cracking, and thermal runaway — all resulting from excessive heating, in particular, from non-uniform heating. Temperature and current flow in a varistor are highly correlated, since the heat is generated by electrical conduction which varies strongly with temperature. Non-uniform Joule heating occurs in varistors as a result of inhomogeneous electrical properties that originate in either the varistor fabrication process or in the statistical fluctuations in properties that generally occur in polycrystalline materials.

The power dissipation characteristics, Joule heating, and energy-handling capability of

ZnO varistors have attracted attention because of their relevance to device failure and service life [3]–[13]. Measurements have been made that directly connect varistor failure to heating, and the various modes of varistor failure have been modeled. Eda [3] has used simple electrical measurements to establish that non-uniform electrical properties cause varistor failures. In his work, a spatial variation of the breakdown voltage in large varistors was measured by placing pairs of small electrodes at various positions. Eda attributed this variation in the breakdown voltage to a broad grain-size distribution. Weak spots with low-breakdown voltage and high-current conduction were attributed to a lower number of grain boundaries present in current paths for materials containing large grains. (Recent evidence suggests that weak spots in small varistors also can result from a reduction in the barrier potentials as well as in the number of grains — i.e., grain boundaries may be present but electrically ineffective [14]–[18]). These highly conducting regions heat up more rapidly than their surroundings under high currents and generate thermal stresses that cause cracking. High current can also cause local heating sufficient to melt the Bi_2O_3 phase and, thereby produce so-called "puncture" failures. Alternatively, the temperature of a varistor may become so hot that a thermal runaway failure ensues due to the material's negative temperature coefficient of resistivity [4]–[7].

Non-uniform electrical conduction in varistors has been observed at a higher spatial resolution by using scanning contacts and galvanic electroplating [8], but such measurements are well suited only to low-current densities and they provide essentially no time resolution. Mizukoshi et al. observed non-uniform heating in varistors by infrared (IR) imaging [9] using an IR camera to image hot zones in station-class arresters, and they quantified the relationships between non-uniformity and varistor failures. But the spatial resolution and temperature resolution of these experimental methods were quite limited. Moreover, the IR images were obtained long after the electrical pulses, so no time resolution of the heating process was observed. In a recent study [10], a new, high-speed IR camera was utilized to monitor more precisely the heating transients occurring in varistors during the electrical breakdown process. Time resolved measurements with high spatial and high temperature resolution provided detailed information on varistor heating on both a macroscopic and a microscopic scale during and after the application of electrical transients. On a macroscopic scale, large arrester blocks exhibited non-uniform heating that is related to their ceramic processing. On an intermediate size scale, non-uniform heating in small varistor disks was found to be most intense along highly conducting current paths that result from the statistical fluctuations in electrical properties which occur in polycrystalline materials. On a microscopic scale, the heat generation in varistors was observed to be localized at the grain boundaries.

In principle, non-uniform heating on each of the above size scales may lead to a varistor failure. The important question how heating on these vastly different time and size scales relates to puncture or cracking has been addressed by applying transport theory and by using computer simulations. The results obtained from these calculations are in excellent agreement with the experimental data. It is shown that on a scale of the grain size the heat transfer is simply too fast to permit temperature differences that could cause a varistor failure. Equally interesting results have been obtained for the case of current localization on an intermediate size scale because of microstructural disorder. Recent results from computer simulations [15]–[18] show that current flowing through the disordered grain microstructure of varistors localizes almost completely in a few very narrow paths. Given these results, it may seem strange that puncture or cracking failure does not always occur! We have resolved this issue by showing that current localization of microstructural origin is important only in small varistors. The degree of current localization is indeed very high in varistors of the a size that can be easily simulated on a computer, but it becomes much less significant in larger varistors. On the other hand, puncture and cracking failures of large varistor blocks are caused by non-uniform current flow and non-uniform heating on a macroscopic scale that can be attributed to imperfect processing. Conditions under which these failures can occur are described in detail in references [6] and [7].

MACROSCOPIC SCALE

Experimental data on two different kinds of large (4.2 cm diameter) varistor blocks are provided in Ref. [10]. One kind of varistors were fabricated with a 1 cm thickness, and the other were a 1 cm thick pieces cut from commercial blocks made with the same diameter but a 4.6 cm length. Both kinds of blocks were heated by a 10 A current pulse with a 1 ms duration and the time resolved thermal images of the surface temperature were recorded with an IR camera. For the case of the varistors fabricated with a 1 cm thickness, the circumference of the blocks represented the hottest region, and the temperature distribution was roughly radially symmetric. Localized heating in the blocks interior, as reported by Mizukoshi et al. [9], was not observed. However, for the 1 cm thick pieces cut from commercial blocks, nonuniform heating localized in hot spots were clearly visible, and the IR images were similar to those reported Mizukoshi et al. These results are consistent with the greater difficulty of fabricating uniform large blocks, particularly blocks with a high aspect ratio.

Thermal images of the varistors were analyzed to determine the temperature distribution in the block as a function of time. Temperature profiles along a disk diameter before, during, and after the current pulse were measured. A few of these profiles for a varistor block fabricated with a 1 cm thickness are shown in Fig. 1(a). An initial temperature profile was taken from an image made during the current pulse. It represents the temperature of the block at a time half way through the pulse, or at 0.5 ms. During the current pulse, the temperature was rising very fast, particularly at the periphery of the block. But after the pulse, the temperature changed relatively slowly on the centimeter size scale of the block.

The temperature rises most at the periphery of the block because most of the current flow is taking place there. This inference was checked using Eda's approach of placing pairs of small electrodes at various positions on the block. A 3 to 4% decrease in the breakdown voltage from the center to the edge was measured directly. These measured values of the breakdown voltage have been expressed as an approximate analytical function of the distance from the center of the disk, and used to simulate its thermal behavior by solving numerically the coupled equations describing the heat transfer and the non-uniform flow of the nonlinear currents. We have assumed that the disk can dissipate heat only radially by convection through the sidewall, and used realistic temperature-dependent data for thermal properties of the material and for the varistor I-V characteristic. At temperatures close to room temperature, we have used: the heat capacity $c = 0.6$ J/(gK), the diffusivity $D = 0.1$ cm^2/s, and the density $\rho = 5.5$ g/cm^3. Details of the model are described in Ref. [6]. For the case of a 1 ms, 10 A current pulse, the temperature profiles of the disk have been calculated for the times corresponding to those in the measurements. The results are presented in Fig. 1(b). They are in very good agreement with the experimental data in Fig. 1(a).

Because the temperature distribution shown in Fig. 1(a) has radial symmetry of the

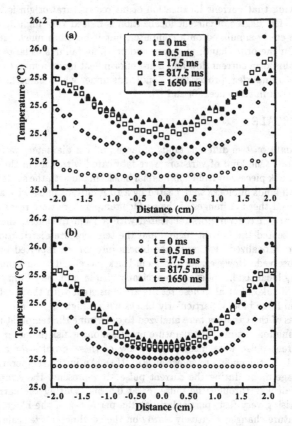

Figure 1: Profiles of the temperature along a varistor diameter before, during, and after a current pulse: (a) measured and (b) simulated.

block, it is logically attributed to the fabrication process. The temperature increases more at the edge of the block because there is a higher current that results from a lower break-down voltage that can be attributed to a larger grain size. The implied grain size increases monotonically from the center to the edge of the block. Friction in the die used for powder compaction causes density variations that can result in such a systematic variation in the grain size. Moreover, grain growth during sintering can be more rapid in the outer part of the block than in the center of the block. It is difficult to fabricate large ceramic parts with uniform properties, but uniformity is especially important in varistor manufacture because any nonuniformity gets amplified in the conduction process by the electrical nonlinearity. The IR images of varistors cut from large blocks shows that less symmetry is obtained in longer blocks. Current localization tends to be the greatest in highly nonlinear materials, and thus, the production of large varistors requires particularly close control of pressing and sintering operations. These conclusions are consistent with the measured [12] and simulated

[6, 7] data on the energy handling capability of varistor blocks used in distribution-class and station-class arresters.

INTERMEDIATE SCALE

It is significantly easier to fabricate varistor disks with uniform properties when they are small. However, small varistors exhibit non-uniform current conduction that is of a statistical origin instead of a processing origin. Property variations due to statistical disorder are more significant in small varistors than in large varistors because statistical fluctuations are in general more important in small systems. In small varistors there is a tendency for the current to flow along those conducting paths that have the lowest breakdown voltages because, for example, they are comprised of the largest grains [15]–[18]. Non-uniform heating of small varistors due to current flow along several localized highly conducting paths was observed in Ref. [10]. Since the number of ZnO grains between the electrodes in the studied samples was only about 40, a variation of 3 – 4 grains can change the current flow in a path by an order of magnitude relative to surrounding paths. Thus, conducting paths with low breakdown voltages carry most of the current and become hotter.

If the number of grains between the varistor electrodes is large enough to apply simple statistics, the difference in the importance of current localization due to statistical disorder in large and small varistors is easy to explain. A grain size distribution characterized by a mean diameter $\overline{d}$ and a variance σ_d will lead to a differing number of grain boundaries along the various conducting paths that will be approximately Gaussian distributed with a mean $\overline{N}$ and a variance σ_N given by:

$$\overline{N} = L/\overline{d} \qquad \text{and} \qquad \sigma_N = \overline{N}^{1/2}\sigma_d/\overline{d}, \qquad (1)$$

where L is the distance between the electrodes. If it is assumed that the breakdown voltage of a current path is just the sum of the breakdown voltage V_b of N identical grain boundaries, then the distribution of the breakdown voltage of the various paths is described by a mean $\overline{V}$ and a variance σ_V that are given by:

$$\overline{V} = \overline{N}V_b \qquad \text{and} \qquad \sigma_V = \overline{N}^{1/2}V_b\,\sigma_d/\overline{d}. \qquad (2)$$

While the voltage increases in proportion to the number of grain boundaries, the variance of the voltage increases only as the square root of the number of boundaries because smaller grains tend to compensate for larger grains. (In reality, grain boundaries are not identical and the breakdown voltage V_b also varies [16]–[18]. However, variations in V_b contribute a similar dependence upon $\overline{N}$ to $\overline{V}$ and σ_V and therefore can be included by just assuming a larger value for $\sigma_d/\overline{d}$.) If a varistor is well into the breakdown region where the current I has a nonlinear dependence on the voltage V that can be described by a power law exponent α, then the distribution of the current among the possible paths is described by a mean $\overline{I}$ and a variance σ_I. In the simplest approximation,

$$\overline{I} = I_0\overline{V}^{\alpha} \qquad \text{and} \qquad \sigma_I = \alpha\overline{I}\,\overline{N}^{-1/2}\sigma_d/\overline{d}, \qquad (3)$$

The above classification of the current will be accurate only if the distributions are relatively narrow and approximately Gaussian, but the results agree with the predictions of a more realistic varistor model, at least in a rough way. We examined the computer predictions of a "brick model" [1] type of grain structure in which the sizes of the grains (i.e., length in the direction of current flow) were generated by a computer program as random numbers from a log-normal distribution with a given mean and variance. An identical but realistic I-V characteristic was assigned to each of the grain boundaries [6]. Small varistors were modeled as two-dimensional arrays of rectangular grains with $N_w = 100$, 200, or 300 grains along the direction perpendicular to the current. All grains were given the same width, but their lengths were randomly generated. The grains were stacked in adjacent columns until a prescribed length corresponding to the thickness of the varistor was reached. Each of the grain boundaries was assumed to have an I-V characteristic with a maximum coefficient of nonlinearity $\alpha = 50$. We assumed current flow only in one direction (i.e., along N_w available parallel current paths containing a varying number of grain boundaries). The current distribution in the sample was found by just calculating separately the current in each path. The general predictions of this simple model agree with those of more realistic Voronoi network models [15]. The current localized in the paths with the least number of grain boundaries, or grains. The localization is particularly pronounced in the nonlinear region of the I-V characteristic. And the maximum current localization occurs at the voltage for which the sample has the maximum nonlinearity coefficient. In Fig. 2, values of $\sigma_I/\overline{I}$ found from this model are compared to the predictions of Eq. (3) for different values of $\sigma_d/\overline{d}$. The agreement is good, and it improves as $\overline{N}$ becomes large or σ_d becomes smaller.

Current localization on highly conducting paths will be negligible if the variance of the current distribution is small relative to its mean, or when $\sigma_I/\overline{I} \ll 1$. This condition together with Eq. (3) leads to a criterion for the number of grain boundaries required in a varistor in which current localization due to statistical variations in grain size will be unimportant:

$$\overline{N} \gg \alpha^2(\sigma_d/\overline{d})^2 . \tag{4}$$

If it is assumed, for example, that $\sigma_d/\overline{d} = 0.1$ and $\alpha = 50$, then $\overline{N} \gg 25$ should result in minimal current localization due to grain size variations. (Note the strong influence of nonlinearity exponent.)

Large arrester blocks contain thousands of barriers in each path, so current localization due to statistical disorder of the grain sizes is too small to cause a failure by puncture or thermal cracking. Based on the results of a computer simulation of non-uniform Joule heating in disordered networks, Vojta and Clarke [13] have recently proposed that puncture failures have microstructural origin. However, our present results indicate that their conclusions are not valid for large (high voltage) varistor blocks. In large varistors, destructive failures are caused by macroscopic nonuniformities due to processing (e.g., compaction and sintering) [6, 7]. Varistors for electronic applications are another matter. Because they may have fewer than 10 barriers in a current path, current localization due to grain size variations can be severe.

A formula that goes beyond inequality (4) to describe the influence of varistor size on current localization due to statistical disorder can be obtained if it is assumed that the breakdown voltages of the various conducting paths have a Gaussian distribution with a mean and variance given by Eq. (2). Then, well into the breakdown region, the current distribution on these paths is

$$p(i) = \frac{i^{-(1-1/\alpha)}}{\alpha \sigma_V \sqrt{2\pi}} \exp\left[-\frac{\overline{V}^2}{2\sigma_V^2}\left(1 - i^{1/\alpha}\right)^2\right], \tag{5}$$

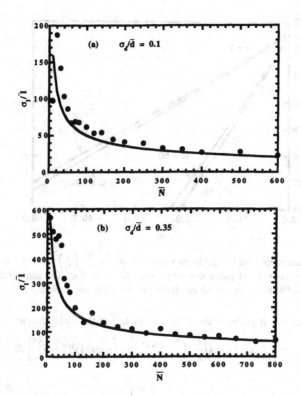

Figure 2: Relative variance $\sigma_I/\overline{I}$ calculated from Eq. (3) (solid lines) and simulated within the disordered "brick model" (full circles) as a function of the average number of grains between the electrodes $\overline{N}$, for $\sigma_d/\overline{d} =$ (a) 0.1 and (b) 0.35.

where i is *defined* as $i = I/\overline{I} = I/(I_0\overline{V}^\alpha)$. The current distribution $p(i)$ is asymmetric, and, depending on the values of α and $\sigma_V/\overline{V}$, can have a long tail for large currents. The presence of such a long tail corresponds to (and describes analytically) the current localization phenomenon. A current path gets hot because of current localization and Joule heating only if it carries more current than an average path by some factor which can be denoted as n:

$$I_{hp} \geq n\langle I \rangle , \qquad (6)$$

where n is a rather arbitrary number between about 2 and 5 depending upon the choice of criterion for "hot". $\langle I \rangle$ in Eq. (6) is the actual expectation value of the distribution (5),

$$\langle I \rangle = I_0\overline{V}^\alpha\langle i \rangle = I_0\overline{V}^\alpha \int_0^\infty i\, p(i)\, di , \qquad (7)$$

which in general is not equal to $\overline{I}$ from the rough approximation in Eq. (3). The fractional number of hot paths f_{hp} in a varistor sample for a given factor n is calculated as the integral

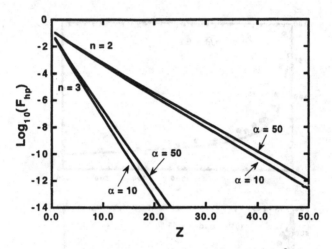

Figure 3: Fractional number of hot paths as a function of $z = \frac{\overline{N}}{2\alpha^2} \left(\frac{\overline{d}}{\sigma_d}\right)^2$. Since z itself depends on α, the fractional number of hot paths is a very strong function of α. The slightly different curves for $\alpha = 10$ and $\alpha = 50$ show that the single dependance on z is not perfect.

describing the number of paths in the high current tail of the distribution:

$$f_{hp} = \int_{n\langle i\rangle}^{\infty} p(i)\, di = \sqrt{\frac{z}{\pi}} \int_{n\langle i\rangle}^{\infty} i^{1/\alpha - 1} \exp\left[-z\alpha^2 \left(i^{1/\alpha} - 1\right)^2\right] di \,, \tag{8}$$

where

$$z = \frac{\overline{N}}{2\alpha^2} \left(\frac{\overline{d}}{\sigma_d}\right)^2 \tag{9}$$

Computed numerically logarithm of the fractional number of hot paths as a function of the parameter z is presented in Fig. 3 for $n = 2$ and 3, and for $\alpha = 10$ and 50. It is seen, for example, that for varistors with $z > 5$ only less than one in about 10^4 paths conducts current 3 times ($n = 3$) higher than the average, which is still not the case of a very hot path. The choice of z as an independent variable in Eq. (8) and in Fig. 3 leads to an approximate scaling of the results, so that they appear as only weakly dependent on α. This, however, does not mean that f_{hp} is not sensitive to the values of α. On the contrary, since z itself is a strongly varying function of α [Eq. (9)], so is f_{hp}.

The predictions of Eq. (8) for small and large varistors contrast remarkably. A typical varistor for electronic applications is a small disk of about 1 cm diameter and 0.1 cm thickness. Typically, the grains of such a varistor are described by $\overline{d} = 100$ μm and $\sigma_d/\overline{d} = 0.1$, and the nonlinearity coefficient might be $\alpha = 25$. If the hot paths are assumed to be those paths that carry a current $I > 2\langle I\rangle$ (i.e., $n = 2$), then the parameter z of Eq. (9) is 0.8, and $f_{hp} \approx 0.1$. Hence, current localization is predicted to be a problem. Commercial varistors for electronic applications have rather low nonlinearity coefficient, although it is not difficult to produce small varistors having an α of 50 or more. However, increase of α, according to Eq. (9), would lead to smaller values of z, and [by Eq. (8)] to an even higher probability of hot paths, thus making the varistor even more prone to failure.

On the other hand, a distribution-class arrester block is typically a disk of about 4 cm diameter and 4 cm thickness. The smaller grains of such a varistor block can be described by $\bar{d} = 20$ μm and $\sigma_d/\bar{d} = 0.1$, and the nonlinearity coefficient of such a varistor might be $\alpha = 50$. Again, if the hot paths are assumed to be those paths carrying a current $I > 2\langle I \rangle$ (which is a very conservative definition), then the parameter z of Eq. (9) is about 40, and $f_{hp} \approx 10^{-10}$. So there will be no hot paths due to statistical disorder in such blocks, even though the total number of available current paths exceeds a million. Clearly, current localization resulting from statistical disorder is only important in small varistors.

MICROSCOPIC SCALE

In Ref. [10], varistor heating was examined at the microscopic level in thin slices cut from low voltage varistors that have grain sizes over 100 μm. These slices were polished to about 100 μm thick, so that electrical conduction was essentially confined to a single two-dimensional layer of grains. The thermal images of the slices taken during 1 ms long high current density pulses showed distinct conducting paths very similar to those obtained from computer simulations of the current flow in a varistor Voronoi network [15]-[17]. This striking similarity substantiate the theoretical calculations and confirm modeling predictions. The paths in the thermal images consisted of chains of localized hot spots that mark the grain boundaries — occurring where the potential is dropped across Schottky-type barriers and the heat is generated. Local temperatures at the grain boundaries were found to rise well above that of the interiors of the grains during 1 ms current pulses of about 300 A/cm². The images of hot spots were similar to photographs of the electroluminescence from grain boundaries [19]-[21]. These observations clearly showed that varistor heating is localized at the grain boundaries during electrical breakdown.

The fine spatial resolution of the IR camera allows the temperature profile along individual conducting paths to be resolved. Figure 4(a) shows one such profile measured along the line through the centers of two distinct hot spots in the IR image of a slice cut from low voltage varistor. These spots were chosen for the purpose of clarity, although the size of the grain between them (about 200 μm), was larger than the average grain size. The profile shows two clear peaks corresponding to grain boundaries, and the difference between the temperature at the grain boundaries and the lowest temperature in the grain interiors is about 15–20°C. In order to interpret the processes that lead to the observed temperature profile, a simple model of the heating was employed. Since the important heat transfer processes take place along conducting paths, a one-dimensional heat diffusion equation of the following form was used:

$$ cD \frac{\partial^2 T(x,t)}{\partial x^2} - c \frac{\partial T(x,t)}{\partial t} + Q(x,t) - H T(x,t) = 0 \ , \qquad (10) $$

where $c = 3.3$ J/(cm³K) is the heat capacity, $D = 0.1$ cm²/s is the thermal diffusivity of the material, and the term $-H T(x,t)$ describes the loss of heat from the grains in a path. The time-dependent temperature $T(x,t)$ at the distance x from the grain boundary is measured relative to the sample temperature at $t = 0$, taken as 20°C. The grain size d is taken as 200 μm, in accordance with the particular measurement shown in Fig. 4(a). In the prebreakdown and in the nonlinear region of the varistor I-V characteristic, almost all of the voltage drop appears across the grain boundaries, so that vast majority of the energy goes into heating the grain boundaries. Therefore, for simplicity, it is assumed that the heat is generated

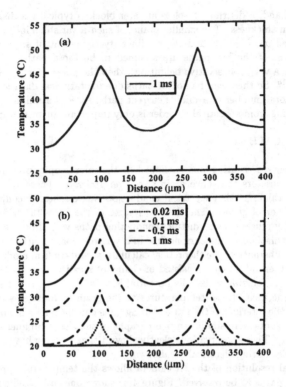

Figure 4: (a) Temperature profile along the line between two bright spots in an IR image of a thin varistor slice. (b) Time-dependent temperature profiles obtained from the solution of the corresponding heat transfer equation.

only at the grain boundaries. Since the grain boundary region is very narrow (~ 1000 Å) compared to the grain size, it is reasonable to approximate the spatial dependence of the power input $Q(x,t)$ in Eq. (10) by the Dirac delta function. The time dependence of $Q(x,t)$ is taken as a constant pulse that starts at $t = 0$, so that $Q(x,t) = q\,\Theta(t)\,\delta(x)$, where $\Theta(t)$ is the unit step function. The input power q at the surface of the grain boundaries can be estimated as follows. The voltage across the sample is 80 V, and the total current is about 1.5 A. However, the current flows mainly along a number of conducting paths. Judging from the IR image of the whole slice, about 10% of the total current flew along the particular path considered here. Finally, taking into account the length ($l = 5.5$ mm), the width ($w = 5$ mm), and the thickness ($h \approx 0.1$ mm) of the sample, we obtain a power input q of about 2200 W/cm². Heat diffuses not only along the line x, as is explicitly described by Eq. (10), but it dissipates in directions perpendicular to the current flow to the grains adjacent to the conducting path, to the substrate beneath the varistor sample, and to some extent to the air above the sample. All of this heat dissipation is described by the single sink term $-H\,T(x,t)$ in Eq. (10), which assumes that the heat loss at position x and time

t is proportional to the temperature rise along the conducting path $T(x,t)$. The parameter H can be estimated simply by assuming that the heat dissipates by conduction to adjacent grains and to a substrate with thermal properties that are not much different from those of zinc oxide. In this particular case, $H = 2cD/(hd) + cD/d^2 \approx 4100$ W/(cm^{3}K), where the first term corresponds to the heat dissipation to adjacent grains, and the second term describes the amount of heat dissipated to the substrate. Since there is also a little heat escaping to the air, we just assume $H = 5000$ W/(cm^{3}K). The boundary conditions that have to be satisfied by the solution of Eq. (10) are

$$T(-d/2, t) = T(d/2, t) \quad \text{and} \quad \left. \frac{\partial T(x,t)}{\partial x} \right|_{x=d/2} = \left. \frac{\partial T(x,t)}{\partial x} \right|_{x=-d/2} = 0 . \qquad (11)$$

These conditions ensure that the temperature profile is symmetric around the grain boundary, and that there is no heat transfer along the x direction at the center of the grain. Eq. (10) can be solved analytically using Laplace transforms, and the solution can be written as

$$T(x,t) = \frac{q}{Hd}(1 - e^{-Ht/c}) + \frac{2q}{cd} \sum_{n=1}^{\infty} \frac{(1 - e^{-A_n t})}{A_n} \cos(2\pi nx/d) \qquad (12)$$

where

$$A_n = \frac{H}{c} + \frac{4\pi^2 n^2 D}{d^2} . \qquad (13)$$

The series in Eq. (12) converges relatively fast and can be calculated numerically. The resulting temperature profiles for $t = 0.02$ ms, 0.1 ms, 0.5 ms, and 1 ms are shown in Fig. 4(b). The final temperature profile at the end of a 1 ms current pulse, shown in Fig. 4(b) as the solid curve, is in excellent agreement with that in Fig. 4(a), measured in the experiment. The calculated data also show that at the beginning of the current pulse the temperature rises fast only at the grain boundaries and in the immediate neighborhood of the grain boundaries. But as the pulse continues and the heat diffuses into the grain interiors, the temperature of the grain boundaries and grain interiors increase at nearly the same rate. This is consistent with the experimental observations and has some significant implications. Varistors exhibit a rather surprising increase in their energy handling capability at very high currents. This was first noticed by Sakshaug, Burke, and Kresge [11], who attributed this increase to a change on a microscopic scale (i.e., the scale of the grain size) in the temperature distribution that occurs at high current. In contrast to currents in the prebreakdown or the nonlinear region of the I-V characteristic, where the voltage drop appears almost exclusively across the grain boundaries, the current is so high when a varistor operates in the upturn region that the voltage drop inside the grains becomes significant, and the heating becomes more uniformly distributed. This, according to the interpretation of Ref. [11], should lead to reduced temperature difference between grain interiors and grain boundaries, thereby reducing the thermal shock and leading to higher energy handling capability. However, as discussed above, the difference between the temperature at the grain boundary and the grain interior quickly reaches an essentially steady-state value, and it never becomes very large — maybe only about 35°C for severe current pulses. This means that the heat transfer is too fast to permit temperature differences on a scale of the grain size that could cause failure by generating thermal stresses. We conclude that varistor failures are not caused by temperature differences on the microscopic scale.

231

SUMMARY AND CONCLUSIONS

The heating of ZnO varistors by electrical pulses over a range of spatial scales that extends from the microscopic to the macroscopic has been discussed. Non-uniform heating due to current localization was shown to have different origins and different consequences for device failure in large and small varistors. On a macroscopic scale, heating has been examined in arrester blocks of a distribution-class size, and non-uniform temperature distributions have been attributed to inhomogeneities which originate during ceramic processing.

In small varistors for electronic applications, non-uniform heating due to current localization appears on a finer spatial scale. Localized conducting paths became important when the number of grains in the electrical paths decreased to the point where statistical variations in the electrical breakdown voltage became significant. It is found that influence of statistical disorder due to grain size and barrier hight variations changes with varistor size in a manner that can be described with relatively simple mathematics.

At the microscopic level, the heating is shown to take place at grain boundaries. Theoretical modeling has confirmed the experimental observation that local temperature increases of up to 20 °C above the grain interiors during varistor breakdown. These results imply that current localization on the size scale of the grains does not cause failures.

ACKNOWLEDGEMENTS

I thank G.D. Mahan, F. Modine and H. Wang for valuable discussions. This work was supported by the U.S. Department of Energy through the Assistant Secretary for Energy Efficiency and Renewable Energy, Office of Transportation Technologies, as part of the High Temperature Materials Laboratory User Program and by the Office of Basic Energy Science under contract DE-AC05-96OR22464, managed by Lockheed Martin Energy Research Corporation.

References

[1] L.M. Levinson and H.R. Philipp, *Ceram. Bull.* **65**, 639 (1986).

[2] T.K. Gupta, *J. Am. Ceram. Soc.* **73**, 1817 (1990).

[3] K. Eda, *J. Appl. Phys.* **56**, 2948 (1984).

[4] M.V. Lat, *IEEE Trans. Power Appar. Syst.* **102**, 2194 (1983).

[5] S. Horiguchi, F. Ichikawa, A. Mizukoshi, K. Kurita and S. Shirakawa, *IEEE Trans. Power Delivery* **3**, 1666 (1988).

[6] M. Bartkowiak, M.G. Comber and G.D. Mahan, *J. Appl. Phys.* **79**, 8629 (1996).

[7] M. Bartkowiak, M.G. Comber and G.D. Mahan, accepted for publication in *IEEE Trans. Power Delivery*.

[8] G. Hohenberger, G. Tomandl, R. Ebert, and T. Taube, *J. Am. Ceram. Soc.* **74**, 2067 (1991).

[9] A. Mizukoshi, J. Ozawa, S. Shirakawa, and K. Nakano, *IEEE Trans. Power Appar. Syst.* **102**, 1384 (1983).

[10] H. Wang, M. Bartkowiak, F.A. Modine, R.B. Dinwiddie, L.A. Boatner, and G.D. Mahan, accepted for publication in *J. Am. Ceram. Soc.*

[11] E.C. Sakshaug, J.J. Burke, and J.S. Kresge, *IEEE Trans. Power Delivery* **4**, 2076 (1989).

[12] K.G. Ringler, P. Kirkby, C.C. Erven, M.V. Lat, and T.A. Malkiewicz, *IEEE Trans. Power Delivery* **12**, 203 (1997).

[13] A. Vojta and D.R. Clarke, *J. Appl. Phys.* **81**, 985 (1997).

[14] Z.-C. Cao, R.-J. Wu, and R.-S. Song, *Mater. Sci. and Engr. B* **22**, 261 (1994).

[15] M. Bartkowiak and G.D. Mahan, *Phys. Rev. B* **51** 10825 (1995).

[16] M. Bartkowiak, G.D. Mahan, F.A. Modine, and M.A. Alim, *J. Appl. Phys.* **79**, 273 (1996).

[17] M. Bartkowiak, G.D. Mahan, F.A. Modine, M.A. Alim, R.J. Lauf, and A.D. McMillan, *J. Appl. Phys.* **80**, 6516 (1996).

[18] C.-W. Nan and D. Clarke, *J. Am. Ceram. Soc.* **79**, 3189 (1996).

[19] G.E. Pike, S.R. Kurtz, P.L. Gourley, H.R. Phillips, and L.M. Levinson, *J. Appl. Phys.* **57**, 1552 (1985).

[20] M. Rossinelli, G. Blatter, and F. Greuter, *British Ceram. Proc.* **36**, 1 (1985).

[21] F. Greuter, G. Blatter, M. Rossinelli, and F. Stucki, *Ceramic Trans.* **3**, 31 (1989).

CURRENT FLOW AND STRUCTURAL INHOMOGENEITIES IN NONLINEAR MATERIALS

F. GREUTER, T. CHRISTEN, and J. GLATZ-REICHENBACH
ABB Corporate Research Ltd., CH-5405 Baden-Dättwil, Switzerland

ABSTRACT

Nonlinear current flow in real 3-d composite materials is far more complex than suggested by usual brickstone models. We present experimental results on controlled seeding of structural inhomogeneities in varistor ceramics. In particular, we discuss the influence of mesoscopic imperfections on the characteristic electrical parameters. A main effect is a change in the statistical distribution of the grain size. As a consequence, the breakdown voltage drops very sensitively due to the disorder, while the shape of the I-V curve is hardly affected. We discuss also the extremely nonuniform filamentary current pattern, being illustrated by electroluminescence studies.

1) INTRODUCTION

Among the materials with nonlinear electrical properties, resistors with strongly positive (PTC) or negative (NTC) temperature coefficients or with a highly nonlinear voltage dependence (VDR, varistors) are the best studied cases. All of them have reached important technical applications, as, e.g., current-limiters, voltage-limiters, and sensors. Due to an interplay between nonlinearity and spatial inhomogeneity, the current flow patterns can be extremely inhomogeneous. While in PTC's a blocking hot plane develops perpendicular to the current flow, in NTC's and varistors current filaments develop along the field direction. Usually, high power densities occur locally, which intrinsically limits most technical applications (see [1-4], and references therein). Figure 1 illustrates by infrared imaging both transverse (PTC) and longitudinal (VDR) patterns, which are triggered by macroscopic inhomogeneities. In an ideal, smart material these two effects would be combined to support each other. An example are varistor composites with built-in PTC-type current limitation under thermal overstress [5].

In the following, we concentrate on the current flow in *varistor* materials and its relation to the structural homogeneity. In real varistor ceramics important inhomogeneities occur on *microscopic, mesoscopic, and macroscopic* length scales. *Microscopic* inhomogeneities are imperfections on the length scale of the grain boundary, like inclusions of a few nm-thick pockets of Bi_2O_3, amorphous grain boundary layers, and triple point phases [6-8]. On the *macroscopic* level, the imperfections are related to the production process like dimensional tolerances, handling and other defects, which can trigger local overstresses in the material [1,9]. Microscopic and macroscopic inhomogeneities and their effects have been intensively studied and reviewed [1-3,6-13]. However, the inhomogeneities on the mesoscopic level and their influence on the current transport are still poorly understood. The *mesoscopic inhomogeneities* of interest are grain size distributions, pores and clusters of the second phases (spinel and Bi_2O_3). They all affect the pattern of the current flow through the real microstructure by creating areas of easy or hindered current flow in the highly nonlinear matrix. Recently, much theoretical effort was put into the computer-modelling of simple or random networks of nonlinear resistors. The results provided an interesting insight in how current paths develop and are affected by the disorder on the grain boundary level and with respect to the grain size distribution [2,3,14]. The filamentary current flow in the highly nonlinear part of the I-V-characteristics is well reproduced and predicts a high sensitivity of the breakdown voltage to the structural or electrical disorder.

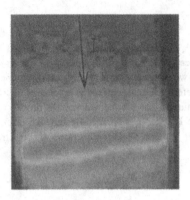 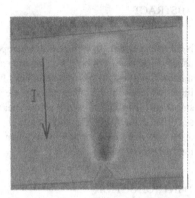

Figure 1: Infrared thermal images of PTC-resistor and ZnO-varistor in their nonlinear current- or voltage limiting state. Left: Polymer-composite PTC [4] with hot plane caused by current limitation. Right: Hot channel in ZnO-ceramic varistor caused by (surface) contacts with protrusion.

Existing simulation work, however, does not include spatially distributed second phases. Depending on the overall composition of the varistor material and the ceramic process quality, these insulating clusters can significantly reduce the available cross section for current conduction and lengthen the current paths. Also, the computer modelling has to be limited to small 2-dimensional networks (<40 x 40) to remain practical. In reality, the third dimension offers additional opportunities for the current to pass the insulating inclusions, which is more difficult in 2- and impossible in 1-dimensions. From the experimental side, usually several parameters are changed simultaneously when changing the microstructure and it is difficult to separate in detail the individual contributions associated with doping, structure, defect equilibrium, etc.

Below, we report on a way of *controlled seeding of microstructural inhomogeneities*, on its influence on the *overall electrical characteristics* and on the *local current flow patterns*.

2) EXPERIMENTS

For this study a multicomponent model composition was used containing the traditional ZnO dopants: Bi_2O_3, Sb_2O_3, CoO, NiO, Mn_2O_3, Cr_2O_3 etc. [10-13]. The powders were prepared by adding all dopants in the form of water soluble salts, whereby a stable cocktail containing all dopants was dispersed with fine ZnO ($\emptyset \approx 0.5$ μm) and spray dried [15]. All ZnO particles are coated with an atomically homogeneous doping layer, resulting in best achievable homogeneity. Samples prepared this way are referred to as *liquid doped*. Sintering and electrical characterization were conventional [10-13,15].

Controlled seeding of structural inhomogeneities was achieved by substituting one of the spinel-forming dopants in the cocktail (e.g., Ni) by oxide particles (NiO) of variable size and concentration, instead of the soluble salt. The composition, of course, remains unchanged. The oxide particles with diameters of interest were classified by a multiple sedimentation procedure. The NiO-particles used here were in the range of 5-11 μm with $d_{50} = 8$ μm. All samples were sintered together under identical conditions. For the electroluminescence studies, samples were cut and polished and thermally etched by a 700-790° C/2h heat treatment with 80° C/h heating and cooling rates. Such a heat treatment is needed to restore the electrical properties of the grain boundaries, which might be damaged (electrically shorted) by the polishing process. Surface contacts were applied and the luminescence was photographed under the light microscope with high-sensitivity films ($\geq$1600 ASA). The details of the luminescence pictures are unfortunatelly partly lost in the present black and white reproduction.

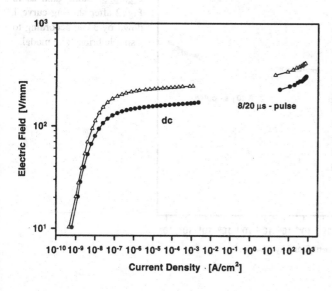

Figure 2: DC- and impulse characteristics of the model varistor ceramics with (circles) and without (triangles) mesoscopic inhomogeneity. Field strength and current density are calculated by normalizing to the sample dimensions.

3) RESULTS

a) Electrical properties

Figure 2 shows the I-V-characteristics of the fully liquid doped ceramics and when all the Ni-salt is replaced by NiO (all other additives still being added as liquid dopants). The breakdown field E_B (defined at 0.1mA/cm^2) is observed to drop by 34% upon introduction of the mesoscopic inhomogeneity. The overall shape of the curve is not much changed, except slight changes in the

237

ohmic region at low fields and in the high-current region. For the characterization of the high-current region normally the 8/20 µs lightning impulse field strength E_p (defined at 250 A/cm^2) and the protection ratio $R_v = E_p/E_B$ are used. R_v is observed to increase slightly by 4-5 %, which is not much from a physics point of view, but is of technological interest. The nonlinearity coefficients α_B, α_L defined as logarithmic derivatives of the I-V characteristics (at 0.1 mA/cm^2 and 1 µA/cm^2) remain almost unchanged or increase slightly. Also the power losses, which are normalized to the measuring voltage and the cross section, are not affected by inhomogeneities. It is interesting that the two DC-curves fit to each other completely by shifting the liquid doped curve down by 32 % and to the left by 13 %. This is equivalent to a nominal decrease of the electrically effective sample thickness h (or the length of current paths) by a factor 0.68 and an apparent increase of the cross section by 14 % relative to the homogeneous sample. Figure 3 shows the two curves after shifting parallel to the field axis only, as in the most simple brickstone model for variable grain size.

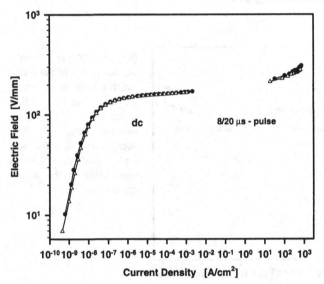

<u>Figure 3</u>: Same data as in Fig. 2 after shifting curve 1 down by 34%, according to a simple brickstone model.

By substituting only part of the spinel-forming Ni-dopant and adding the rest as a liquid doped Ni-salt, the mesoscopic inhomogeneity can be introduced in a systematic way. The response of the electrical properties is shown in Figure 4.

The breakdown field E_B decreases linearly with increasing disorder, whereas the protection ratio $R_v = E_p/E_B$ increases. On the other hand, the nonlinearity coefficients α and the normalized power losses P_V (not shown) are not affected by the degree of inhomogeneity achieved here (see below). The low-current nonlinearity α_L shows a slightly increasing trend. This can be understood by the small shift of the DC-curve to the left (see above), thereby moving into regions of higher α-values for the fixed current density of 1 μA/cm^2.

Note that only little of an inhomogeneity in the unsintered state can lead to the above large effects, in particular regarding the field strength. The state of 100% NiO corresponds to about 0.3 volume-% fraction of coarse particles with $d_{50} \approx 8$ μm in an otherwise homogeneously doped ZnO matrix with all the other dopants (Bi-, Sb-, Co-, Mn-, Al-, ...salts; percent to ppm range). In other words, *1 coarse NiO-particle out of 10^5 ZnO-particles* (∅ = 0.5 μm) is sufficient to create enough inhomogeneity in order to strongly influence the electrical behaviour!

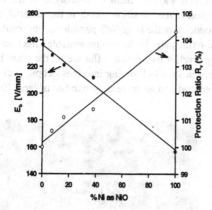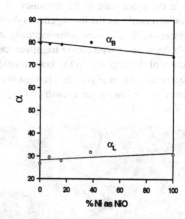

Figure 4: Changes in characteristic electrical quantities due to a systematic introduction of mesoscopic inhomogeneity at maintained composition by substituting part of the liquid dopant Ni-acetate by coarse NiO: 100% Ni = x% NiO + (100 - x)% Ni-salt.

In the next section we show that it is the mesoscopic structural homogeneity (insulating clusters of second phases) which is is modified, but not the chemical homogeneity (dopant distribution in active part of matrix). These inhomogeneities can also be introduced by other spinel-forming dopants (e.g. Sb), but not with the liquid-phase forming main dopant Bi$_2$O$_3$, which has no effect even at particle sizes as large as 20-30 μm. The high sensitivity to specific dopants is related to the liquid phase sintering-process and the complex phase-formation sequence [16].

b) Microstructure

The very different distribution of the second phases (Bi$_2$O$_3$ + spinel clusters) for the two extreme cases of 0 % (all liquid doped) and 100 % NiO is shown in Figure 5. For the fully liquid doped sample the clusters of 2nd phases are small and rather homogeneously distributed throughout the microstructure, whereas in the other limit they form large, mostly spherical islands of insulating material besides the small pockets at the triple points. Note that both cases contain the same total volume of 2nd phases, as the composition is the same. Obviously the large clusters are seeded by addition of the coarse NiO-particles. Their number density, but not their size, decreases with decreasing NiO-fraction in Figure 4. Part of the Ni enters the ZnO-lattice as a substitutional dopant, whereas the rest acts as one of the spinel forming dopants (= Sb and transition metals Co, Cr, Mn, Ni [10,11,16]). From point analysis by SEM/EDX on a large number of spinel grains one finds that they all show the same chemical composition, independent of whether the spinel is trapped inside a ZnO-grain, at a small triple point or within a large cluster. This must be due to the (large volume) of liquid phase formed during sintering, where the liquid Bi_2O_3 dissolves the dopants and is an excellent transport medium to achieve chemical homogeneity. The structural homogeneity can not be restored by the liquid phase, once it is present in the green state of the ceramics. In a similar way, the SEM/EDX-analysis of the ZnO grains as well confirms the homogeneous distribution of the transition metals in the material, and no reminders of the chemical inhomogeneity introduced by the large NiO-particles is observed.

While the seeding with coarse transition metal-oxide particles has no measurable effect on the distribution of the dopants on the grain-boundary level, it has a clear influence on the grain size distribution, as seen in Figure 6. Through the uneven spinel-clustering into mesoscopic pockets, the control of the ZnO-grain growth [10,11] is reduced and an increasing number of large ZnO-grains is observed.

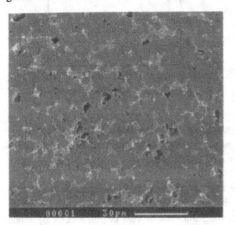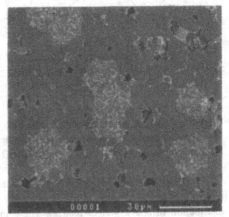

Figure 5: Microstructure of homogeneous, liquid doped ceramics (left) and with seeded inhomogeneity (right). In the backscattering imaging mode the Bi_2O_3 and spinel phases show up white and light grey, respectively.

To a good approximation the liquid doped material shows a log-normal distribution, whereas the inhomogeneous sample deviates from such a distribution. The curves in Figure 6 can be quantified by: i) the most frequent grain size $\emptyset_p$ of ~ 8 µm and ~10 µm, ii) the average grain size $\emptyset_a \approx 10$ µm and ≈ 13 µm and iii) the largest observed grain diameter $\emptyset_m = 24$ µm and 40 µm, respectively. By comparing microstructures of commercial varistors of different origins, both types of distributions as in Fig. 6 can be observed, i.e. rather symmetric and highly asymmetric ones, indicating different levels of process technology. As the size of the NiO seeding grains is reduced, the diameter of the Bi_2O_3 spinel-clusters decreases and the breakdown fields approach the value of the homogeneous (0 % NiO) material.

Bearing in mind Emtage's [17] results on the relation between electrical properties of the varistor and the fractional variance σ of the grain size distribution, we mention that we find $\sigma = 0.4$ and $\sigma = 0.46$ for the liquid doped and the strongly inhomogeous case, respectively. Clearly, σ is rather large and indicates that simple uniform brickstone models may not be appropriate descriptions. In other words, a shift in breakdown voltage cannot simply be interpreted by changes in the average grain size but must include the effects of a broad size distribution [2,3,14].

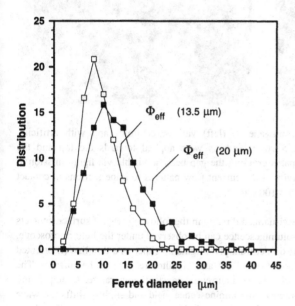

Figure 6: ZnO-grain size distribution (Ferret dia= meters) of homogeneous (empty squares), and in= homogeneous (full squares) ceramics. The average grain sizes are 10 μm and 13 μm, and the mean square deviations are 4 μm and 6 μm, respectively. $\varnothing_{eff}$ is the effective, electrically active grain size.

c) Electroluminescence

Below the high-current region, current transport through the ceramics is completely governed by the grain boundaries. Here, each ZnO/ZnO interface acts as a micro-varistor with a breakdown voltage $V_B \approx 3.2$ V for a good material [2,7,10-14]. Due to the nature of the double Schottky barriers formed at each grain boundary, the applied voltage V drops within the narrow depletion region (≈ 200 nm). Hence, locally very high electric fields [12,13] are created - high enough to produce hot electrons, which are able to ionize defect centers deep in the band gap or even the valence states (minority carriers, triggering breakdown). Impact ionization leads to electroluminescence in the visible (main intensity at $h\nu = 1.78$ eV, Co-centers) and the near UV-region [13,19,20].

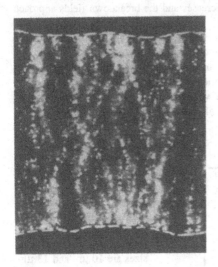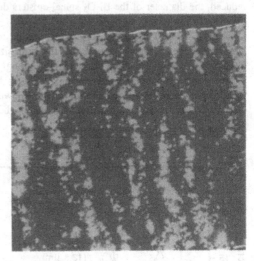

Figure 7: Electroluminescence of homogeneous (left) varistor ceramics and with artificial, mesoscopic inhomogeneities (right). Surface contacts were applied towards the top and the bottom of the photographs. Current paths crossing the gap become visible via light emission at each grain boundary. Note the sensitivity of the current flow patterns to imperfections in contact geometry and microstructure. ⊢———⊣ = 100 μm.

Individual current filaments can be visualized directly in the following way. If surface contacts are applied to the ceramics, the electroluminescence can be observed under the light microscope. Each grain-boundary plane, being part of a current path, lights up either as a thin line (if oriented perpendicular to the surface) or a more smeared out area (if tilted relative to the surface). The total light intensity increases linearly with the total current at constant temperature. A temperature increase, e.g. caused by selfheating, reduces the luminescence yield and slightly shifts the wave lengths. Due to a strong optical absorption the observed light stems from a narrow surface layer only. For the present contact geometry the main current flow at breakdown is concentrated in the surface region. Figure 7 illustrates such luminescence patterns for the fully liquid doped and the 100 % NiO-doped materials. Current filaments can be seen running from one electrode to the other, when the total current is high enough (>50 μA/cm for the naked eye). The filamentary structure is rather stable and does not move around with time or upon switching on and off the power source, showing that there are (structurally) preferred paths. At this stage no systematic attempt was made to quantify the current distribution of the filaments made visible by the grain boundary luminescence.

As a tendency the following differences and similarities appear among the two types of samples:

i) The fully liquid doped sample has a more homogeneous appearance, which is partly due to the smaller grain size and partly due to the higher structural homogeneity, both resulting in more filaments per width of the electrode.

ii) In the homogeneous sample there are more continuous filaments bridging both electrodes, whereas for the 100 % NiO-sample more filaments are observed to disappear from the electrode (into the third dimension) or eventually join with another filament. With increasing disorder the current paths have to avoid the insulating clusters and are forced to escape sideways. Large bends however are very unlikely. For the same total current there are fewer, but more intense filaments in the inhomogeneous material.

iii) Both cases are sensitive to imperfect contact geometries even if they have deviations of just a few grain sizes. The homogeneous case might be more sensitive than the disordered one.

iv) By comparing carefully the location of the luminescence and the corresponding microstructure [13,19], we find that all ZnO/ZnO boundaries are varistor-active along a visible current path.

v) Emitting grain boundaries beneath the surface have a blurred and diffuse appearance and boundary planes ideally oriented perpendicular to the surface are rather seldom.

4) DISCUSSION

For an understanding of the above experimental findings there are two major contributions to be separately quantified:

i) the second phases forming current blocking pockets at the triple points and in larger clusters of $\varnothing \approx 20$–$60\ \mu m$ throughout the ceramics

ii) the reduced control of grain growth for the inhomogeneous powder process, resulting in a broader grain size distribution

All the discussions must be seen on the background of the pronounced filamentary current flow in the nonlinear region of the I-V characteristics. Many of the details seen in electroluminescence pictures (Fig. 7) are well reproduced by nonlinear resistor network simulations [2,3,14], although they are restricted to the purely 2-dimensional case, where a filament is not able to bypass an inhomogeneity by diving into the third dimension.

From Figure 5 it is evident that even for the homogeneous material the Bi_2O_3- and spinel-phases take away a good fraction of the geometrical cross section A_{geom} accessible for current transport. In a first approximation, an electrically effective cross section A_{eff} can be derived from comparing the measured high frequency capacitance to the one calculated from the height ϕ_B of the potential barrier and the shallow donor density N_o (determined by IR-reflectance). The barrier ϕ_B can be derived iteratively from I-V-measurements, where the cross section A_{eff} enters the formula only weakly [13,20]. The discrepancy between calculated and measured capacitance is then assigned to the difference between A_{geom} and A_{eff}. For most commercial, Bi_2O_3-based varistor materials we find a ratio $A_{eff}/A_{geom} \approx 0.3$–$0.4$, which is compatible with micrographs. Only for Pr-doped varistors (Fuji) a significantly higher effective cross section $A_{eff}/A_{geom} \approx 0.6$ is observed, in agreement with the microstructure being almost free of 2nd phase. An effective cross section can also be derived from measurements of the upturn characteristics, considering the constriction resistance imposed by the grain boundaries [21]. This gives similar typical values $A_{eff}/A_{geom} \approx 0.35$.

If now an additional mesoscopic inhomogeneity is introduced by the large Bi_2O_3/spinel clusters, as shown in Figs. 4 and 5, the available cross section for current transport and the length of the current paths are further modified. To quantify the influence of such large inhomogeneities we have introduced circular insulating islands in a ideally homogeneous varistor matrix and calculated the current distribution by a Finite Element tool (ACE) capable to handle high nonlinearities. Assuming a regular, face centred lattice with the diameter of the circles to be equal to the separation between two circles, a regular "Swiss cheese" is simulated for an assumed I-V characteristics of the matrix as in Fig. 2. Although this represents an extreme situation, the influence on the overall I-V-curve, i.e.the shape, is surprisingly small. The finite element calculations show that both ohmic regions at low and high current densities are shifted to lower values by ~46 % and the breakdown field is increased by less than ~1 %. In order to achieve a larger influence on the breakdown voltage, the size of the insulating island has to be increased much more, to counteract the tendency to form straight filaments and to enforce a meander-type shape of these filaments. The shift of the simulated I-V-curve however is in the opposite direction to the experiment in Figure 2. Therefore, there must be an additional origin. By collecting the 2nd phases in large clusters the average dimension of the remaining small triple point clusters has to decrease. The effective cross section A_{eff} will then increase and with this the observed current density of the voltage controlled material will go up. This effect will counteract in reality the calculated shift towards lower current densities caused by the large islands. The overall I-V characteristics, in particular the breakdown field, seems to be rather robust against insulating, mesoscopic inhomogeneities. Obviously, it needs a much higher cluster density than in Fig. 5 to disturb the straight filamentary current flow (see Fig. 7) and to remarkably shift the I-V curve. Hence the main origin of the phenomena in Figs. 2-4 comes from the disturbed grain size distribution upon introduction of mesoscopic inhomogeneities.

From Figure 6 the average grain size is seen to increase by a factor of about 1.3 when going from the fully liquid doped to the 100 % NiO- doped case. From electrical measurements we can also derive an electrically, effective grain size, since the breakdown voltage per boundary a in an good varistor is ~3.2 ± 0.1 V. Figure 7 and simulations [2,3,14] tell us that at breakdown the current paths are rather straight and contain good barriers only . Variations in the junction cross-section A along a path do hardly affect the voltage drop per boundary (e.g. $\Delta U < 0.2$ V for A_i / A_j = 100). Then the effective grain size becomes ≈ 3.2 V / E_B, or about 13.5 µm and 20 µm for the two extreme cases in Fig. 2. On the grain size distribution curve in Fifure 6 theses values are displaced from the maximum towards larger diameters, as expected. Obviously, the effective grain size is even larger than the mean grain size. This can be explained by the dominating influence of large grains and the need to find a continuous, connecting path from one electrode to the other. This condition of connectivity requires that from the distribution of grain sizes a certain selection is being made. In a simple brickstone model the variation of the grain size distribution results in a shift of the I-V curve parallel to the field axis, as it is shown in Figs. 3 and 4. This shift already explains the main difference between the homogeneous and inhomogeneous sample. In a lowest order approximation, the ratio of the shifted breakdown fields (237/156) is expected to be inverse proportional to the ratio of the average grain sizes (10/13). This alone however only predicts a 23% reduction in the field strength, whereas experimentally we observe -34%. More sophisticated theories also include the higher moments of the grain size distribution [14,17]. With increasing second moment σ, these references predict a rapid drop of ~30 % of the breakdown voltage, which then saturates at this level. In our case (Fig. 6) the increase in the variance σ amounts to an additional reduction of -10% in the field strength, in good agreement with the experiment.

Only an additional small shift of ~15 % along the current density axis is needed to get the full overlap of the two curves. This small horizontal shift must be the combined result from the introduction of the large Bi_2O_3/spinel clusters (reducing the available cross section) and the accompanying increase of the average ZnO/ZnO junction area (via reduced average size of the triple point phases and the larger grain size).

Similar arguments may apply to qualitatively understand the small changes in the high current region (Figs.2-4). As we still are in the nonlinear region (α >10), we have to expect filamentary current flow and are sensitive to the effectively available cross section. A second contribution comes from the finite resistivity of the ZnO grains. At these current levels, the I-V curve is governed partly by the grain boundaries and partly by the voltage drop over the grains (including restriction resistance). For low field varistors with large grains the latter contribution is stronger and higher protection ratios R_v must be expected, as seen in Fig. 4.

By comparing under the microscope a large number of electroluminescence and geometrical pictures, we observe that all grain boundaries connected by a current filament actually are also electrically active. This contradicts several reports in the literature, which claim a good percentage of low ohmic, inactive boundaries from microprobe measurements (see references in [2,3,14]). From our own experience, we are tempted to assign such observations to damages caused by preparing for the microprobe measurements, knowing the high sensitivity of the grain boundaries, e.g. to small changes in the monomolecular adsorption layers of oxygen [13].

5) CONCLUSIONS

We reported on physical phenomena occurring in current transport in nonlinear varistor materials at various degrees of mesoscopic disorder. By controlled seeding with spinel forming coarse oxide particles, the structural homogeneity of a chemically doped material can be modified without disturbing the doping homogeneity. While the breakdown field of the varistor is strongly affected, the shape of the I-V characteristic is almost unchanged over the entire current range. By electroluminescence techniques, it is possible to visualize the current flow, which is rather insensitive to the mesoscopic inhomogeneities itself, but which is sensitive to the consequent change in the grain size distribution. Both the grain size distribution and the filamentary current flow in the nonlinear region must be considered for understanding the overall electrical properties of varistor ceramics and the effect of mesoscopic inhomogeneities.

Acknowledgement: We thank V. Schmid, P. Kluge-Weiss, D. Minichiello, and S. Mainardi for their expertise and support.

REFERENCES

1. K. Eda, J. Appl. Phys. **56** (10) p 2948 (1984).
2. M. Bartkowiak, M.G. Comber, G.D. Mahan, J. Appl. Phys. **79** (11) p 8629 (1996) and IEEE Trans. Power Delivery 97 (in press).
3. A. Vojta, D.R. Carke, J. Appl. Phys. **81** (2) p 985 (1997), Computational Mater. Sci. **6** p 51 (1996).
4. R. Strümpler, J. Skindhøj, J. Glatz-Reichenbach, IEEE Trans PD (1997) in print; R. Strümpler, G. Maidorn, J. Rhyner, J. Appl. Phys. **81** (10), p 6786 (1997); R. Strümpler, J. Glatz-Reichenbach, F. Greuter, MRS Symp. Proc. **411**, p 393 (1996).
5. R. Strümpler, P. Kluge-Weiss, F. Greuter, Advances in Sci. and Techn. **10**, p 15 (1995).
6. D.R. Clarke, J. Appl. Phys. **49** p 2407 (1978).
7. E. Olsson, L.K.L. Falk, G.L. Dunlop, R. Österlund, J. Mater. Sci. **20**, 4091 (1985) and J. Appl. Phys. **66** p 4317 (1989) and E. Olsson PhD Thesis U. of Göteborg (1988).
8. H. Wang, Y.M. Chiang, J. Amer. Ceram. Soc. (1997) in press; H. Wang PhD Thesis MIT (1996).
9. H.F. Nied, Ceram. Trans. **3**, p 274 (1988).
10. L.M. Levinson and H.R. Philipp, Am. Ceram. Soc. Bull. **65** p 639 (1986).
11. T.K. Gupta, J. Am. Ceram. Soc., **73** p 1817 (1990).
12. G.D. Mahan, L.M. Levonson, H.R. Philipp, J. Appl. Phys. **50** p 2799 (1979).
13. F. Greuter, G. Blatter, Semicon. Sci. and Techn. **5** (2) p 111 (1990).
14. M. Bartkowiak, G.D. Mahan, Phys. Rev. B **51** p 10825 (1995). M. Bartkowiak, G.D. Mahan, F.A. Modine, M.A. Alim, J. Appl. Phys. **79** (1) p 273 (1996), ibid **80** (11), p 6516 (1996), Jpn. J. Appl. Phys. **35** L414 (1996), Proc. 8th EPS-APS Intern. Conf. Physics Computing PC96, Krakow, p 289 (1996), and this proceeding.
15. M. Osman, R.S. Perkins, F. Schmückle, Europ. Patent 0 200-126 (1985).
16. M. Inada, Jpn. J. Appl. Phys **17**, p1, (1978).
17. P. R. Emtage, J. Appl. Phys. **50**, p 6833 (1979).
18. G.E. Pike, S.R. Kurtz, P.L. Gourley, H.R. Philipp, L.M. Levison, J. Appl. Phys. **57**, p 5512 (1985).
19. F. Greuter, M. Rossinelli, G. Blatter, patent applic. 3921/85-0 (1985); Br. Ceram. Proc. **36**, p 1 (1985); ibid **41**, p 177 (1989).
20. F. Greuter, G. Blatter, M. Rossinelli, F. Schmückle, Mat. Sci. Forum **10-12**, p 235 (1986).
21. L.M. Levinson, H.R. Philipp, J. Appl. Phys. **47**, p 3116 (1976).

HIGH POWER SWITCHING BEHAVIOR IN CONDUCTOR-FILLED POLYMER COMPOSITES

ANIL R. DUGGAL and LIONEL M. LEVINSON
General Electric Corporate Research and Development, Niskayuna, NY 12309

ABSTRACT

It has generally been assumed that the switching properties of conductor-filled polymer composites are based on a positive temperature coefficient of resistance (PTCR) effect where, at a certain switch temperature, the material resistivity increases by orders of magnitude. Here we present studies of the electrical switching behavior at high current densities which demonstrate that, in the high power regime, the observed switching is not based on the PTCR effect. Instead, we show that this type of switching appears to be a general feature in conductor-filled polymer composite materials and a qualitative model for the switching phenomenon is proposed. These results suggest that conductor-filled polymer composite materials can provide a new non-mechanical way of rapidly limiting high power short circuit currents. This should have broad applications in the circuit protection industry.

INTRODUCTION

It has been known for over 30 years that certain conductor-filled polymer composites exhibit reversible switching from a low to a high resistance when electrical currents above a certain magnitude are run through them [1-4]. These materials exhibit a positive termperature coefficient of resistance (PTCR) effect where, at a certain temperature, the material resistivity increases by orders of magnitude. The most studied polymer-composite system which exhibits this effect consists of polyethylene (PE) loaded with carbon black (CB). The temperature dependence of the resistivity of a particular PE/CB PTCR material is shown in Figure 1. At temperatures below the ~130C melting point of PE, the resistivity does not depend strongly on temperature; at 130C, the resistivity rises by orders of magnitude. This anomalous resistivity increase is believed to be due to an increase in CB particle separation resulting from the discontinuous PE expansion upon melting [5-6].

The PTCR property of these materials explains their reversible switching as a function of current and forms the basis for their use as current limiting devices in electrical circuits. In overload conditions, the joule-heating due to the excess current causes the material temperature to rise above the PTCR transition temperature thereby causing the material to switch to its high resistance state. In this state, the overload current is limited to an acceptable value. When the overload condition is cleared, the device cools to below its switch temperature and normal circuit operation is resumed [7].

Until recently, current limiting devices based on the PTCR effect have all been designed for low power circuit applications. Recently, a current-limiting device utilizing a PE/CB PTCR material was designed for much higher power circuit applications where the system voltage is on the order of 450V and the maximum current during a short-circuit is on the order of 100 kA [8]. As a result of material damage, this device is only designed for ~5 current limiting operations but this still provides a substantial performance advantage when compared to conventional methods of high power current limiting such as single-use fuses. If the current limiting performance and number of operations under high power conditions could be increased further, these devices could revolutionize the circuit protection industry. Hence there is substantial interest in better understanding their high power switching properties.

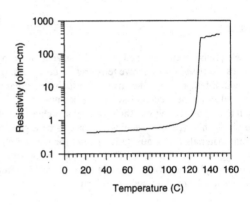

Fig. 1. Resistivity vs temperature of a polyethylene/carbon black (PE/CB) composite.

EXPERIMENTAL

The switching properties of polymer composite materials under high power conditions were probed using power amplifiers which allowed the application of arbitrary voltage waveforms with voltage and current limits of 400V and 200A respectively. The sample testing configuration consisted of a fixture where material samples 3/4" in diameter and 0.05" in thickness were placed between two 1/4" diameter nickel-plated copper electrodes with an applied pressure of ~200 PSI. Elevated temperature experiments were performed simply by placing the fixture into an oven with electrical feedthroughs.

The PE/CB PTCR material used in these experiments was obtained from a commercial current limiting device[8]. The room temperature resistivity and carbon loading of this material were measured to be 0.3 ohm-cm and 53 wt% respectively. The temperature dependence of the resistivity of this material measured at atmospheric pressure is shown in Figure 1. Note that the material exhibits a change in resistivity of greater than 2 orders of magnitude at ~130C.

The silver-filled silicone material that was tested was fabricated by mixing a vinyl silicone organopolysiloxane fluid with 80 wt% of silver particles. The mixture was poured into a mold and then cured in a Carver press at 150C for 30 min at 5000 pounds pressure.

RESULTS

Figure 2 shows typical current and voltage traces when a 3.8V-380V-3.8V voltage step sequence is applied to the PE/CB material. Prior to this voltage pulse, the sample resistance was ~0.4 ohm. During the first millisecond, when only the first voltage step is applied, the material resistance does not change from this initial value. During the second voltage step, the current increases to an amplifier limited value of ~200A (600A/cm^2) and then decreases to a value below ~20A. This resistance increase of ~50X marks a switching event. As the applied voltage is decreased to its final value, the resistance decreases to a value of ~0.8 ohm. A few seconds after the termination of this applied voltage pulse, the resistance of this device stabilized to a value of ~0.3 ohm. On successive pulsing with the same voltage step sequence, the same type of resistance switching and then recovery behavior is obtained.

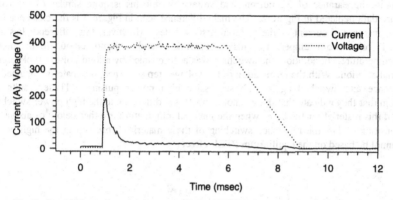

Fig. 2. Current through and voltage across the PE/CB composite material during an applied high current density electrical pulse. The ambient temperature is 25C.

In contrast to the switching behavior of conventional low power devices, the switching event is accompanied by a loud noise, visible light generation, and material ablation. The material ablation leads to some loss of material in the region between the two electrodes. Typically, with the voltage pulse sequence depicted in figure 2, ~10 mg of material is lost and the distance between the two electrodes is decreased by ~0.01". The amount of material ablated scales approximately linearly with the amount of energy absorbed. Visual examination indicates that the material ablation occurs only from one or both of the material/contact interfaces and not from the bulk of the material. The material at these two interface regions is pitted and scarred but the damage is localized to a thin (<100 micon) region near the surface. Time-lapse photography during switching show that a bright light - presumably from arcing or ablating material - is also localized to the material/electrode contact regions.

The localization of damage and light emission suggests that the resistance switching event is confined to the material/electrode interface regions rather than the bulk of the material. Further evidence for this was achieved by repeatedly pulsing a sample while monitoring the sample thickness before and after each pulse. After 7 pulsing events, the material thickness had reduced from 0.05" to 0.02". However, no significant differences were observed in either the time-dependence of the resistance or in the final resistance achieved while in the switched state in spite of the fact that the thickness of the material changed by more than a factor of two. If the switching had a significant bulk component, one would expect to see a significant difference - at least in the switched resistance achieved - with this degree of change in material length. It thus seems likely that the switching is localized to a thin region at the material/electrode interfaces.

In order to determine the influence of the PTCR effect on the observed high power density resistance switching, experiments were performed with ambient temperatures above the PTCR switch transition temperature. A pressure-contacted PE/CB material was placed in an oven and its resistance was monitored as the temperature was increased to 150C. The resistance exhibited a PTCR effect at ~130C but with a factor of 2X change in resistance rather than the typical ~100X change observed with no static pressure (e.g. Figure 1). This reduced change in resistance under pressure has been observed previously[9] and we speculate that it is simply due to a reduction in the magnitude of expansion of the PE at the melt transition under pressure. Figure 3 shows the results when a high power density pulse is applied to the material at 150C. Note that the qualitative appearance of the current and voltage waveforms is quite similar to the room temperature data depicted in Figure 2. The main difference seen in Figure 3 is the existence of a momentary increase in current as the material first switches. However, this difference has also been seen in other room temperature pulsing experiments and so is not unique to the higher material temperature. In addition, the switching was accompanied by a loud noise, visible arcing and material ablation. With the termination of the voltage step sequence, the material regained its low resistance and switched again with successive high power pulsing. These results are surprising in that they indicate that there is not a substantial difference in the high power switching behavior of this material for the cases when the material temperature is either below or above the PTCR transition. Thus the resistance switching of these materials observed in the high power regime cannot be based on the equilibrium PTCR effect.

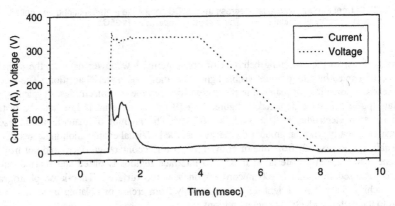

Fig. 3. Current through and voltage across the PE/CB composite material during an applied high current density electrical pulse. The ambient temperature is 150C.

These results open the possibility that other classes of conductor-filled composite materials may also exhibit switching. This was tested by fabricating materials using matrix polymers and conducting fillers that are significantly different from the CB (semimetal) filled PE (thermoplastic). In particular, silver-silicone (metal in elastomer) and commercially available nickel-epoxy (metal in thermoset) materials were fabricated and subjected to high power pulse tests. The resulting current and voltage traces for both materials are shown in Figures 4. Note that both materials exhibit switching from a low to high resistance with current and voltage traces that are qualitatively similar to the PE/CB curves in Figures 2 and 3. Again, the switching was confined to the contact interfaces, accompanied by a loud noise, visible light generation, and material ablation and yet afterwards the materials regained their initial low resistance and were capable of switching again. This suggests that repeatable high power resistance switching is a general feature of conductor-filled polymer materials.

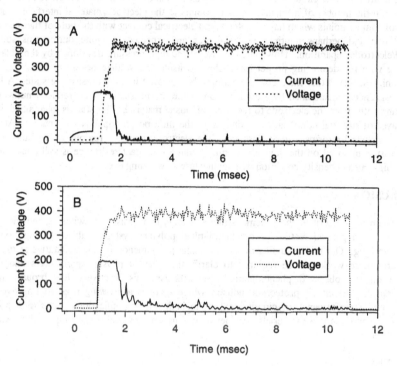

Fig. 4. Current through and voltage across a silver/silicone composite material (A) and a nickel/epoxy composite material (B) during a high current density pulse.

These experiments clearly indicate that the observed resistance switching phenomenon under high power electrical pulsing does not rely on the equilibrium PTCR effect. It is instructive to note the common features in the materials and in the switching behavior described above. All of the materials consist of a metal or semi-metal filler in an organic binder. Note that with this class of materials, the binder is expected to pyrolize or ablate and, in so doing release gas, at temperatures that are far below the melting or vaporization temperatures of the fillers. In all cases, the switching is confined to the electrode/material interface region where there is an "excess" contact resistance.

On the basis of these observations, we propose the following qualitative model for the observed switching. During the initial stages of a high current density input, the joule heating of the material is essentially adiabatic. Thus, the temperature increases much faster at the thin electrode/material contact regions than in the rest of the sample since these regions exhibit an extra contact resistance. Switching is instigated when the temperature at the contact regions rises above the temperatures where pyrolysis and ablation of the binder material occurs. Material is ablated, and the ensuing gas release causes separation of conducting filler particles in the contact region as well as a pressure expansion which leads to a partial or complete physical separation of one or both of the electrodes from the material. The contact resistance thus increases even further and essentially all of the circuit energy gets localized to one or both of the contact regions. This reduces both the number of mechanical contact points at the electrode/material interface and the number of contact points which have uninterrupted electrical contact with the bulk of the material. At least one and perhaps multiple electrical arcs are expected to ignite during this sudden material/electrode separation. However, because of the high pressure developed by the ablating gas along with the deionizing properties of the gas itself, the voltage necessary to sustain the arc(s) is high so that either the arc(s) are extinguished or exist in a high resistance state. A high device resistance is maintained by a combination of the arc resistance and a reduced number of conducting paths from the electrode to the bulk composite material. This state is maintained, with a thin layer of material continually ablating, until the high power input is terminated. At this point, the material reforms a good contact at the electrode interface regions as a result of the static pressure imposed on the device. Assuming that the contact resistance is still appreciable, the next high current density condition will again cause a switching event.

CONCLUSION

The results described above demonstrate a novel type of resistance switching behavior that appears to be a general feature of conductor-filled polymer systems subject to high power electrical pulsing. Our understanding of this switching phenomenon is only qualitative at present and clearly more work will be required to clarify the relevant physical parameters and their influence on the observed phenomenon. Nevertheless, this discovery has broad-ranging implications for the circuit protection industry where there is a growing need to improve the current-limiting performance of circuit breakers in high power applications. The high power switching observed in these materials provides a new, non-mechanical way to rapidly limit short-circuit currents.

REFERENCES

1. F. Kohler, US Patent 3,243,753, I (29 March 1966).
2. K. Ohe and Y. Naito, Jap. J. Appl. Phys., 10, 99, (1971).
3. J. Meyer, Polym. Eng. Sci., 13, 462 (1973).
4. F. Bueche, J. Polym. Sci., 11, 1319, (1973).
5. P. Sheng, E. K Sichel, and J. J. Gittleman, Phys. Rev. Lett., 40, 1197 (1978).
6. R. D. Sherman, L. M. Middleman, and S. M. Jacobs, Polym. Eng. Sci., 23, 36 (1983).
7. F. A. Doljack, IEEE Trans. Comp., Hybrids, Manuf. Tech., CHMT-4, 372 (1981).
8. "PROLIM Current Limiter", Asea Brown Boveri Control, Vasteras, Sweden.
9. J. Meyer, Polym. Eng. Sci. 13, 462 (1973).

COMPUTER SIMULATION OF ZnO VARISTORS FAILURES

M. BARTKOWIAK*, G.D. MAHAN*, M.G. COMBER**, M.A. ALIM[†]
*Department of Physics and Astronomy, University of Tennessee, Knoxville, TN 37996-1200, and Solid State Division, Oak Ridge National Laboratory, Oak Ridge, TN 37831-6030, bartkowiakm@ornl.gov
**Hubbell Power Systems, 210 North Allen Street, Centralia, MO 65240-1395
[†]Hubbell Incorporated, The Ohio Brass Company, 8711 Wadsworth Road, Wadsworth, OH 44281

ABSTRACT

A simple thermo-mechanical model is applied to evaluate the influence of the nonuniformity of ZnO varistor disks used in surge arresters on their energy handling capability. By solving heat transfer equations for a varistor disk with nonuniform electrical properties, we compute the time dependence of the temperature profile and the distribution of thermal stresses. The model can identify the energy handling limitations of ZnO varistors imposed by three different failure modes: puncture, thermal runaway, and cracking. It conforms to the available failure data, and explains the observation that energy handling improves at high current densities.

INTRODUCTION

Zinc oxide varistors are multi-component ceramic devices produced by sintering ZnO powder together with small amounts of other oxides. Highly nonlinear current-voltage (I-V) characteristics of ZnO varistors are used in electrical surge arresters. They protect electrical equipment from damage by limiting overvoltages and dissipating the associated energy. Therefore, the energy handling capability is crucial. It is defined as the amount of energy that a varistor can absorb before it fails.

There are three main failure modes of varistor elements: thermal runaway, puncture, and cracking. The leakage current, and consequently the Joule heating of a varistor, increase with temperature. Thus, if the temperature is raised above the thermal stability temperature T_s, power input may exceed heat dissipation, and thermal runaway occurs. In puncture, a small hole results from melting of the ceramic where high current is concentrated [1]. Nonuniform heating can also cause thermal stresses higher than the failure stress of the material and can lead to cracking [1, 2]. Currents in the nonlinear region of the I-V characteristics tend to concentrate into narrow paths. This current localization has been detected by applying small spot electrodes on the surfaces of varistors, by using infrared cameras [3, 4] and by electroplating techniques [5].

Measurements of the energy handling capability of varistors have been reported [1, 6, 7], but the nature of the failures is not well understood. Puncture has been studied by Eda [1], who showed that at high currents, a hot spot may reach about 800 °C and cause local melting. However, two important factors were omitted by Eda: (1) the influence of the upturn in the I-V characteristics, and (2) that thermal stresses may cause cracking before a puncture can take place.

ZnO varistors exhibit a complex dependence of the energy handling capability upon pulse magnitude and duration. Initially the energy handling capability decreases with increasing current, but, as first pointed out by Sakshaug et al. in [6], it *increases* again if the current becomes very high and the pulse duration becomes very short. It is the main purpose of the present paper to provide a fundamental explanation of the phenomenon. The explanation is of added interest because Ringler et al. [7] have recently shown that the energy absorption capability of varistors used in station-class arresters increases almost 4 times as the current level increases from 0.8 A_{peak} to 35 kA_{peak}. These tests are simulated in the present study, and the results are compared with the experimental data. The model is also used to evaluate the influence of the nonuniformity of varistor disks on their energy handling capability for current surges of various magnitudes and durations.

THERMO-MECHANICAL MODEL OF VARISTOR DISKS

Details of the thermo-mechanical model used in the present study have been presented elsewhere [8]. Here, only its brief description is presented in order to settle the terminology and define the basic concepts.

The behavior varistor disks is simulated by solving the coupled nonlinear equations of current conduction and heat diffusion to obtain the time and spatial dependence of the current density and of the temperature. The temperature dependence of the thermal conductivity [9] is taken into account in the heat diffusion equation. The boundary conditions include the assumption that the disks can dissipate heat only radially by convection to air through the sidewall, and that the ambient temperature is 20 °C. A model temperature-dependent I-V characteristic typical for high-voltage varistors is used in the simulations [8]. At $T = 293$ K, the breakdown field (i.e., the field at 1 mA/cm^2) is $F_b \approx 1870$ V/cm, the prebreakdown resistivity $\rho_{pb} = 5 \times 10^{11}$ Ω cm, and the resistivity in the upturn region (i.e., the resistivity of the grains) is $\rho_{up} = 1$ Ω cm. The coefficient of nonlinearity has the maximal value $\alpha_m \approx 50$. Nonuniformity of the varistor disk is simulated by assuming that a hot spot of diameter $2R_{hs}$ extends axially through the block at its center, so that the heat diffusion problem remains cylindrical symmetric. The breakdown voltage at the hot spot F_{bhs} is assumed $p\%$ lower than that for the rest of the varistor block, i.e., $p = (F_b - F_{bhs})/F_b$. The quantity p (expressed in percents) will be called *the hot spot intensity*. As follows from [1], typical varistor disks have hot spots with $p = 5\%$, but the intensity of hot spots in highly nonuniform disks can be as high as 10%. The small difference between the electrical properties of the hot spot, and those of its surroundings, causes high nonuniformity in the spatial distribution of the current, and thereby in the input power density due to Joule heating.

The temperature profile for the disk and its time evolution, is used to calculate the distribution of thermal stresses and to identify the cracking failure mode. To determine the maximum energy that can be absorbed by a varistor disk before it cracks, it is necessary to know the strength of the material. Its value strongly depends on the details of the ceramic processing used, on the presence of preexisting flaws, and on other factors, so that it may vary considerably for individual disks. Here, a simple assumption that the tensile strength of the varistor ceramic is $S_{ft} = 20$ kpsi [2] is used. Two types of failures are considered: (1) cracking at tension, when thermal tensile stress exceeds S_{ft}; and (2) puncture, when the temperature at the hot spot exceeds 800 °C [1].

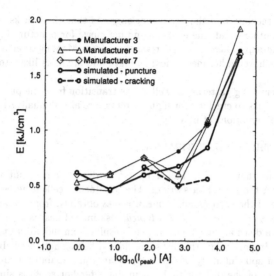

Figure 1: The simulated and mean measured [7] energy handling capabilities of station-class-arrester disks as a function of the peak test current.

SIMULATION OF THE EXPERIMENTAL DATA

Recently, Ringler et al. [7] measured the energy absorption capability of commercial varistor disks used in station-class arresters, at current levels ranging from 0.8 A_{peak} to 35 kA_{peak}. Here, the model described in the preceding section is used to simulate these experiments. A typical disk, 23 mm high and 63 mm diameter, is assumed to have a 1.2 cm diameter hot spot ($R_{hs} = 0.6$ cm), and the upturn resistivity $\rho_{up} = 1\ \Omega$ cm. The computed energy handling capabilities for puncture and cracking as functions of the peak test current are shown in Fig. 1, together with the measured mean values of the total energy to destroy varistor disks from three different manufactures. The same symbols and code numbers for manufactures as in [7] are used.

At low currents corresponding to 0.8 A_{peak} and 7 A_{peak}, the only possible failure mode (besides a thermal runaway) is puncture. This is consistent with the results of [7], where at these test currents the failure resulted in a single hole through the bulk ceramic, and no cracking or fragmentation failures were observed. At 600 A_{peak}, the tensile tangential stresses around the hot spot become higher than the strength of the varistor ceramic before the temperature at the hot spot reaches 800 °C. Therefore, the disks become likely to crack along lines branching out from the hot spot. Indeed, the most significant external damage was found in [7] for the currents at 600 A_{peak}. The simulations indicate that cracking remains the most likely failure mode also at 4 kA_{peak}. However, for very high current pulses of 35 kA_{peak}, the simulated thermal stresses never become high enough to cause cracking, and the disks fail either due to a puncture or due to overheating. This is again consistent with the data reported in [7].

Varistors from Manufacturers 5 and 7 exhibit an unexpected dip in energy absorption capability at 600 A_{peak}. This agrees surprisingly well with the simulated energy handling characteristic for cracking, and the agreement suggests that these disks crack on failure at 600 A_{peak}. On the other hand, the energy handling curve for puncture in Fig. 1 is in good agreement with the experimental data for varistors from Manufacturer 3. Apparently, the latter varistor disks have higher mechanical strength and are more likely to exhibit puncture than cracking.

The predicted cracking patterns, as well as the transition from the puncture failure mode at low current densities to cracking for high currents, are also in qualitative agreement with the experimental observations of Eda [1].

INFLUENCE OF THE NONUNIFORMITY

The assumptions that the disk had a 1.2 cm in diameter hot spot with the intensity $p = 5\%$ was used in the preceding section. The obtained excellent agreement between the simulated results and the corresponding *mean values* obtained from a statistical analysis of the experimental data [7] indicates that a *typical* disk indeed has such a hot spot. However, as shown in [7], the disks are not identical, and results from individual measurements may vary considerably. In particular, the disks can have different hot spots. In this section, the influence of the hot spot intensity on the energy handling capability is discussed. Thermo-mechanical behavior of the disks with 1 cm in diameter hot spots is simulated, assuming three different intensities: $p = 2.5\%$, 5%, and 7.5%. The energy handling characteristics of disks with these hot spots, obtained for the case of dc currents, are shown in the left panels of Fig. 2.

Of course, the disks with hot spots of higher intensity have lower energy handling capability. The minimum energy handling decreases from about 1 kJ/cm^3 for $p = 2.5\%$, to 450 J/cm^3 for $p = 5\%$, and 200 J/cm^3 for $p = 7.5\%$. Also, the region of the current densities for which the disks exhibit punctures extends to lower currents as the intensity increases. Simulations for the case of $p = 2.5\%$ show that the thermal stresses never become high enough to cause cracking, independently of the magnitude of the applied current. This shows that disks with hot spots of low intensities do not crack, and can only fail by puncture. Cracking caused by tensile thermal stresses in tangential direction is the dominating failure mode at high current densities for $p = 5\%$ and $p = 7.5\%$. Moreover, for higher p, the disks become more likely to crack also at lower current densities.

The size of the hot spot is another parameter that can significantly deviate from the typical value of about 1 cm. To evaluate its influence on the energy handling capability, station-class varistor disks having hot spots with $p = 5\%$, and three different diameters: 1 mm, 1 cm, and 2 cm are simulated. Their energy handling characteristics, obtained for a dc current flow, are shown in the right panels of Fig. 2. The value of the minimum energy handling only weakly depends on the size of the hot spot, and varies from about 650 J/cm^3 for the disk with a 1 mm hot spot, through 450 J/cm^3 when $2R_{hs} = 1$ cm, to 480 J/cm^3 for the disk with a 2 cm in diameter hot spot. These results indicate that the minimum energy handling is not a monotonic function of R_{hs}, and that it becomes lowest for disks with hot spots of an intermediate size. In all three simulated cases, the minimal E corresponds to a puncture, and moves towards lower current densities as the size of the hot spot increases. As follows from the data in Fig. 2, the disks are more likely to crack on failure when they have large hot spots. The disk with a 1 mm hot spot never cracks, whereas cracking is the dominating failure mode even at 1 A/cm^2 for the disk with $2R_{hs} = 2$ cm.

256

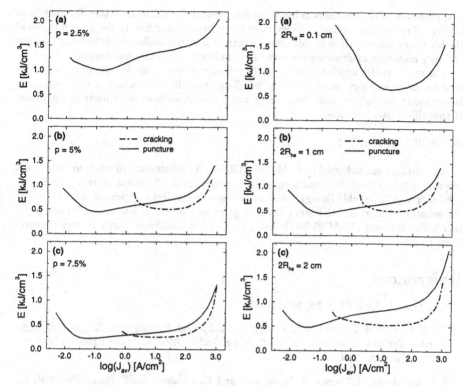

Figure 2: Energy handling characteristics of station-class varistor disks having 1 cm diameter hot spots with different intensities (left panels); and having hot spots of the intensity $p = 5\%$ and three different sizes (right panels)

CONCLUSIONS

A simple theoretical model can identify the energy handling limitations of ZnO varistors imposed by three different failure modes: puncture, thermal runaway, and cracking. The energy handling depends upon which of these failure modes is predominant for a current pulse. Each failure mode can be limiting, depending on the disk shape, its electrical uniformity, and the current magnitude.

The model conforms to the available failure data, and explains the observation of Sakshaug et al. [6] and Ringler et al. [7] that energy handling improves at high current densities. It can also be used to simulate thermo-mechanical behavior and estimate the energy handling capability of various types of varistor disks without performing destructive experiments.

Cracking and puncture are caused by a localization of the current, which causes local heating leading to nonuniform thermal expansion and thermal stresses. Puncture is most likely in varistor disks with low geometrical aspect ratio and when the current density has intermediate values. Cracking dominates at higher current densities and for disks with high aspect ratio. Puncture and cracking do not occur when the current is small, because the time evolution of the nonuniform heating is slow enough for the temperature distribution to flatten. They are also unlikely at very large currents corresponding to the upturn region of the I-V characteristic, since in this case the current becomes uniformly distributed. For low and very high current densities the most likely failure mode is thermal runaway.

The model is also applied to evaluate the influence of the nonuniformity of varistor disks used in surge arresters on their energy handling capability. Puncture is the dominating failure mode for slightly nonuniform disks, but cracking becomes more likely as the degree of nonuniformities increases.

ACKNOWLEDGEMENTS

The authors are indebted to F. Modine (ORNL) for stimulating discussions. This work was supported by the U.S. Department of Energy through the Assistant Secretary for Energy Efficiency and Renewable Energy, Office of Transportation Technologies, as part of the High Temperature Materials Laboratory User Program and by the Office of Basic Energy Science under contract DE-AC05-96OR22464, managed by Lockheed Martin Energy Research Corporation.

References

[1] K. Eda, *J. Appl. Phys.* **56**, 2948 (1984).

[2] H.F. Ellis, R.M. Reckard, H.R. Philipp, and H.F. Nied, Fundamental Research on Metal Oxide Varistor Technology, *EPRI Report EL-6960* (Electric Power Research Institute, Palo Alto, CA, 1990).

[3] A. Mizukoshi, J. Ozawa, S. Shirakawa, and K. Nakano, *IEEE Trans. Power Appar. Syst.* **102**, 1384 (1983).

[4] H. Wang, M. Bartkowiak, F.A. Modine, R.B. Dinwiddie, L.A. Boatner, and G.D. Mahan, accepted for publication in *J. Am. Ceram. Soc.*

[5] G. Hohenberger, G. Tomandl, R. Ebert, and T. Taube, *J. Am. Ceram. Soc.* **74**, 2067 (1991).

[6] E.C. Sakshaug, J.J. Burke, and J.S. Kresge, *IEEE Trans. Power Delivery* **4**, 2076 (1989).

[7] K.G. Ringler, P. Kirkby, C.C. Erven, M.V. Lat, and T.A. Malkiewicz, *IEEE Trans. Power Delivery* **12**, 203 (1997).

[8] M. Bartkowiak, M.G. Comber, and G.D. Mahan, *J. Appl. Phys.* **79**, 8629 (1996).

[9] Y.S. Touloukian, R.W. Powell, C.Y. Ho, and P.G. Klemens, Thermophysical Properties of Matter, Plenum, New York - Washington, 1970.

Part VI

Ionic and Mixed Conductors

NON-DEBYE AND CPA BEHAVIORS OF IONIC MATERIALS

J. C. Wang
Energy Division, Oak Ridge National Laboratory, Oak Ridge, Tennessee 37831-6185

ABSTRACT

Non-Debye and constant-phase-angle (CPA) behaviors associated with the bulk and interfacial processes involving ionic materials are discussed in terms of complex impedance, admittance, and dielectric spectra. The yielding of a CPA and/or a broad non-Debye dielectric loss peak in a spectrum from fractal, pore, and ion-hopping models are compared and reviewed. The observed wide frequency ranges of the CPA behavior suggest that the fractal and pore models, which require a wide range of special structures down to very fine scales, may not be realistic. The ion-hopping model treats the bulk and interfacial processes as a chemical reaction having a thermally-activated Arrhenius form. Because of thermal fluctuations, the activation energies for ion hopping (e.g., in a potential double-well) have a double-exponential distribution which yields a non-Debye dielectric loss peak and a CPA spectrum over a wide frequency range above the loss peak. The distribution also has a special temperature dependence which may explain the invariance of dielectric spectral shapes with temperature, an observation by Joscher. The construction of CPA elements (in a generalized Warburg impedance form) using three distinct types of resistor-capacitor networks are presented and used to aid the discussion.

INTRODUCTION

Non-Debye and constant-phase-angle (CPA) behaviors associated with the bulk and interfacial processes involving ionic materials have been widely observed, for example, with impedance measurements [1]. A complex impedance with a CPA has the form

$$Z(\omega) = A_Z(j\omega)^{-n} = A_Z(\cos\frac{n\pi}{2} - j\sin\frac{n\pi}{2})\,\omega^{-n} \ , \tag{1}$$

where $j = (-1)^{\frac{1}{2}}$, $\omega = 2\pi \times frequency$, and $0 < n < 1$. $Z(\omega)$ shown in Eq. (1) is called a CPA impedance because the $\ln[\text{Re}(Z)]$ and $\ln[-\text{Im}(Z)]$ vs $\ln[\omega]$ graphs form a pair of parallel lines (Fig. 1) and the ratio $\text{Im}(Z)/\text{Re}(Z)$, which is equal to $\tan[(-n\pi)/2]$, is independent of ω. A complex impedance spectrum such as that shown in Eq. (1) and Fig. 1 can also be expressed in forms of complex admittance and dielectric functions. In this paper, all three forms of representing a spectrum (i.e., the complex impedance, admittance, and dielectric function) are used interchangeably and which form to use in a given case depends on the system under consideration. In the admittance form, Eq. (1) can be written as

$$Y(\omega) = \frac{1}{Z(\omega)} = A_Y(j\omega)^n \ , \tag{2}$$

where $A_Y = 1/A_Z$. In the dielectric function form, Eqs. (1) and (2) can be written as

Mat. Res. Soc. Symp. Proc. Vol. 500 © 1998 Materials Research Society

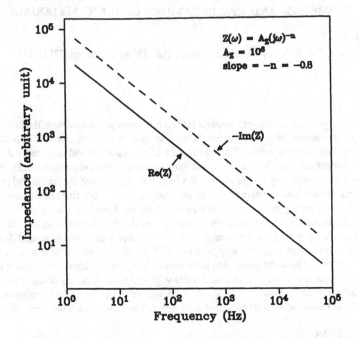

$$Z(\omega) = A_z(j\omega)^{-n}$$
$$A_z = 10^6$$
$$\text{slope} = -n = -0.8$$

Fig. 1. The $\ln[\text{Re}(Z)]$ and $\ln[-\text{Im}(Z)]$ vs $\ln[\omega]$ graphs of a CPA impedance showing that they form a pair of parallel lines.

$$\varepsilon(\omega) - \varepsilon_\infty = \frac{Y(\omega)}{j\omega} = A_Y(j\omega)^{n-1} \quad , \tag{3}$$

where ε_∞ is the high-frequency dielectric constant.

Equivalent circuits are often used to help understanding the physical process associated with an observed spectrum. To help understanding the CPA behavior, it is useful to have some systematic methods of constructing equivalent circuits that can yield a CPA element. Three types of such equivalent circuits in forms of resistor-capacitor (rc) networks are shown in Fig. 2. The resistors and capacitors in the networks can be discrete elements, in which case only an approximation to a CPA element is obtained, or continuous distributions satisfying some mathematical relations. For example, if the resistors in Fig. 2(a), from left to right, have values

$$r_k = r_o(a/M)^k \quad , \quad k = 0, 1, 2, \dots \tag{4}$$

and the capacitors have values

$$c_k = c_o M^k \quad , \quad k = 0, 1, 2, \dots \quad , \tag{5}$$

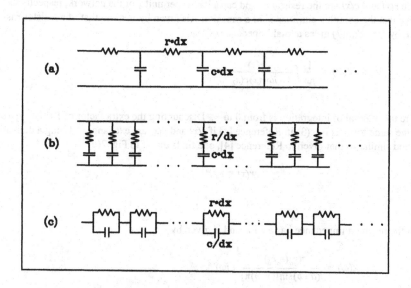

Fig. 2. Three types of rc networks that can yield a CPA element.

where r_o, c_o, a, and M are constants with $a > M$, then the network in Fig. 2(a) will give a CPA impedance with [2]

$$n = 1 - \frac{\ln(M)}{\ln(a)} \tag{6}$$

This result has been used in Reference [2] to verify the derivation of a fractal model [3] for a blocking electrode/electrolyte interface.

Methods of constructing CPA elements using the rc networks shown in Fig. 2(a) and 2(b) have been reported in References [2] and [4], respectively. For completion, the corresponding method for the network shown in Fig. 2(c) is derived in this paper. These networks can be useful in discussing the non-Debye and CPA behaviors of ionic materials.

CONSTRUCTION OF CPA ELEMENTS WITH DISTRIBUTION OF PARALLEL rc-PAIRS

In Fig. 2(c), the impedance associated with dx at x, which is treated as a parameter, is given by

$$dZ(\omega) = \frac{1}{\dfrac{1}{r(x)dx} + j\omega\dfrac{c(x)}{dx}} = \frac{r(x)}{1 + j\omega r(x)c(x)} dx \quad , \tag{7}$$

where $r(x)$ and $c(x)$ are the resistance and capacitance per unit x of the network, respectively. Notice that the complex admittance of a capacitor c is given by $j\omega c$ and that of a resistor r is given by $1/r$. Eq. (7) gives a total impedance of

$$Z(\omega) = \frac{1}{j\omega} \int \frac{j\omega r(x)}{1+j\omega r(x)c(x)} \, dx \quad , \tag{8}$$

where the interval of integration is from 0 to ∞. Discounting the extra factor of $1/(j\omega)$, Eq. (8) has the same form as Eq. (3) in Reference [4] if $r(x)$ and $c(x)$ are exchanged. Using a derivation method similar to that given in Reference [4], one finds that if in Fig. 2(c)

$$r(x) = r_o x^a$$

and

$$c(x) = c_o x^b \quad ,$$

then the resultant impedance has a CPA and is given by

$$Z(\omega) = \frac{\pi r_o (r_o c_o)^{-n}}{(a+b)\sin[(1-n)\pi]} \, (j\omega)^{-n} \quad , \tag{9}$$

where

$$n = \frac{a+1}{a+b} \quad , \quad \text{with the restriction } 0 < n < 1 \quad .$$

Utilizing a transformation method similar to those used in References [2] and [4], other CPA-element producing $r(x)$ and $c(x)$ pairs can be obtained. For example, if in Fig. 2(c)

$$r(x) = r_o e^{ax}$$

and

$$c(x) = c_o e^{bx} \quad ,$$

then the resultant impedance is given by

$$Z(\omega) = \frac{\pi r_o (r_o c_o)^{-n}}{(a+b)\sin[(1-n)\pi]} \, (j\omega)^{-n} \quad , \quad , \tag{10}$$

where

$$n = \frac{a}{a+b} \quad , \quad \text{with the restriction } 0 < n < 1 \quad .$$

264

COMPARISON OF FRACTAL AND PORE MODELS FOR ELECTROLYTE/ BLOCKING ELECTRODE INTERFACE

It is often observed that the impedance of the interface between an electrolyte and a blocking electrode can be represented by a CPA element. According to a pore model [5], this impedance can be attributed to pores with some special shapes and the value of the exponent n in Eq. (1) is related to the detailed shape of the pores. Fractal models [3,6] have also been proposed to explain this CPA behavior. In these fractal models, the micro structure of an interface is represented by electrolyte-filled grooves (or pores) on the electrode. In a model developed by Liu [3], each groove subdivides into smaller grooves from one stage to the next in a way similar to the Cantor-bar problem as the electrolyte intrudes deeper into the electrode [Fig. 3(a)]. In a model proposed by Nyikos and Pajkossy [6], the electrode contains a distribution of grooves of various sizes and, inside each groove, there are grooves of smaller sizes. The side view of the interfacial boundary has the pattern of a generalized Koch curve (Fig. 4).

In Liu's model, the fractal dimension of a generalized a Cantor-bar is given by

$$D = \frac{\ln N}{\ln a} , \tag{11}$$

where N is the number of new grooves formed in each old groove [e.g., $N = 2$ in Fig. 3 (a)] and a is the cross-section reduction factor from one stage to the next. According to Liu [3], the fractal dimension defined in Eq. (11) and the exponent of the resultant CPA impedance is related by

$$n = 1 - D . \tag{12}$$

Notice that the fractal dimension D defined by Eq. (11) is that of a generalized Cantor-bar. Its

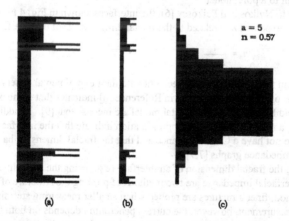

$$a = 5$$
$$n = 0.57$$

(a)   (b)   (c)

Fig. 3. (a) and (b): Liu's fractal interfaces with $D=0.43$. At each new stage, the width of the grooves (shaded areas) is reduced by a factor $a=5$. (c): An equivalent pore (groove) constructed from (b). In all three cases, the frequency exponent n is the same. (After Reference [7])

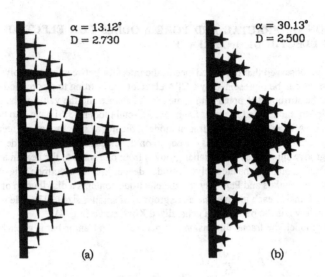

$\alpha = 13.12°$
$D = 2.730$

$\alpha = 30.13°$
$D = 2.500$

(a)   (b)

**Fig. 4. Interfaces in the fractal model of Nyikos and Pajkossy.
The interface with grooves (shaded areas) of smaller angle α has a
greater fractal dimension D.** (After Reference [7])

exact meaning for the actual interface is not clear. For example, measuring the fractal dimension of the interface shown in Fig. 3(b) with a coast-line method may yield a fractal dimension quite different from that given by Eq. (11) [7]. As demonstrated in Reference [7], Liu's Cantor-bar interface shown in Fig. 3(b) can be transformed into a pore as shown in Fig. 3(c) and his fractal model is equivalent to a pore model.

According to Nyikos and Pajkossy [6], the interfaces shown in Fig. 4 have a CPA impedance and the exponent n is related to the fractal dimension D of the interface by

$$n = \frac{1}{D-1} \tag{13}$$

These authors claim that Eq. (13) is in agreement with their experimental observation [8]. However, a review of the derivation given in Reference [6] indicates that some of the equations used by Nyikos and Pajkossy for their fractal model are inconsistent [9]. In addition, numerical calculations using a rc transmission-line approximation indicate that the interfaces like those shown in Fig. 4 do not have a CPA impedance and that the fractal dimension has little effect on the shape of the impedance graphs [7,9].

Measuring the fractal dimension of an interface (e.g., using the coast-line method) and measuring the interfacial impedance are two methods of probing the structure of the interface. In the coastline method, finer structures are probed when smaller measuring step sizes are used. In the impedance measurement, however, the current penetration depends on both the pore size and the location of the pore. For example, in Fig. 4(a), at some frequencies the current may reach everywhere in the smallest groove near the top of the figure, but does not penetrate the smallest grooves deep inside the largest groove at the center of the figure. In other words, the current

266

does not probe the structures of the same size equally as the fractal dimension measurement does. This does not prove that n and D are not related, but because of this difference in measuring methods, there should be no simple relation between them. However, if the fractal dimension is defined in a way consistent with the current penetration concept as in Liu's model, a simple relation such as Eq. (12) can exist. It is interesting to note that when Liu's model is generalized, there is also no universal relation in which n is simply a function of D [10].

NON-DEBYE DIELECTRIC RESPONSE AND DISTRIBUTION OF ACTIVATION ENERGIES

The Debye dielectric function has the form [1,11]

$$\varepsilon(\omega) - \varepsilon_\infty = \frac{\varepsilon_0 - \varepsilon_\infty}{1 + j\omega\tau} , \tag{14}$$

where ε_0 and ε_∞ are the low- and high-frequency dielectric constants, respectively, and τ is a time constant. For the convenience of discussion, we shall express Eq. (14) and other dielectric functions discussed in this paper in the form

$$\frac{\varepsilon(\omega) - \varepsilon_\infty}{\varepsilon_0 - \varepsilon_\infty} \equiv \varepsilon'(\omega) - j\varepsilon''(\omega) , \tag{15}$$

here $\varepsilon'(\omega)$ and $\varepsilon''(\omega)$ represent the dispersion and loss characteristic, respectively, of the dielectric response of ionic materials. As can be seen from Eqs. (2), (3), and (14), the Debye dielectric function corresponds to an admittance in the form $Y(\omega) = 1/r + j\omega c$, which can be represented by an equivalent circuit consisting of a resistor and a capacitor in series [like those shown in Fig. 2(b)]. This equivalent circuit can be obtained from a physical model for a charged particle hopping back and forth in a potential double-well [12]. The activation energy E is equal to the difference between the potential energy of the system when the hopping particle is at the saddle point and the potential energy when it is at the bottom of the well. A possible example is given by an O^{2-} vacancy in stabilized zirconia.

As illustrated in Fig. 5, $\varepsilon''(\omega)$ for the Debye dielectric function is symmetric about its maximum value at $\omega\tau = 1$. Many ionic materials, however, display a non-Debye behavior characterized by a broad asymmetric loss peak and by a CPA behavior at frequencies above the loss peak. This non-Debye dielectric response can be described quite well by an empirical expression proposed by Havriliak and Negame (H-N) [13]:

$$\frac{\varepsilon(\omega) - \varepsilon_\infty}{\varepsilon_0 - \varepsilon_\infty} = \frac{1}{[1 + (j\omega\tau')^{1-\alpha}]^\beta} , \tag{16}$$

where τ' is a constant, $0 \leq \alpha < 1$, and $0 < \beta \leq 1$. Equation (16) becomes the Cole-Cole function if $\beta = 1$, becomes the Davidson-Cole function if $\alpha = 0$, and becomes the Debye function if $\alpha = 0$

and $\beta = 1$. A schematic comparison of a Debye and an H-N complex dielectric functions is given in Fig. 5.

Although many models have been proposed, the origin of the non-Debye dielectric response is not well understood. It is sometimes assumed that the Debye function corresponds to a relaxation process with a single time constant [τ in Eq. (14)] while a non-Debye dielectric function such as the H-N function represents a superposition of many Debye functions with various relaxation times [14]. If the relaxation is thermally activated and the time constant can be written as

$$\tau = \tau_o \, e^{\frac{E}{kT}} \,, \tag{17}$$

where k is the Boltzmann constant and T the temperature, then the non-Debye dielectric function becomes a superposition of Debye functions with various activation energies:

$$\frac{\varepsilon(\omega) - \varepsilon_\infty}{\varepsilon_0 - \varepsilon_\infty} = \int \frac{1}{1 + j\omega\tau_o \, e^{E/kT}} \, G(E) \, dE \,, \tag{18}$$

where $G(E)$ is the distribution of activation energies. Because each Debye dielectric function can be related to the admittance of a resistor and a capacitor in series, a possible equivalent circuit for a dielectric function represented by Eq. (18) is that shown in Fig. 2(b) with proper $r(x)$ and $c(x)$ distributions.

One interesting distribution of activation energies for a non-Debye dielectric response is the double-exponential $G(E)$ given by [15,16]

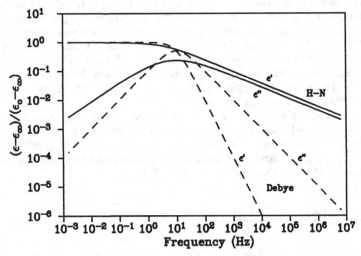

Fig. 5. A schematic comparison of the Debye dielectric function and a non-Debye dielectric function represented by the empirical H-N function.

$$G(E) = \frac{ab}{(a+b)kT} e^{-a(E_o-E)/kT} \quad , \qquad \text{for } E \le E_o \; ,$$

$$= \frac{ab}{(a+b)kT} e^{-b(E-E_o)/kT} \quad , \qquad \text{for } E \ge E_o \tag{19}$$

where a, b, and E_o are constants. As demonstrated in Reference [16], the double-exponential $G(E)$ can yield dielectric functions very close to the H-N function. Since the latter can represent the non-Debye dielectric response of many materials, this implies that the double-exponential $G(E)$ can also represent the non-Debye dielectric response of these materials. A discussion of how such a distribution of activation energies can be obtained from a physical model is given in Reference [16]. In essence, the potential energy of a hopping particle in a potential double-well can be described by the Boltzmann's factor [e.g., $(a/kT)exp(-aV/kT)$, with V being the potential energy and a an arbitrary parameter] at the bottom of the potential well and at the saddle point. By defining the activation energy for each successful ion jump as the difference of the two potential energies, the distribution of activation energies shown by Eq. (19) can be obtained.

Several $G(E)$'s that can be obtained from Eq. (19) are shown in Fig. 6(a)-6(d). Their corresponding dielectric functions calculated numerically with Eq. (18) are shown in Fig. 6(a')-6(d'). As mentioned before, the Debye model corresponds to a process with a single relaxation time or a single activation energy [Fig. 6(a)]. Its dielectric function has fixed slopes as shown in Fig. 6(a'). If an exponential decay in $G(E)$ for $E \le E_o$ is introduced [Fig. 6(b)], a CPA behavior of the dielectric function appears above the loss peak [Fig. 6(b')], but the slopes of $\varepsilon'(\omega)$ and $\varepsilon''(\omega)$ below the loss peak are identical to those of the Debye model [Fig. 6(a')]. If now we extrapolate the $G(E)$ in Fig. 6(b) from $E \le E_o$ to $E \ge E_o$, so that it increases exponentially with E indefinitely, that is, changing Eq. (19) to

$$G(E) = \frac{ab}{(a+b)kT} e^{-a(E_o-E)/kT} \quad , \qquad \text{for all } E \; , \tag{20}$$

then the CPA behavior shown in Fig. 6(b') will extend to all frequencies. The zero value of $G(E)$ above E_o in Fig. 6(b), therefore, is responsible for the termination of the CPA behavior at low frequencies and for the appearance of the loss peak in Fig. 6(b').

If an exponential decay in $G(E)$ for $E \ge E_o$ is introduced [Fig. 6(c)], only the slope of $\varepsilon''(\omega)$ below the loss peak is altered [see Fig. 6(a') and 6(c')]. The trend is in the direction of broadening the Debye loss peak.

The combined result of Fig. 6(b) and 6(c) is the double-exponential $G(E)$ shown in Fig. 6(d). Its corresponding dielectric function [Fig. 6(d')], as mentioned before, has all the main features of the H-N function and can describe many experimental results quite well. It is interesting to note the simple relations among the slopes in Fig. 6(d) and 6(d'). They suggest that the exponential decays in $G(E)$ are important features that can be associated with the experimental dielectric response.

The possible physical origin of the double-exponential $G(E)$ can be summarized as the following: Identical potential double-wells seen by hopping particles at $T = 0$ will no longer be identical at finite T. In the diffusion theory [17], an averaging process is used so that these wells remain identical statistically at finite T. The appearance of a single activation energy in the Debye model can be attributed to this averaging process. The distribution-of-activation-energies approach represents an attempt to remove this averaging process.

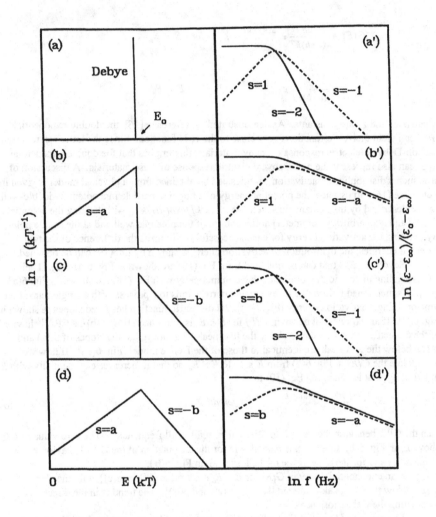

Fig. 6. Several $G(E)$'s specified by Eq. (19) and their corresponding $\varepsilon'(\omega)$'s and $\varepsilon''(\omega)$'s, showing the simple relations among the slopes. Notice that in order to get the indicated slopes for the $G(E)$ graphs, the natural logarithm is used and the unit for energy is chosen to be kT. (After Reference [16])

The double-exponential $G(E)$ given by Eq. (19) has an unusual temperature dependence, i.e., factor $1/kT$ in the exponent. A consequence of this dependence is that the distribution of activation energies broadens as T increases. As demonstrated in Reference [16], this broadening of $G(E)$ with T yields spectral shapes of $\varepsilon'(\omega)$ and $\varepsilon''(\omega)$ independent of temperature, which is consistent with a large amount of experimental data. As summarized by Jonscher [1], for many materials, the log-log graphs of the dielectric function have almost an identical shape at different temperature. For such a material, the experimental data may be satisfactorily normalized into a master curve, indicating the spectral shapes remain invariant with temperature.

DISCUSSION

The rc networks giving CPA elements can be helpful for understanding the non-Debye and CPA behaviors of inoic materials. Fig. 2(a) is more relevant to the current penetration concept of the pore and fractal models for electrolyte/blocking electrode interfaces. When an ac signal is applied from the lefthand side of the network (or transmission line), how far it can penetrate depends on the frequency used [2]: the lower the frequency, the greater the penetration depth. If the network shown in Fig. 2(a) has a finite length, the CPA behavior may cease to exist at low frequencies [2,5]. This implies that the fractal and pore models discussed in this paper may require an unrealistically wide range of special features in order to produce a CPA behavior over a sizable frequency range.

The rc network shown in Fig. 2(c) could be used to see whether under some conditions the change of material properties near an electrolyte/electrode interface can yield a CPA element. For example, there is experimental evidence that the resistivity of single-crystal NaCl near a electrode is about 25 times greater than that of the bulk [18]. According to Eq. (10), if both $r(x)$ and $c(x)$ increase exponentially toward the electrode such that $a = b = 1$, then the interface impedance will have a CPA with $n = 0.5$, like that of a Warburg impedance [19]. However, it should be stressed that the exact meaning of such a result is unclear.

The rc network shown in Fig. 2(b) can be very helpful for discussing the non-Debye dielectric response in terms of distribution of activation energies [4,16]. This is because each rc pair can represent a Debye response and can be associated, through a physical model [12], to a charged particle hopping back and forth in a potential double-well. The transformation method derived in Reference [4] can be used to identify the distributions of $r(x)$ and $c(x)$ (and therefore the distributions of activation of energies) which can yield a CPA behavior.

As mentioned above, the model for the non-Debye dielectric response represents an attempt to remove the averaging nature of the diffusion theory. This seems to be a reasonable thing to do because the averaging procedure was not really justified in the theory and the dielectric response function exhibits effects of all transition rates, not the averaged rate. However, a justification for the use of the arbitrary constants a and b in Eq. (19) is needed.

The $G(E)$ considered for the non-Debye dielectric response [Eq. (19)] represents the fluctuating barrier heights induced by thermal agitations. Even for a crystal with identical potential double-wells, $G(E)$ should not be a delta function of E. This may explain why the Debye dielectric response, which corresponds to a single activation energy, is not commonly observed in solids [1]. As mentioned above, the temperature dependence of $G(E)$ shown in Eq. (19) is consistent with the invariance of the spectral shapes of the loss characteristics with temperature, an observation based on extensive experimental data [1].

The CPA elements and non-Debye dielectric functions discussed in this paper may be applied to analyze and interpret experimental data. An example is given in References [20] and [21] for the analysis of the complex dielectric response data of a yttria-stabilized zirconia sample.

ACKNOWLEDGMENT

This work was sponsored by the U.S. Department of Energy under contract No. DE-AC05-96OR22464 with Lockheed Martin Energy Research Corp.

REFERENCES

1. A. K. Jonscher, <u>Dielectric Relaxation in Solids</u>, Chelsea Dielectrics Press, London (1983).

2. J. C. Wang, J. Electrochem. Soc.**134**, p. 1915 (1987).

3. S. H. Liu, Phys. Rev. **55**, p. 529 (1985).

4. J. C. Wang, Solid State Ionics **39**, p. 277 (1990).

5. J. C. Wang and J. B. Bates, Solid State Ionics **18/19**, p. 224 (1986).

6. L. Nyikos and T. Pajkossy, Electrochim. Acta **30**, p. 1533 (1985).

7. J. C. Wang, Solid State Ionics **28-30**, p. 1436 (1988).

8. T. Pajkossy and L. Nyikos, J. Electrochem. Soc. **133**, p. 2061 (1986).

9. J. C. Wang, Electrochim. Acta **34**, p. 987 (1989).

10. T. Kaplan, L. J. Gray, and S. H. Liu, Phys. Rev. **B35**, p. 5379 (1987).

11. P. Debye, <u>Polar Molecules</u>, Dover, New York (1929).

12. J. C. Anderson, <u>Dielectrics</u>, Reinhold, New York, p. 67 (1964).

13. S. Havriliak and S. Negami, Polymer **8**, p. 161 (1967).

14. C. J. F. Bottcher and P. Bordewijk, <u>Theory of Electric Polarization</u>, Elsevier, Amsterdam (1978).

15. J. C. Wang and J. B. Bates, Mat. Res. Soc. Symp. Proc. **135**, p. 57 (1989).

16. J. C. Wang and J. B. Bates, Solid State Ionics **50**, p. 75 (1992).

17. G. H. Vineyard, J. Phys. Chem. Solids **3**, p. 121 (1957).

18. J. M. Wimmer and N. M. Tallan, J. Appl. Phys. **37**, p. 3728 (1966).

19. E. Warburg, Ann. Phys. **6**, p. 125 (1901).

20. J. B. Bates and J. C. Wang, Solid State Inoics **28-30**, p. 115 (1988).

21. J. C. Wang, Electrochim. Acta **38**, p. 2111 (1993).

IMPEDANCE SPECTROSCOPY OF NANOCRYSTALLINE Y-STABILIZED TETRAGONAL ZIRCONIA

P. MONDAL and H. HAHN
Department of Materials Science, Thin Film Division, Darmstadt University of Technology, Petersenstr. 23, D-64287 Darmstadt, Germany

ABSTRACT

Impedance Spectroscopy, X-Ray Diffraction and High Resolution Scanning Electron Microscopy have been used to study the effect of extremely fine grain size on electrical properties such as dc-conductivity and activation energies in nanocrystalline Y-stabilized tetragonal zirconia. The samples were prepared from powders produced by the Inert Gas condensation method. X-Ray Diffraction was used to characterize phase and average grain size of the sample. In addition, grain size and microstructure of the sample was examined using High Resolution Scanning Electron Microscopy. With Impedance Spectroscopy relaxations of O^{2-}-ions in the lattice and in the grain boundaries could be resolved. The dc-conductivities of the lattice and the grain boundaries were deduced from the data. The activation energies for ac- and dc-conductivity of the lattice and grain boundary relaxation are reported.

INTRODUCTION

In conventional polycrystalline materials with grain sizes larger than 1 μm the fraction of atoms in the grain boundaries of the order of 10^{-4} or less is nearly negligible. Nanocrystalline structures exhibit extremely fine grain size and consequently large fractions of grain boundaries. Hence, the properties of these materials are expected to be dominated or at least modified by the properties of the grain boundaries [1]. It has been suggested that high angle grain boundaries in nanocrystalline materials reveal lower atomic density than in microcrystalline materials due to the different rigid body translations required to maximise the atomic density in the nanocrystalline material [1, 2], causing additional property changes.

Because of its nearly pure ionic conductivity microcrystalline Y-stabilized zirconia (YSZ) is a well known material for oxygen sensors and solid oxide fuel cells for many years [3, 4]. Conduction takes place by diffusion of O^{2-}-ions via O^{2-}-vacancies introduced by the substitution of Zr^{4+} by Y^{3+} in the zirconia lattice. Impedance Spectroscopy is therefore a good method to investigate the proposed structural changes in the grain boundaries of the nanocrystalline material by the possibility of measuring the relaxation dynamics of the O^{2-}-ions in lattice and grain boundaries in different temperature and frequency ranges [5].

EXPERIMENT

The ZrO_2- and Y_2O_3-powders used for the preparation of the nanocrystalline YSZ were produced by the Inert Gas Condensation Method [6]. As precursors ZrO and Y were used. The prepared powders were characterized by X-ray Diffraction (XRD) concerning the

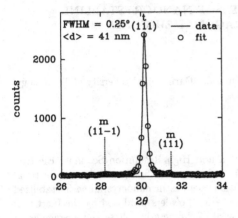

Figure 1: Left: XRD-pattern of a nanocrystalline Y-stabilized zirconia sample with 87 % of the theoretical density after sintering at 1000 °C; right: HRSEM micrograph of a fracture surface of the same sample.

phase and average grain size, which can be deduced from the peak broadening using the Scherrer formula [7]. In addition the specific surface area of the powders was measured by nitrogen adsorption using a method introduced by S. Brunauer, P. Emmet and E. Teller (BET) [8] from which an average particle size can be estimated assuming that the powder consists of spherical monosized particles. Powder mixtures with a Y_2O_3-content of 3 mol% were prepared by dispersing appropriate amounts of powders in water with an ultrasonic treatment. The mixed powders were dried for several days and subsequently pressed into pellets by applying uniaxial pressures of 700 MPa, which yielded green bodies with 49 to 52 % of the theoretical density. The green bodies were sintered under vacuum at 1000 °C for one hour and were oxidised completely in flowing O_2 during the cooling period. Phases present and average grain size of the ceramic samples were characterized by XRD, the microstructure was examined by High Resolution Scanning Electron Microscopy (HRSEM).

The specimens were contacted with silver or gold paste as thin as possible to produce porous contacts which were checked by an optical microscope. The samples were mounted in the pseudo four probe method in a furnace with which measurements from room temperature up to 1100 K were carried out. The impedance of the samples was measured in fixed temperature intervals with the auto balance bridges HP 4284 and HP4285 in the frequency range of 20 Hz to 3 MHz.

RESULTS

Analysis of different charges of powders prepared by the Inert Gas Condensation method with XRD revealed average grain sizes of 5 to 10 nm for the zirconia and 7 to 20 nm for the yttria. The specific surface areas of the powders were determined with the BET method to 150 to 200 m^2/g and 60 to 150 m^2/g for ZrO_2 and Y_2O_3, respectively. From these specific surface areas average particle sizes of 7 to 9 nm for zirconia and 8 to 20 nm for yttria can be calculated. The excellent agreement of the average grain sizes and the average particle

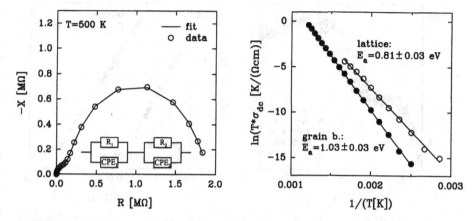

Figure 2: Left: Nyquist-plot of the investigated nanocrystalline Y-stabilized zirconia sample; right: Arrhenius-plot of the dc-conductivities of lattice and grain boundaries.

sizes indicates that the produced powders are not or only weakly agglomerated.

Fig. 1 (left) shows the XRD pattern of a sintered sample with 87 % of the theoretical density. The sample reveals nearly fully tetragonal structure and no indications of a separate yttria phase pointing to a complete interdiffusion of yttria and zirconia. The FWHM of the tetragonal peak is 0.25°. Taking the instrumental broadening into account, an average grain size of 41 ± 5 nm is calculated using the Scherrer formula.

A scanning electron micrograph of a fracture surface of the same sample is shown in Fig. 1 (right). The sample reveals a homogeneous distribution of grain sizes between 30 and 60 nm. Smaller structures visible on the grains result from gold clusters which had to be sputtered on the sample to avoid charging effects. Pores with diameters between 80 and 150 nm are visible on the micrograph. The average grain size calculated from the XRD corresponds to the observed grain size range indicating the significance of the used methods.

The frequency dependent impedance data taken at different temperatures reveal clearly three different relaxation processes which can be attributed to the diffusion of O^{2-}-ions by a vacancy mechanism in the grain interior (lattice process), between neighbouring grains (grain boundary process) and at the contact [5]. The contact process could be identified by measuring samples with different contacts (gold and silver). No discernable differences were observed in the data taken during heating and cooling, which could have been the result of a possible diffusion of gold or silver into the samples during the measurement. Lattice and grain boundary relaxation could be distinguished because the lattice relaxation did not change from sample to sample like the grain boundary process. Fig. 2 (left) shows the imaginary part versus the real part of the impedance (Nyquist-plot) of the sample at 500 K. The lattice relaxation (small semicircle on the left side) and the grain boundary relaxation (large semicircle on the right side) are overlapping and a simple extrapolation of the semicircles to the intersection with the real axis, which corresponds to the dc-resistances of both processes, is not possible. Therefore, the response of an equivalent circuit also shown in the plot was fitted to the data to determine the two dc-resistances R_1 and R_2. The fit result is depicted in Fig. 2 as a solid line. The necessity to use constant phase elments instead

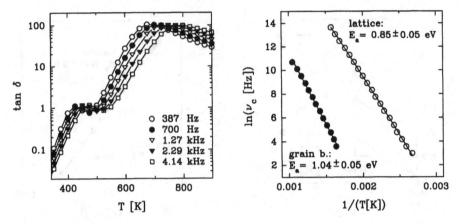

Figure 3: Left: Dielectric loss angle versus temperature of the nanocrystalline Y-stabilized zirconia sample; right: Arrhenius-plot of the characteristic relaxation rates deduced from the tan δ vs. T-plot.

of simple capacitors (impedance of a constant phase element: $Z_{CPE} = A(i\omega)^{-b}$ where ω is the angular frequency and A and $b \in [0, 1]$ are frequency independent constants) indicates a distribution of relaxation times for both the lattice and the grain boundary relaxation, which can be attributed to distributions of activation energies [9, 10, 11].

In Fig. 2 (right) the dc-conductivities of the lattice and grain boundary processes are plotted in Arrhenius representation. Both processes reveal straight lines indicating thermally activated diffusion of the O^{2-}-ions in the lattice and via the grain boundaries. The activation energies are determined as 0.81 ± 0.03 eV and 1.03 ± 0.03 eV for the lattice and the grain boundary dc-conductivity, respectively.

The logarithm of the dielectric loss angle tan δ is shown in Fig. 3 (left) versus the temperature for several frequencies. Relaxations appear in this representation as peaks shifting to higher temperatures with increasing frequency. In analogy to the Nyquist plot in Fig. 2 the relaxation on the lower temperature side corresponds to the lattice and that on the higher temperature side to the grain boundary process. From the peak shift with increasing frequency the temperature dependence of the characteristic relaxation rates ν_c can be deduced. From the linear relation in the Arrhenius representation (Fig. 3 (right)) the activation energies for ac-conductivity can be determined for the lattice process to 0.85 ± 0.05 eV and for the grain boundary process to 1.04 ± 0.05 eV. The attempt frequencies determined from the tan δ - plot are $4.5 \pm 0.5 \cdot 10^{12}$ Hz and $1.8 \pm 0.5 \cdot 10^{10}$ Hz for the lattice and the grain boundary process, respectively. The excellent agreement of the activation energies determined for dc- and ac-conductivity indicate that the observed relaxations are pure conductivity relaxations and that there are no other relaxations with lower activation energies than those observed for dc-conductivity [12].

To compare our dc-conductivities for the present nanocrystalline 3 mol% YSZ with an average grain size of 41 nm with that obtained by other authors for the microcrystalline material all results are summerized in an Arrhenius representation in Fig. 4 [13, 14]. For the

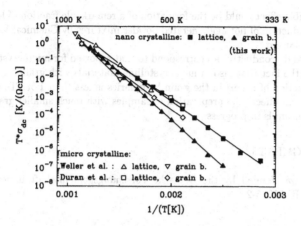

Figure 4: Comparison of the dc-conductivites of the investigated sample with the reported results for microcrystalline samples.

nanocrystalline sample thermally activated behavior could be proved over a conductivity range of seven orders of magnitude which is a wider range than for the literature results. Within the experimental uncertainty there is a good agreement of the activation energies (Weller et al.[13]: 0.92 ± 0.03 eV (lattice), 1.17 ± 0.03 eV (grain boundaries); Duran et al.[14]: 0.82 eV (lattice), 0.98 eV (grain boundaries)). In addition the absolute dc-conductivities are in the same range. The consistency of the activation energies of the nanocrystalline sample with those reported for microcrystalline samples reveal that the same thermally activated processes are responsible for the transport of the O^{2-}-ions what indicates that no structural changes in the lattice or in the grain boundaries occur in the nanocrystalline sample compared to microcrystalline material. In order to compare the dc-conductivities the influence of the porosity in the nanocrystalline samples has to be considered using the "effective medium approximation" [15, 16]. For a sample with 87 % of the theoretical density a factor of about 1.4 can be assesed for typical geometry factors. An increase of the conductivity by this factor would not result in a visible change in the logarithmic plot of Fig. 4.

CONCLUSION

The impedance of a nanocrystalline Y-stabilized zirconia sample with an extremely fine average grain size of 41 ± 5 nm and a density of 87 ± 2 % of the single crystal density has been reported in the frequency range 20 Hz to 3 MHz for temperatures between room temperature and 1100 K.

The relaxation processes of O^{2-}-ions in the lattice and in the grain boundaries could be resolved and the activation energies of these processes have been determined for ac- and dc-conductivity. The activation energies ascertained are consistent with those reported for microcrystalline samples, which implies that lattice and grain boundaries in the nanocrystalline material reveal the same structure as in the microcrystalline material. This result might be a consequence of the agglomeration of the powder mixtures during the dispersion process. A possible effect could be the formation of a remarkable fraction of low angle grain boundaries. Production of not agglomerated powder mixtures by Chemical Vapor Synthesis [6] is therefore planned.

The deduced dc-conductivities correspond to that reported for microcrystalline samples which is due to the fact that also in nanocrystalline materials with grain sizes in the range of 40 nm the fraction of atoms in the grain boundaries is less than 1 %. To study possible grainsize dependent effects the preparation of samples with much smaller grain sizes in the range of 10 to 20 nm is in progress.

ACKNOWLEDGMENTS

This work was funded by the Deutsche Forschungsgemeinschaft under contract no. Ri 510/5-1 and Ri 510/5-2.

REFERENCES

[1] H. Gleiter, Prog. Mat. Sci. **33**, 223 (1998).

[2] M. J. Weins, H. Gleiter and B. Chalmers, J. Appl. Phys. **42**, 2639 (1971).

[3] B. C. H. Steele, in Electro Ceramics, edited by B. C. H. Steele (London, 1991).

[4] C. Wagner, Proc. Int. Commun. Electrochem. Thermo. Kinetics (CITCE), (Butterworth Scientific Publ., London, 1957).

[5] J. E. Bauerle, J. Phys. Chem. Solids **30**, 2657 (1969).

[6] H. Hahn, Nanostruct. Mat. **9**, 3 (1997).

[7] P. Scherrer, Göttinger Nachrichten **2**, 98 (1918).

[8] S. Brunauer, P. Emmet and E. Teller, J. Am. Chem. Soc. **60**, 309 (1938).

[9] K. S. Cole, R. H. Cole, J. Chem. Phys. **9**, 341 (1941).

[10] J. R. Macdonald, Solid State Ionics **13**, 147 (1984).

[11] J. R. Macdonald, J. Appl. Phys. **58**, 1955 (1985).

[12] J. C. Dyre, J. Appl. Phys. **64**, 2456 (1988).

[13] M. Weller and H. Schubert, J. Am. Ceram. Soc. **69**, 573 (1986).

[14] P. Duran, P. Recio, J. R. Jurado, C. Pasual, M. T. Hernandez, C. Moure, J. Mat. Science **24**, 717 (1989).

[15] A. D. Brailsford and D. K. Hohnke, Sol. State Comun. **11**, 133 (1983).

[16] D. S. McLachlan, M. Blaszkiewicz and R. E. Newnham, J. Am. Ceram. Soc. **73**, 2187 (1990).

Microstructural Correlation with Electrical Properties for Y_2O_3 Doped CeO_2 Thin Films

Chunyan Tian and Siu-Wai Chan

*School of Engineering and Applied Science, Materials Science Division,
Columbia University, New York, NY 10027*

Abstract

High quality textured 0.58% Y_2O_3 doped CeO_2 films with (001), (111)/(001) and (110) were prepared using an e-beam deposition technique on substrates of (001) $LaAlO_3$, r-cut sapphire, and fused silica, respectively. The composition and stoichiometry of the films were verified by Rutherford backscattering spectroscopy analysis. Both x-ray diffraction and transmission electron microscopy analyses gave consistent microstructural information. Complex impedance measurements have been performed to study the electrical properties of these films as a function of temperature. The conductivities of the films were dominated by grain boundaries of high conductivities as compared to that of the bulk ceramic of the same dopant concentration. The activation energies for the film conductivities were only slightly higher than that for the bulk lattice conductivities, but much lower than that for the bulk grain boundary conductivity. These results have been discussed in terms of the differences of the grain size and grain boundary microstructures between the films and the bulk ceramics.

Introduction

CeO_2 based electrolytes have been shown to have an exciting prospect when used in oxygen sensors and intermediate temperature (500~700°C) solid oxide fuel cells (SOFC) as compared to the YSZ based devices which operate at 1000°C or higher.[1] Furthermore, thin layers of solid electrolytes can minimize the ohmic loss in the electrolytes so as to lower the operating temperature and maintain the required power output.[2,3,4] Therefore, it is highly desirable to produce thin layers of these materials for solid oxide device applications.

There is a well known, so-called "grain boundary effect"[5], which can lower the effective DC conductivity by 2-3 orders of magnitude[6]. Several studies of this phenomenon have been done for bulk materials, but few deals with thin films. Since the ratio of grain boundary area to lattice volume in thin films is larger than that of bulk ceramics, grain boundaries will have a larger impact on film properties. Thus, it is important to understand the "grain boundary effect" in thin film electrolytes. In addition, the advantage of the microstructural engineering of thin films make it possible to correlate the microstructure with their electrical properties to optimize the film properties for device applications. Here, we prepared yttria doped ceria thin films using an electron-beam deposition technique and investigated the relation between the microstructure and the electrical properties of CeO_2 thin films.

Experimental

Films were grown using the electron-beam deposition technique. The film deposition parameters were described previously.[7] During deposition, oxygen was introduced to fully oxidize the films. After deposition, films were annealed and cooled to room temperature under an oxygen partial pressure (Po2) of 10^{-3} Torr. The Ce, Y, and oxygen contents of the films were determined by Rutherford Backscattering Spectroscopy (RBS) with 2.275 MeV He^{2+}. The microstructures of the as-deposited films were studied using x-ray diffraction and transmission

279

electron microscopy (TEM). TEM samples of the films on r-cut sapphire and LaAlO₃ substrates were prepared by mechanically polishing, dimpler thinning, and ion milling to electron transparency[8]. For films on quartz substrates, the substrates were chemically etched off using hydrofluoric acid.

Complex impedance measurements were performed from room temperature to 500°C with the frequency ranging from 0.1Hz to 10MHz using a Solartron 1260 Impedance/Gain-Phase analyzer. The sample configuration used for the thin film electrical measurement is shown in Fig. 1.

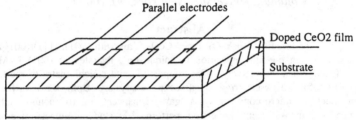

Fig. 1. Geometry of the thin film samples for electrical measurements

Results and Discussion

Figure 2 shows the RBS spectrum of the Y₂O₃ doped CeO₂ film on r-cut sapphire. The actual dopant concentration of the as-deposited film is 0.58% Y₂O₃, although the evaporation target used was 4% Y₂O₃ doped CeO₂.

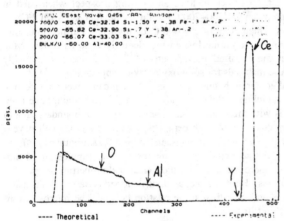

Fig. 2. Rutherford backscattering spectrum of a 1400 Å thick Y₂O₃ doped CeO₂ film

Figure 3 shows x-ray diffraction results of films with the following textures on different substrates: (100) texture on (001) LaAlO₃, (110) texture on fused silica, (100) and (111) mixed texture on r-cut sapphire. The microstructures of the films were investigated with TEM. Bright field TEM images with the imbedded electron diffraction patterns are shown in Fig.4. Many gaps were observed along grain boundaries. These gaps came either from incomplete coalescence during film deposition process or from ion milling during TEM sample preparation due to preferential ion erosion along grain boundaries.

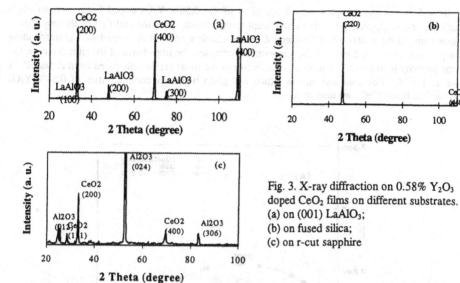

Fig. 3. X-ray diffraction on 0.58% Y_2O_3 doped CeO_2 films on different substrates.
(a) on (001) $LaAlO_3$;
(b) on fused silica;
(c) on r-cut sapphire

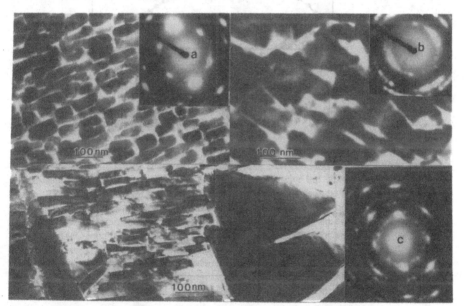

Fig. 4. TEM micrographs of the as-deposited CeO_2 films on different substrates.
(a) on (001) $LaAlO_3$, (001) oriented film with brick-shape grains of 40nm
(b) on fused silica, (110) textured film of average grain size about 100nm
(c) on r-cut sapphire, (100) oriented rectangular grains of 50nm and (111) oriented triangular grains of 0.5μm in size

Figure 5 shows the complex impedance plots of the 0.58% Y_2O_3 doped CeO_2 film on (001) .aAlO3 and r-cut sapphire measured at different temperatures. Z' is the real part of the impedance, orresponding to the resistance of film, and Z" is the imaginary part of the impedance, corresponding ɔ the polarization of the film. As the temperature increases, the impedance of the film decrease, i.e. ıe conductivity increases. The ionic conductivities of the films can be calculated from Z' and Z", as hown in Fig. 6. For comparison, the lattice and grain boundary conductivities of 0.58% Y_2O_3 .oped bulk ceramic CeO_2 are also included.

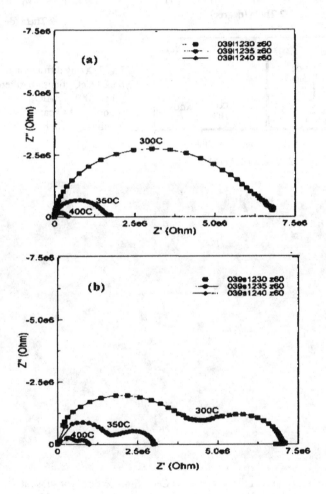

Fig. 5. Complex impedance diagrams of 0.58% Y_2O_3 doped CeO_2 thin films at different temperatures.   (a) on (001) LaAlO3,   (b) on r-cut sapphire

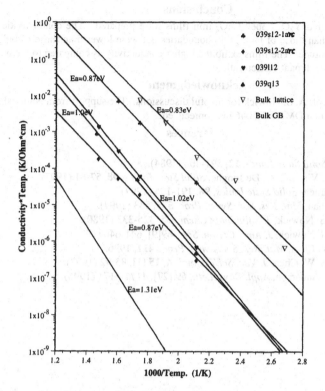

Fig. 6. Product of temperature and conductivity as a function of reciprocal temperatures for 0.58% Y_2O_3 doped CeO_2 films on different substrates

The conductivities of the films on (001) $LaAlO_3$ and fused silica fall between the conductivities of bulk lattice and grain boundary. The film conductivities are about 3 orders of magnitudes higher than the bulk grain boundary conductivity but lower than that of the bulk lattice. These results suggest that the conductivities of the films can be attributed to that of grain boundaries of high conductivities, since nothing related to the lattice conductivity can contribute to the observed difference. This possibility is further supported by the conductivities of the films on (001) $LaAlO_3$ and fused silica being close to that obtained from the second arc (low frequency arc) in the complex impedance of the film on r-cut sapphire. That second arc was identified as a grain boundary arc. The difference between the conductivities of the films and bulk grain boundary may be explained by a grain-size-dependent-grain-boundary-impurity-segregation phenomenon, where the average boundary impurity concentration decreases with increasing grain boundary area.[9] However, it also should be noted that the activation energies of the films were slightly higher than that of bulk lattice but much smaller than that of bulk grain boundary, which cannot be explained only by the size-dependent impurity segregation. This effect may involve the microstructures of the grain boundary, as shown in Fig. 4. The films exhibited highly aligned grains at least in one crystallographic direction contrast to the completely random grain orientations in the bulk.

Conclusions

Highly textured Y_2O_3 doped CeO_2 thin films were prepared. The conductivities of the films were higher than those of bulk grain boundaries but were lower than those of bulk lattice by 2 orders of magnitude. The films exhibited higher conductivity as compared to the bulk grain boundary of the same dopant concentration.

Acknowledgment

We thank Prof. A. S. Nowick for helpful discussions. The support from National Science Foundation under grant DMR-93-50464 is appreciated.

References

[1] B. C. H. Steele, *Solid State Ionics*, **12**, 391-406 (1984).
[2] S. de Souza, S. J. Visco. L. C. De Jonghe, *Solid State Ionics*, **98**, 57-61 (1997)
[3] T. Tsai. S. A. Barnett, *Solid State Ionics*, **98**, 191-196 (1997).
[4] C. Tian, S. W. Chan, *Mat. Res. Soc. Symp. Proc.* **575**, 533 (1997).
[5] D. Y. Wang, A. S. Nowick, *J. Solid State Chem.*, **35**, 325-333 (1980).
[6] R. Gerhardt, A. S. Nowick, *J. Ame. Ceram. Soc.*, **69**[9], 641-646 (1986).
[7] C. Tian and S. W. Chan, *Mat. Res. Soc. Symp. Proc.*, **411**, 1996.
[8] C. Tian, Y. Du, S. W. Chan, *J. Vac. Sci.Technol. A*, **15** [1], 85-92 (1997).
[9] I. Kosacki, H. U. Anderson, *Appl. Phys. Lett.*, **69** [27], 4171-4173 (1996).

CORRELATION BETWEEN ELECTRICAL PROPERTIES AND COMPOSITION / MICROSTRUCTURE OF Si-C-N CERAMICS

C. HALUSCHKA *, C. ENGEL*, R. RIEDEL*, H.-J. KLEEBE **, R. FRANKE***
* Technische Universität Darmstadt, Fachbereich Materialwissenschaft, Fachgebiet Disperse Feststoffe, Darmstadt, Germany
** Universität Bayreuth, Institut für Materialforschung, Bayreuth, Germany
*** Universität Bonn, Physikalisches Institut, Bonn, Germany

ABSTRACT

In this paper we report on the measurement of electrical properties of multielement ceramics in the ternary Si-C-N system using the impedance spectroscopy. The results were correlated to the chemical composition, the hybridization state and the microstructural characteristics investigated by chemical analysis, X-Ray absorption near edge spectroscopy (XANES), Raman Spectroscopy, high resolution transmission electron microscopy (HRTEM) and X-Ray powder diffraction (XRD).

INTRODUCTION

Ceramics obtained by thermolytical conversion of suitable polymers have attracted high attention in the last few years due to their excellent mechanical properties and their high thermal stability [1-3]. Ceramics in the ternary Si-C-N system described here were derived from the commercially available polyhydridomethylsilazane NCP 200[1]. The precursor crosslinked at 350 °C and pyrolyzed at 1000 °C leads to an amorphous ceramic with the composition of $Si_{1.7}C_{1.0}N_{1.5}$ [4-5]. The chemical composition of these materials as well as their structural properties can be controlled by varying the conditions of either the polymer-to-ceramic transformation or the final heat-treatment. At temperatures higher than 1400 °C the ternary silicon carbonitride separates into the binary phases SiC and Si_3N_4 [6].

Also the electrical properties depend very strongly on the synthesis conditions. It will be shown that the electrical properties correlate with composition and microstructure. Therefore, investigation of the electrical behavior can be used as a nondestructive method to characterize these materials.

EXPERIMENT

The impedance was measured in the frequency range between 20 Hz and 30 MHz using autobalancing bridges HP 4284 A and HP 4285 A, Hewlett Packard, USA. These measurements continuously covered the temperature range from 100 K to 1100 K. D.c.- and a.c.-conductivity as well as the permittivity were determined by this technique.

The chemical composition was evaluated by EC 12 for carbon and TC 436 for nitrogen and oxygen by hot gas extraction method, both Leco Corporation, USA. The hydrogen content was determined using the ^{15}N-method provided at the University of Frankfurt, Germany. The energy of the incident nitrogen ions was in the range between 6 - 8 MeV. XANES measurements were performed at the Bonn electron stretcher and accelerator,

---

[1] Nichimen Corp., Japan

Mat. Res. Soc. Symp. Proc. Vol. 500 © 1998 Materials Research Society

Germany. The C_K-spectra were obtained by using a high-energy toroidal grating monochromator in the electron yield mode. The energy range of the incident radiation was 270 - 330 eV, the resolution about 10^{-3}. Raman spectra at room temperature were recorded by a RFA 106 Raman accessory, Bruker, Germany. The Laser excitation was supplied by a Nd:YAG laser, operating with an average power of about 100 mW. Microstructural characterization was performed by transmission electron microscopy using a JEOL 4000 EX microscope, operating at 400 kV. The point resolution was 0.17 nm. For analytic characterization an electron energy-loss spectrometer (PEELS, Gatan 666) fitted to the microscope was used. Under optimized working conditions, an energy resolution of 0.80 eV was achieved. Finally, x-ray powder diffractometry was performed on a D 5000 powder diffractometer, Siemens, Germany.

RESULTS

Fig. 1 represents the electrical conductivity $\sigma_{d.c.}$ measured at room temperature depending on the temperature and time of the heat-treatment in N_2. $\sigma_{d.c.}$ increases with temperature by several orders of magnitude. Related to the phase composition, 3 temperature regions can be distinguished.

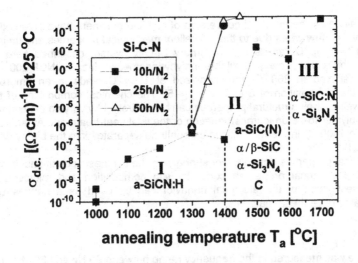

Fig. 1   Electrical conductivity (d.c., T = 25 °C) of Si-C-N ceramics depending on the time and temperature of the heat-treatment in N_2.

In region I there exists only amorphous silicon carbonitride, which contains residual hydrogen (a-SiCN:H). It is assumed that the change of hybridization of the carbon causes the high increase of $\sigma_{d.c.}$ in the temperature range up to 1300 °C. XANES-spectroscopy at the C_K-edge shows an increasing content of sp^2-hybridized carbon (π*-resonance at 285 eV characteristic for C=C bonds [7]) with increasing temperature (Fig. 2). In addition, the amount of C-H bonds (characteristic resonance at 288 eV [7]) is decreasing. The H-content measured by the ^{15}N-method showed a decrease from 5 at.% to

286

1 at.% in the same temperature range. This correlation between H-content, sp^2-hybridization and $\sigma_{d.c.}$ is similar to that reported for amorphous C and SiC [8-10].

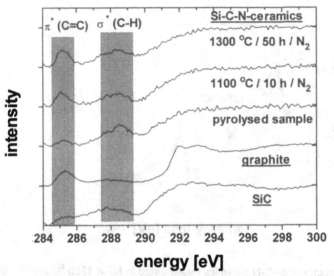

Fig. 2   XANES spectra of Si-C-N ceramics depending on the annealing conditions in comparison to graphite and SiC.

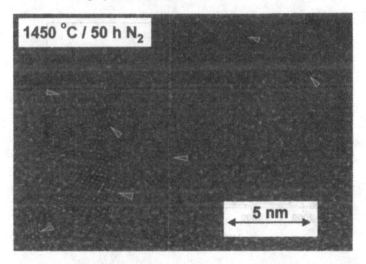

Fig. 3   TEM image of a Si-C-N sample heat-treated at 1450 °C: marked areas are nanocrystalline SiC-particles.

In the temperature region above 1300 °C, a pronounced increase of the conductivity can be measured. While the formation of Si_3N_4 has no influence on $\sigma_{d.c.}$, nanocrystalline SiC-particles (Fig. 3) are assumed to be responsible for the high rise in conductivity due to the formation of conducting percolation paths.

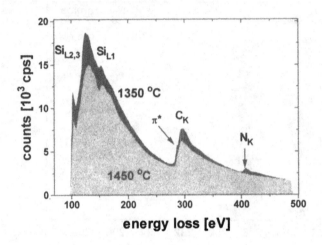

Fig. 4 EELS spectra of Si-C-N ceramics heat-treated in N_2 at 1350 °C and 1450 °C.

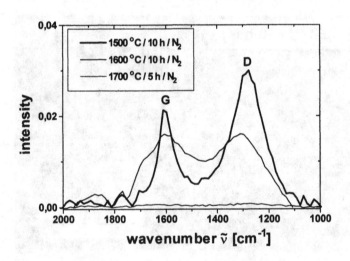

Fig. 5   Raman spectra of Si-C-N ceramics heat-treated in the range between 1500 °C and 1700 °C: G (graphite band), D (defect band).

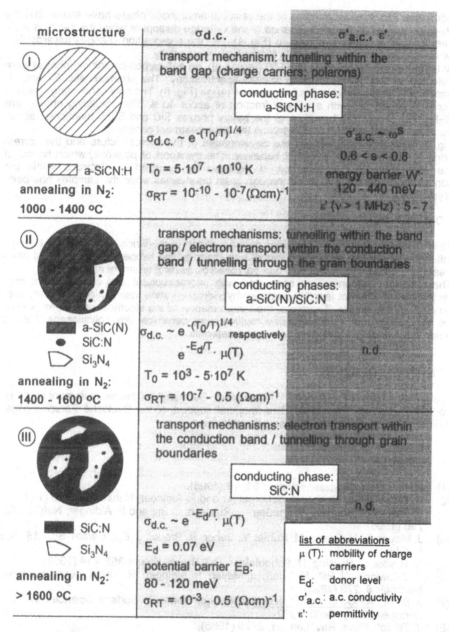

Fig. 6   Correlation between microstructure and electrical properties of Si-C-N ceramics derived from polysilazanes.

Additional EELS-measurements of the residual amorphous phase have shown that the N_K-edge as well as the π^*-resonance of the C_K-edge disappear when the samples were heat treated at higher temperature (Fig. 4), i.e., the composition changes towards SiC, which is thought to be responsible for the high conductivity.

At temperatures higher than 1600 $^\circ$C (region III), no sp^2-carbon containing phases were detected in the ceramics using Raman-spectroscopy. The characteristic G- and D-bands [11] disappeared in this temperature range (Fig. 5). The only conducting phase is nitrogen-doped SiC with a volume fraction of about 40 %. The insulating phases are Si_3N_4 and pores. The formation of the binary phases SiC and Si_3N_4 mentioned above was also proved by XRD depending on the heat-treatment conditions.

Fig. 6 illustrates schematically the development of the microstructure and the corresponding change of the electrical behavior. The transport of polarons, which typical of the amorphous Si-C-N ceramics, is replaced by transport of electrons within the conduction band and tunnelling through grain boundaries when the samples has been crystallized.

CONCLUSIONS

The electrical properties of polymer derived silicon carbonitride ceramics are strongly influenced by the synthesis conditions, especially by the temperature of the final heat-treatment. It was shown that $\sigma_{d.c.}$ can be varied by several orders of magnitude.

The electrical properties are correlated with microstructural and compositional characteristics of different length scales like hybridisation state, distribution and composition of all of the phases. Therefore, the determination of the electrical properties is seen as a powerful tool to characterize multielement ceramics like polysilazane derived amorphous and polycrystalline composite materials.

ACKNOWLEDGEMENTS

We greatfully thank Dr. Balogh (TU Darmstadt, Germany) for the measurement of the H-content using the ^{15}N-method. The financial support of the Deutsche Forschungsgemeinschaft, Bonn, Germany is also acknowledged.

REFERENCES

[1] R. Riedel, Naturwissenschaften **82**, 18 (1995).
[2] R. Riedel, H.-J. Kleebe, H. Schönfelder and F. Aldinger, Nature **374**, 526 (1995).
[3] R. Riedel, A. Kienzle, W. Dreßler, L. Ruwisch, J. Bill and F. Aldinger, Nature **382**, 796 (1996).
[4] J. Mayer, D.-V. Szabó, M. Rühle, M. Seher, R. Riedel, J. Eur. Ceram. Soc. **15**, 703 (1995).
[5] R. Riedel, G. Passing, H. Schönfelder, R.J. Brook, Nature **355**, 714 (1992).
[6] A. Jalowiecki, J. Bill, M. Frieß, J. Mayer, F. Aldinger, R. Riedel, Nanostructured Materials **6**, 279 (1995).
[7] J. Stöhr, NEXAFS Spectroscopy, (Springer Series in Surface Sciences Vol. 25, Springer Verlag, Berlin 1992).
[8] J. Tersoff, Phys. Rev. Lett. **61**, 2879 (1988).
[9] G. Galli, R.M. Martin, R. Car and M. Parrinello, Phys. Rev. Lett. **62**, 555 (1989).
[10] P.C. Kelires, Europhys. Lett. **14** (1), 43 (1991).
[11] J. Robertson, Advances in Physics **35** (4), 317 (1986).

THE USE OF COMPUTER SIMULATIONS TO INTERPRET AND UNDERSTAND ELECTRICAL MEASUREMENTS

Edward J. Garboczi, Building Materials Division, 226/B350, NIST, Gaithersburg, MD 20899.

ABSTRACT

It is very rare to obtain complete 3-D information in the form of images of the microstructure of a material. Most often this information is incomplete because the resolution is inadequate, or is restricted to 2-D, via some kind of micrograph, or is not available at all. In the case of incomplete microstructural information, electrical measurements are then used to try to check a hypothesized microstructure, to see if it can account for the measured electrical response. But even when complete microstructural information is available, if the microstructure is random, then it is not possible to analytically calculate the electrical response of the microstructure. The use of computer simulations, both to generate material shape and topology and numerically solve the electrical equations, is then required. Computer simulations allow the use of more complex hypotheses for the microstructure of a material, as the electrical response can be accurately computed for a wide range of microstructural shapes and topologies.

INTRODUCTION

In materials science, there are really two uses for electrical measurements. The first is to see if a material has the proper electrical properties for a given application. The second use is the general one of trying to discover how the shape, topology, and electrical properties of material phases in a microstructure affect the overall electrical properties. This paper will focus on this second use of electrical measurements.

Sometimes a picture of the microstructure is available, but usually in 2-D and less often in 3-D, and sometimes not at all. When direct information about the microstructure is not available, usually a hypothesis is made about the microstructure, which can hopefully explain the overall electrical behavior seen. This depends on whether or not the electrical response of the hypothesized microstructure can be calculated. Agreement with the measured electrical properties lends support to the hypothesized microstructure. But even when direct images of the microstructure, in 3-D, are available, if randomness or complexity is present there is still a difficulty in knowing how to compute the electrical response. It is not an easy task to go from an image to the electrical response. The general task of going from a structure to an electrical response is especially difficult because the electrical response of a material depends not only on the shape and topology of the phases but also on the material properties of the individual phases.

The reason that this task requires computer simulations is that for many real shapes and topologies, especially in random materials, exact analytical equations cannot be developed. Complex hypotheses about microstructure, which are required for most real materials, are not useful when there is no way of exactly solving a complex hypothesized microstructure for its effective electrical properties. Simple hypotheses have been made that while often useful, cannot capture the full intricacies of a random microstructure. Computer simulations, as will be shown, can, within the limitations of processing power and memory constraints, handle arbitrary shapes, topologies, and electrical properties. It is important to emphasize that the electrical response of a microstructure is indeed a mixture of shape, topology, and electrical properties. Two different topology microstructures can give a similar electrical response depending on the electrical properties of the different phases.

In this paper, "electrical properties" means DC conductivity, with the amplitudes of applied fields small enough so that the material response is always linear. The term "computer simulations" means direct construction of a microstructure in a computer, and then solution of the appropriate electrical equations on this microstructure under some set of boundary conditions. The main kind of simulation that will be discussed in this paper is typified by programs like *dc3d.f*, whose code and operating manual can be found at http://ciks.cbt.nist.gov/garboczi/ [1]. This is a finite-difference program for DC linear 3-D steady-state conduction problems, where the microstructure is described by a digital image. This program is especially adapted for working on 3-D digital images. The problem being solved is given by the equation

$$\nabla \cdot (\sigma \vec{E}) = 0 \qquad (1)$$

where the current flux density is given by

$$\vec{j} = \sigma \vec{E} \qquad (2)$$

so that eq. (1) is the steady state charge conservation equation.

Besides studying random microstructures, one can also be concerned about particle shapes when a second phase is particulate. In the case of very small particles, it may be difficult to directly image them and so their shape must be inferred from property measurements. One can also be concerned about sample and electrode shape. Simple sample shapes are usually preferred, like cylinders, cubes, etc. because when the electrodes are applied uniformly across the ends, the applied field is uniform in the sample. However, it may be experimentally useful to have an "odd" sample and/or electrode shape. The electrical equations will usually not be analytically solvable for a general shape. In this case as well, computer simulations make it possible to understand the response of these kind of shapes. The remainder of this paper will show how computer simulations can be used at the individual particle level, the microstructural level, and the sample level, to compute the overall electrical signal.

INDIVIDUAL PARTICLE LEVEL

When a second phase is particulate, and is dilute, so that its volume concentration is small, say at most 5% or so, then much can be inferred about particle shape and electrical properties from the electrical signal. To be more specific, the case is that a second phase, of some conductivity σ_2 and volume fraction c_2, embedded as isolated particles in a matrix of conductivity σ_1 and volume fraction c_1. We wish to infer the shape and properties of the particles from the overall conductivity, where

σ_1 is known. In this case, much can be done analytically, for certain shapes.

The quantity to be discussed in this case is the *intrinsic conductivity*, $[\sigma]$, defined in the following way [2]:

$$[\sigma] \equiv Lim_{c_2 \to 0} \frac{(\sigma - \sigma_1)}{\sigma_1 c_2} \qquad (3)$$

where σ is the measured conductivity. Equations (1) and (3) can be solved analytically for tri-axial ellipsoidal shapes (which includes spheres) for any value of σ_2. In the limit of σ_2 being zero (insulating particle) or infinity (superconducting particle), other shapes can also be solved analytically [2]. Ref. [2] also shows how many different shapes can be solved numerically, via computer simulation, in the limits of an insulating or superconducting particle. A simple Padé approximant can then be used, along with the simulation or analytical results for the insulating and superconducting limits, to predict the intrinsic conductivity for any value of the relative conductivity σ_2 / σ_1 [3]. The intrinsic conductivity as a function of σ_2 / σ_1 can be also computed via simulation [3].

A simple shape whose intrinsic conductivity cannot be computed analytically is shown in Fig. 1a. This shape is formed from a unit cube but with an m x m (m < 1) square channel cut through the center of each face, leaving only a solid frame. Figure 1b shows a graph of the intrinsic conductivity of this object for m = 0.941, vs. its dimensionless conductivity, $x = \sigma_2/\sigma_1$. In the insulating limit of x, the curve is fairly flat, going to a value of about -1.5. When x becomes very large, in the highly conducting particle limit, the value of the intrinsic conductivity also becomes quite large, around 50. In contrast, for a solid cube or sphere, this limit is only about 3 [2]. Figure 1c shows a linear-log graph of the intrinsic conductivity of the particle, in the superconducting limit, denoted $[\sigma]_\infty$, as a function of the empty air portion of the unit cube from which the particle is made. The air volume fraction of the particle is simply given in terms of m by

$$\textit{Air volume fraction} = m^2 (3 - 2m) \qquad (4)$$

The intrinsic conductivity goes up rapidly with the empty air portion of the particle, becoming very large as the value of m approaches one and the particle shape approaches a wire frame, since the volume fraction of the object, c_2, becomes very small while the particle still has an appreciable effect on the overall conductivity, so that eq. (3) become large.

If one has no idea what the shape of a second phase particle might be, then it is difficult to use the simulation and analytical results, as there are many combinations of shape and σ_2 / σ_1 that can agree with the measured results. However, if one has a hypothesis about the particle shape and electrical properties, then these kind of results can be used as a quantitative test of the hypothesized shape and electrical properties of the second phase. There are many other references in Refs. [2] and [3] that

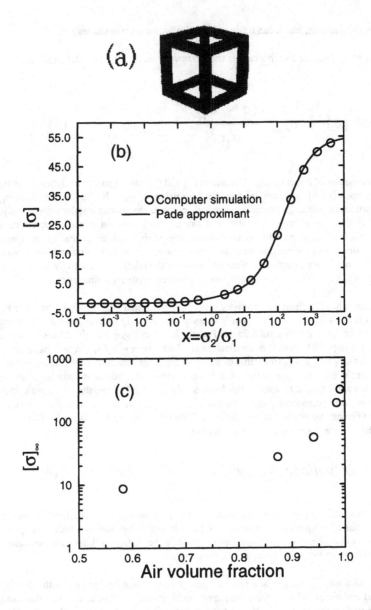

Figure 1: (a) particle shape, (b) intrinsic conductivity of object with m=0.941 vs. ratio of object conductivity to matrix conductivity, and (c) intrinsic conductivity of object in the highly conducting limit, vs. the air content of the object.

detail uses of these results in materials science, and particularly polymer science, where this method of inferring particle properties is commonly employed. One should note that if x is close to 1, then the intrinsic conductivity is rigorously known to be insensitive to particle shape [3]. Also, when x << 1, the intrinsic conductivity is also not very sensitive to particle shape [3], so in these two regimes, determining particle shape from the electrical response would be an even more difficult task.

MICROSTRUCTURAL LEVEL

At the microstructural level, we wish to see if a hypothesized microstructure can produce the measured electrical response. For non-dilute volume fractions or for topological connection of a phase throughout the material (percolation), exact analytical formulas, except in very special cases, are impossible to produce. These special cases include the case when the phases in a microstructure are exactly aligned in series or parallel behavior, or when they follow one of the "ideal" microstructures represented by the Hashin bounds [4]. For random materials, this almost never happens, so that simple hypothesized microstructures, whose electrical responses can be calculated analytically, are not very useful. However, if one produces a complex hypothesized or model microstructure via computer simulation, and then uses computer simulation again to compute the electrical response, progress can be made in understanding the material.

An example of this approach is from cement paste, which is a combination of portland cement and water, that changes from a viscous suspension to a rigid solid via hydration reactions between the solid phase and the water. The solid particles are on the scale of 10-20 μm, and so the pore space, the water-filled space between the particles called the capillary pore space, is also at this scale. The solid phase grows, because the reaction products are less dense than the cement, so that water is consumed as the initial pore space is filled up, resulting in a porosity that decreases with time. The electrical properties come from the fact that the water contains dissolved ions from the cement, making the pore space a conductor. The main reaction product, called CSH, is also porous on the nanometer scale, so that even as it fills up the capillary pore space, it creates a nano-porous network which is also filled with aqueous electrolyte.

Models have been made of this material in 3-D that closely mimic the evolving microstructure. By computing the electrical properties of this model [5-7] one can compare model predictions to experiment and extract information about the microstructure. Figure 2 shows a slice from such a 3-D model, where (a) shows a slice from the simulated microstructure, and (b) shows the electrically conductive phases colored white and gray, and all the insulating phases colored black. There are two conducting phases, the capillary pore space, with a given conductivity (white), and the nano-porous CSH, which is treated as a uniform conductor at the micrometer scale.

At room temperature, the conductivity of the capillary pore space dominates the overall conductivity. It is possible to measure the conductivity of the pure capillary pore space by squeezing out the pore water and measuring its conductivity separately [5]. However, it is not possible to prepare CSH by itself that is equivalent to its form inside the cement paste. There is also no way, in this random material, to analytically extract the two separate conductivities from the overall conductivity.

By using the above model, which gives an accurate description of the microstructure, and comparing

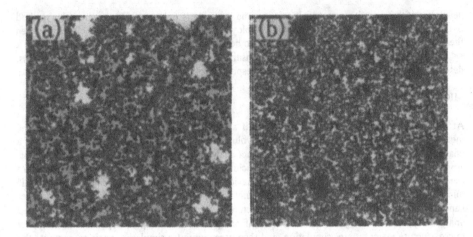

Figure 2: (a) model cement paste microstructure (black = capillary pores, white = unreacted cement, dark gray = CSH, light gray = other insulating reaction product), and (b) map of electrical conductivity (white = capillary pores, gray = CSH, black = insulating phases).

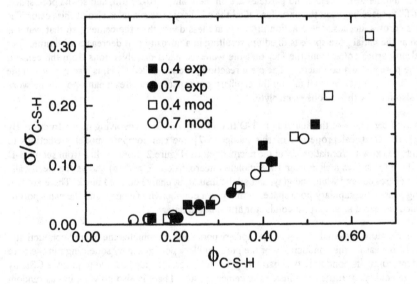

Figure 3: Conductivity at -40°C normalized by the conductivity of the pure CSH phase, for two different cement pastes, mixed with different amounts of cement and water.

to experiment, a fairly accurate value of the conductivity of the CSH phase relative to that of the capillary pore space was obtained [5-7]. In the model, the capillary pore space is given a conductivity of unity, so that the computed conductivity will be relative to the capillary pore space value. Experimentally, the capillary pore space conductivity is measured, as described above, and the overall conductivity normalized by this value. The model is then used to compute the overall conductivity for a range of chosen CSH conductivity values. By matching to experiment, the value of the CSH conductivity at room temperature was found to be about 0.0025-0.01 relative to the capillary pore space [5-7].

When the temperature is lowered below the freezing point of water this situation reverses [8]. By lowering the temperature to -40°C, the water in the larger micrometer scale pores is frozen, while the water in the nanometer scale CSH pores is still mobile. This switches the conductivities of the two phases, even though both are lower than at room temperature, so that now the CSH phase has about 100 times the conductivity of the capillary pore space and has become the dominant conducting phase. However, like at room temperature, the absolute value of the conductivity of the pure CSH is not separately measurable. Using the cement paste model, simulations were run of the frozen cement paste conductivity, normalized by the (unknown) CSH conductivity. This was done by assigning a conductivity of unity to the CSH, and 0.01 to the capillary pore space, and then running the model. These were run for two different combinations of water and cement (ratio of the water mass to the cement mass of 0.4 and 0.7). These results were then compared to experimental data. It was found that by choosing a certain value for the frozen conductivity of the CSH phase, and normalizing the experimental data by this value, the simulation and experimental results agreed very closely. These results are shown in Figure 3. The model also predicted a percolation threshold of about 15% volume fraction for the CSH phase, which is confirmed by Fig. 3. At this temperature, the CSH phase is by far the main conducting phase, so that the conductivity goes approximately to zero at the CSH percolation threshold, which in this case is the signature of the percolation threshold.

This kind of comparison between experiment and model results was also done for electrical measurements at frequencies up to about 10^7 Hz, with some degree of success [9]. See Ref. [9] for details of this work, and Ref. [1] for the details of the computer programs, equivalent to $dc3d.f$, that can be used at finite frequencies.

SAMPLE STRUCTURE LEVEL

In a simple experimental set-up, a cylindrical sample with electrodes covering both ends fully, one can assume that the applied field is uniform over the sample, and it is then easy to convert measured resistances into resistivities or conductivities. Often, however, in electrical experiments, the geometry of the sample or of the applied electrodes is an experimental variable. In this case one cannot simply assume that the applied field is uniform. In these cases, if the material property is known, the effect of the sample and electrode geometry can be predicted via computer simulation, to see quantitatively what effect this geometry has on the results. This use of computer simulation has been carried out recently in a case involving variable electrode geometry.

This case occurred in a study of the effect of electrode geometry on three-point measurements [10,11]. In this kind of measurement, the *working* and *counter* electrodes are used to apply an electrical signal, and the complex voltage is sensed at the *reference* electrode for the purpose of looking at the material-electrode response. The geometry of the working and counter electrodes, and the geometry and position of the reference electrode, was varied across typical experimental geometries to see what effect this might have on properly measuring material-electrode impedances. The samples were cylindrical with a circular cross-section, with the electrodes applied across the ends of the sample. Figure 4 shows two electrode configurations that were investigated numerically, out of the many that were investigated experimentally [10,11]. The electrical field was applied across the CE and WE, and the impedance was measured at the RE. Using a 3-D digital image, the sample and electrode geometry were duplicated, and the impedance response at a number of frequencies was simulated using a finite difference technique applied to the digital image [1,9]. In the computer simulations, a known material-electrode response was used, so that artifacts from the electrode arrangements could be readily seen.

Clear differences were seen experimentally between different electrode arrangements. These artifacts, due to electrode geometry, were confirmed by the computer simulation results. The true material-electrode response, for some arrangements of the electrodes, was not seen because of these distortions. The electrode arrangements, although simple, were such that analytical solutions of eq. (1) were not possible, so that computer simulation played a vital role here in correctly interpreting experimental results. Simple guesses of how a certain electrode configuration would respond turned out to be very misleading. Only computer simulation could be used to sort out the experimental artifacts from the true response.

In some situations, it would be experimentally convenient to have a non-uniform arrangement of electrodes and/or sample shape. This is usually not done because the applied field would not then be uniform. The availability of these computer simulation techniques alleviates this difficulty and makes it possible to be more flexible with electrode and sample shapes.

DISCUSSION

One point that needs further discussion is the electrical signature of a percolation threshold in a material. The connectivity of a phase is important information, and the electrical response is often looked to for obtaining this information. Even after restricting the discussion to two phases, there is still an interplay between two factors: (1) the actual geometrical percolation of either phase, and (2) the electrical conductivity contrast between the two phases, $x = \sigma_2 / \sigma_1$.

Suppose that it is geometrically possible that both phases can become percolated. This will be the case in a material like cement paste, as described above, or in a two-phase polymer blend, or in a random interpenetrating phase ceramic composite [12]. Consider the case when phase 2 is not percolated, but is becoming percolated through some sort of chemical or physical process. Will the percolation of phase 2 be seen in the electrical conductivity of the material? That depends on the value of x. If x is on the order of 100 or more, there will be a sharp upturn in the overall conductivity when phase 2 percolates, thus giving the signature of the percolation threshold (volume fraction of phase 2 at percolation). If x is between 1 and 10, a sharp upturn will not be seen, but only a fairly gradual increase, which will not pick out a percolation threshold. For values between 10 and 100, there may or may not be a sharp increase, depending on the microstructure. For values of x less than

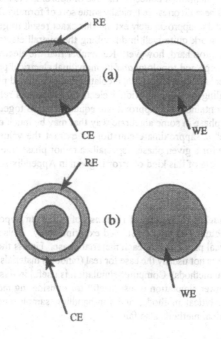

Figure 4: Showing two different experimental electrode arrangements that were investigated via computer simulation [(a) and (b)]. Electrodes are applied to ends of samples, gray = metallic electrode, WE = working electrode, CE = counter electrode, and RE = reference electrode.

one, even for x = 0, nothing will be seen as in 3-D the percolation of one phase does not require the disconnection of another phase. The percolation of phase 2, which is insulating in the case of x = 0, will only decrease the overall conductivity and will not give a sharp downturn. So unless x is large, the geometrical percolation of phase 2 will not have a signature in the conductivity.

This fact is an important point to remember. Suppose one has a material with a second phase in it that one thinks may be becoming percolated. If a percolation threshold is not seen in the electrical conductivity, as denoted by a sharp upturn in the conductivity, phase 2 may be percolated and yet have a low value of x, or it may not be percolated and have a fairly high value of x. Or, if the conductivity is decreasing as the volume fraction of phase 2 increases, it certainly means that x is less than unity, but it does not say anything about the percolation aspects of phase 2.

Another point that should be mentioned concerns the use of approximate analytical expressions like effective medium theories. These theories are typically some sort of formula that is exact in the dilute limit, with statistical ideas used to approximately extend this exact result to general volume fractions. These equations can often work quite well in describing the overall electrical conductivity. This success should not be pushed too hard, however. Remember that the electrical response of a phase is some combination of its shape and topology *and* its individual electrical properties. For example, the above discussion of percolation shows that unpercolated and percolated phases can give the same electrical response, depending on their individual electrical properties relative to the rest of the material in which they are contained. An approximate equation mixes together the geometrical and electrical properties of the phase in some algebraic way that may be much different from reality. If one tries to "deconvolute" an approximate equation to extract the value of volume fraction or electrical property or shape for a given phase, especially a minor phase, the answer one gets can be seriously in error. An example of this kind of error is given in Appendix A of Ref. [13].

CONCLUSION

A calculation must be made of the electrical response of either an hypothesized model for the microstructure, or of an actual 3-D image obtained experimentally. This calculation involves the shape, topology, and electrical properties of each material phase. Unless the phases are arranged in a simple arrangement, which is not usually the case for real (random) materials, this calculation cannot be made exactly by analytical methods. Computer simulation is useful in this case for computing the electrical response. Computer simulation is also useful for considering individual particles whose shape precludes exact analytical methods, and for handling sample and electrode shapes and arrangements where analytical methods also fail.

ACKNOWLEDGEMENTS

I would like to thank my collaborators at NIST, J.F. Douglas and D.P. Bentz, and my collaborators at Northwestern University, T.O. Mason and H.M. Jennings, for their contributions to this general problem, and the NSF Center for Advanced Cement-Based Materials, for partial funding of this work.

REFERENCES

1) E.J. Garboczi, *Finite element and finite difference programs for computing the linear electric and elastic properties of digital images of random materials*, NIST Internal Report, in press (1998); available at http://ciks.cbt.nist.gov/garboczi/, Chapter 2.
2) J.F. Douglas and E.J. Garboczi, Adv. in Chem. Phys. **91**, 85-153 (1995); available at http://ciks.cbt.nist.gov/garboczi/, Chapter 11.
3) E.J. Garboczi and J.F. Douglas, Phys. Rev. E **53**, 6169-6180 (1996).
4) Z. Hashin, J. Appl. Mech. **50**, 481-505 (1983).
5) B.J. Christensen, T.O. Mason, H.M. Jennings, D.P. Bentz, and E.J. Garboczi, "Experimental and computer simulation results for the electrical conductivity of portland cement paste," in *Advanced Cementitious Systems: Mechanisms and Properties*, edited by F.P. Glasser, G.J. McCarthy, J.F. Young, T.O. Mason, and P.L. Pratt (Materials Research Society Symposium Proceedings Vol. 245, Pittsburgh, 1992), pp. 259-264; also at http://ciks.cbt.nist.gov/garboczi/, Chapter 5.

6)   E.J. Garboczi and D.P. Bentz, J. Mater. Sci. **27**, 2083-2092 (1992); also at http://ciks.cbt.nist.gov/garboczi/, Chapter 5.

7)   R.T. Coverdale, B.J. Christensen, T.O. Mason, H.M. Jennings, E.J. Garboczi, and D.P. Bentz, J. Mater. Sci. **30**, 712-719 (1995); also at http://ciks.cbt.nist.gov/garboczi/, Chapter 5.

8)   R.A. Olson, B.J. Christensen, R.T. Coverdale, S.J. Ford, G.M. Moss, H. M. Jennings, T.O. Mason, and E.J. Garboczi, J. Mater. Sci. **30**, 5078-5086 (1995); also at http://ciks.cbt.nist.gov/garboczi/, Chapter 5.

9)   R.T. Coverdale, B.J. Christensen, T.O. Mason, H.M. Jennings, and E.J. Garboczi, J. Mater. Sci. **29**, 4984-4992 (1994); also at http://ciks.cbt.nist.gov/garboczi/, Chapter 5.

10)   G. Hsieh, T.O. Mason, E.J. Garboczi, and L. R. Pederson, Sol. St. Ionics **96**, 153-172 (1997).

11)   G. Hsieh, D.D. Edwards, S.J. Ford, J.-H. Hwang, J. Shane, E.J. Garboczi, and T.O. Mason, *Electrically Based Microstructural Characterization Vol. 411*, edited by R.A. Gerhardt, S.R. Taylor, and E.J. Garboczi (Materials Research Society, Pittsburgh, 1996), pp. 3-12.

12)   D.R. Clarke, J. Amer. Ceram. Soc. **75**, 739-759 (1992).

13)   E.J. Garboczi and D.P. Bentz, "Multi-scale analytical/numerical theory of the diffusivity ofconcrete," Adv. Cem.-Based Mater. **8**, 77-88 (1998).

CHARACTERIZING THE DISPERSION OF CONSTITUENTS IN CONCRETE BY ELECTRICAL RESISTIVITY

PU-WOEI CHEN, XULI FU AND D.D.L. CHUNG,
Composite Materials Research Laboratory, State University of New York at Buffalo
Buffalo, NY 14260-4400

ABSTRACT

Characterization of the degree of dispersion of constituents in cement paste or mortar by DC electrical resistivity measurement was demonstrated for the constituent being much more conducting than cement (i.e., short carbon fibers) and much less conducting than cement (i.e., latex particles). The fiber dispersion is described by the ratio of the measured conductivity to the calculated value. The latex particle dispersion is described by a factor in an electrical conduction model.

INTRODUCTION

The dispersion of a component in a concrete mix is critical to the effectiveness of the component for reinforcing (as in the case of fibers), for permeability reduction (as in the case of a polymer) or for other functions. In spite of the mixing, the dispersion is never perfect, though agglomeration is avoided. The use of microscopy to characterise the degree of dispersion of constituents in concrete is difficult and tedious. Furthermore, it is ineffective for distinguishing between small differences in the degree of dispersion. Although the degree of dispersion affects the mechanical properties, the mechanical properties cannot be used to indicate the degree of dispersion. This is because of the multiplicity of factors other than the degree of dispersion that also affect the mechanical properties. These factors include bond strength, void content, void size, etc. On the other hand, for components that are either much more or much less electrically conducting than the other components, their degrees of dispersion can be indicated by the electrical resistivity, provided that they do not agglomerate. For a component that is much more conducting than the other components, the greater is its degree of dispersion, the lower is the resistivity, provided that they do not agglomerate. For a component that is much less conducting than the other components, the greater is its degree of dispersion, the higher is the resistivity. This is the notion behind the dispersion characterization method described in this paper. Although the contact resistivity of the interface between the component and the cement matrix also affects the resistivity, the effect is negligible when the difference in volume resistivity between the component and the cement matrix is sufficiently large. The fact that the addition of sand to cement paste essentially does not change the volume resistivity of the composite (10^5 Ω.cm) allows the method of dispersion characterization by resistivity measurement to be applicable whether sand is present or not. The further addition of a coarse aggregate increases the resistivity (10^7 Ω.cm), so its presence may complicate this method for characterizing the dispersion of a component which is much less conducting than the cement matrix. A component which is much more conducting than the cement matrix is carbon fibers. A component which is much less conducting than the cement matrix is a polymer, such as latex. This work uses these two components to illustrate the method of dispersion characterization by resistivity measurement.

DISPERSION OF SHORT CARBON FIBERS

The carbon fibers were of diameter 15±3 μm, length ~ 5 mm and resistivity 3 x 10^{-3} Ω.cm, as provided by Ashland Petroleum Co. (Ashland, KY). The matrix was portland cement, Type I, from Lafarge Corporation (Southfield, MI); the mean particle size was from 0.1 to 0.2 μm. The resistivity of the cement paste (with a water/cement ratio of 0.32, corresponding to a slump of 150 mm) at 7 days of curing was 1.49×10^5 Ω.cm. The additional filler was silica fume (Elkem Materials, Inc., Pittsburgh, PA; EMS 965; average particle size 0.15 μm, particle size range 0.03-0.5 μm; 94% SiO_2) used in the amount of 15% of the cement weight. The addition of silica fume to the cement paste (with the water/cement ratio increased to 0.35 in order to maintain the slump at 130 mm) slightly increased the resistivity to 2.32×10^5 Ω.cm when the fibers were absent. This implies that the additional filler and the matrix were about the same in resistivity. However, when the fibers were present, the addition of silica fume decreased the resistivity, as described below.

Because the fiber dispersion is enhanced by a small amount (0.4% of cement weight) of methylcellulose (Dow Chemical Corp., Midland, MI; Methocel A15-LV), methylcellulose was used when the fibers were present, whether or not silica fume was present. The defoamer used with methylcellulose was Colloids 1010 (Colloids, Inc., Marietta, GA) in the amount of 0.13 vol.%. The methylcellulose had no effect on the resistivity when the fibers were absent.

To investigate the effect of the particle size on the effectiveness of a filler in increasing the conductivity, sand (coarser than silica fume) was used in place of silica fume. It was natural sand (100% passing 2.36 mm sieve, 99.91% SiO_2). Its addition to cement paste (without silica fume or fibers) was associated with an increase of the water/cement ratio from 0.32 to 0.475, a resulting increase of the slump from 150 to 180 mm, and essentially no change in the resistivity. Thus, the resistivity of the sand was essentially the same as that of the cement paste matrix. Both types of additional filler were the same as the cement matrix in resistivity.

The water/cement ratio and water-reducing-agent/cement ratio were chosen to increase with increasing fiber volume fraction in order to maintain the slump from 77 to 200 mm. The water reducing agent was TAMOL SN powder (Rohm and Haas), which contained 93%-96% sodium salt of a condensed naphthalenesulfonic acid. Mixing was conducted by using a Hobart mixer with a flat beater. Methylcellulose was dissolved in water and then the fibers and the defoamer were added and stirred by hand for 2 min. Then this mixture, cement, water and the water reducing agent (and sand and/or silica fume, if applicable) were mixed in the mixer for 5 min. After pouring the mix into oiled molds, a vibrator was used to reduce air bubbles.

The volume resistivity at 7 days of curing were measured by the four-probe method, using silver paint for electrical contacts. The resistivity decreased much with increasing fiber volume fraction, whether a second filler was present or not. Fig. 1(A) shows the resistivity and Fig. 1(B) shows the measured conductivity as a fraction of the calculated value obtained from the Rule of Mixtures by assuming that the fibers were continuous and parallel along the axis of the conductivity measurement. This fraction describes the degree of fiber dispersion. At a given fiber volume fraction, it was higher when silica fume was present, whether sand was present or not. When sand was absent, silica fume did not affect the percolation threshold (1 vol.%), but increased the fibers' effectiveness. When sand was present, silica fume diminished the percolation threshold. The addition of sand without silica fume increased the threshold. In spite of the smaller proportion of silica fume (silica fume/cement ratio of 0.15) compared to sand (sand/cement ratio of 1.5), silica fume was much more effective, due to its smaller particle size. Consistent with the improved fiber dispersion was the increased flexural toughness and strength due to the presence of silica fume [1].

304

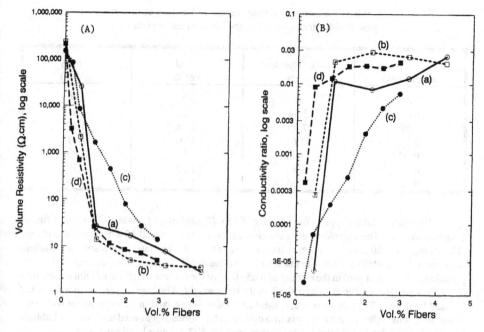

Fig. 1(A)   Variation of the volume electrical resistivity with carbon fiber volume fraction.

(B)   Variation with the carbon fiber volume fraction of the ratio of the measured volume electrical conductivity to the calculated value. The matrix conductivity used in the calculation was the measured conductivity for the case without fibers but containing the corresponding additives.

(a)   Without sand, with methylcellulose, without silica fume,
(b)   Without sand, with methylcellulose, with silcia fume,
(c)   With sand, with methylcellulose, without silica fume,
(d)   With sand, with methylcellulose, with silica fume.

DISPERSION OF LATEX PARTICLES

The addition of a polymer to concrete reduces permeability, improves vibration damping ability [2], increases flexural toughness [1] and increases bond strength between reinforcement (fibers or rebars) and the concrete [3-5]. Among the polymers, latex (styrene butadiene) is particularly common for concrete. Its volume resistivity is 10^{14} Ω.cm, compared to 10^5 Ω.cm for plain cement paste. In spite of the properties of latex-modified cement, the microstructure of this material has not been clarified. Most attention regarding the microstructure has been given to the voids, the content of which decreases with increasing latex/cement ratio. The polymer (latex) is a distinct phase, but its observation by microscopy is difficult, partly due to the water in the cement paste and partly due to the small particle size (0.2 µm) of latex.

305

Table 1. Latex content, void content and density of cement
pastes with various values of the latex/cement weight ratio

Latex*/cement weight ratio	Latex dispersion vol.%	Void vol.%	Density (g/cm³)
0	0	2.32	1.99
0.05	3.64	2.07	1.95
0.10	7.35	1.88	1.90
0.15	11.07	1.70	1.87
0.20	14.91	1.53	1.83
0.25	18.81	1.25	1.79
0.30	22.55	1.10	1.76

Portland cement (Type I) from Lafarge Corp. (Southfield, MI) was used without fine or coarse aggregate. The water/cement ratio was 0.23, except that it was 0.45 when latex was absent. The slump was 160 mm. No water reducing agent was used. The latex was a styrene butadiene dispersion (Latex 460NA, Dow Chemical Corp., Midland, MI; 48 wt.% solid, density 1.01 g/cm³, particle size 0.19-0.21 µm) in the amount of 0.05 - 0.30 of the cement weight, used with an antifoam (Dow Corning 2210) in the amount of 0.5% of the latex weight. The latex and antifoam were mixed by hand for 1 min. Then this mixture, cement and water were mixed in a Hobart mixer with a flat beater for 5 min. After pouring the mix into oiled molds, a vibrator was used to reduce air bubbles. Specimens were demolded after 1 day and then cured at 40% relative humidity for 28 days.

The air content was measured on cement pastes using ASTM C185-91a, modified for the absence of sand and the presence of latex. This method also gave the latex volume fraction and the density. The volume electrical resistivity was measured by the four-probe method, using silver paint for the electrical contacts. Table 1 shows that the density decreases with increasing latex/cement ratio, in spite of the decrease in void content. This is because the density of the latex dispersion (1.01 g/cm³) is lower than that of cement (3.15 g/cm³). Fig. 2 shows that the volume resistivity of cement paste increases with latex/cement ratio.

Assuming that latex and cement phases are in parallel (Fig. 3(a)) and neglecting the conductivity of latex compared to that of cement, the resistivity of latex-modified cement $\rho_{||}$ is ρ_c/V_c, where ρ_c and V_c are the resistivity and volume fraction of the cement phase respectively. Assuming that latex and cement phases are in series (Fig. 3(b)) and neglecting the conductivity of latex, the resistivity of latex-modified cement ρ is ∞. In reality, the situation is between these two extremes. Fig. 3(c) is an electrical conduction model of the real situation. In this model, the latex and cement phases are partly in parallel (e.g., sections 1, 3 and 4 of Fig. 3(c)) and partly in series (e.g., sections 2 and 5 of Fig. 3(c)). The sections which are in series do not contribute to conduction, so the effective volume fraction of the cement phase is $(1-f) V_c$, where f is the volume fraction of the cement phase which belongs to the sections that are in the series configuration. The resistivity of the latex-modified cement ρ is $\rho_c/(1-f)V_c$. The quantity f describes the degree of latex dispersion. For a given polymer, f increases with the polymer content. The value of f is obtained, if ρ, ρ_c and V_c are known; V_c is obtained by deducting the latex and void volume fractions (Table 1) from unity.

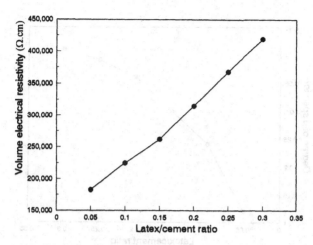

Fig. 2   Effect of the latex/cement ratio on the volume electrical resistivity.

Figure 4 shows that f increases more rapidly with the latex/cement ratio at low values of this ratio than at high values of this ratio. This is because a high ratio corresponds to a situation in which some of the sections of Fig. 3(c) (such as Section 5) have more than one horizontal band of latex. A section does not contribute to conduction as long as it has at least one band. Therefore, f underestimates the true degree of dispersion when the ratio is high, particularly > 0.2.

CONCLUSION

Characterization of the dispersion of constituents in cement paste or mortar by DC resistivity measurement was demonstrated for the constituent being much more conducting than cement (i.e., short carbon fibers) and much less conducting than cement (i.e., latex particles). The degree of fiber dispersion is described by the conductivity ratio, which is the ratio of the measured conductivity to the calculated value obtained from the Rule of Mixtures by assuming that the fibers were continuous and parallel along the axis of conductivity measurement. The addition of a discontinous filler (silica fume) to a composite with short carbon fibers enhanced fiber dispersion when the fiber volume fraction was less than 3.2%. The maximum fiber volume fraction for silica fume to be effective for

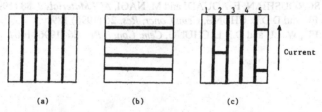

(a)   (b)   (c)

Fig. 3   Electrical conduction model of in latex-modified cement paste. (a) Latex and cement phases in parallel. (b) Latex and cement phases in series. (c) Latex and cement phases partly in series and partly in parallel. Dark bands : latex phase. Remaining areas : cement phase.

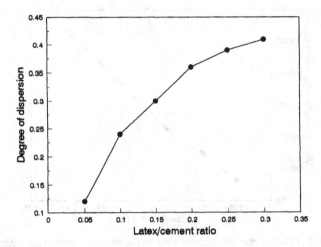

Fig. 4   Degree of dispersion (f) of latex as a function of the latex/cement ratio.

enhancing fiber dispersion was only 0.5% when the additional filler was sand (much coarser than silica fume). The silica fume addition did not affect the percolation threshold, but the sand addition increased it. The degree of dispersion of latex particles in latex-modified cement paste was assessed by measurement of the resistivity and modeling it in terms of latex and cement phases that were partly in series and partly in parallel. The assessment was best at low values of the latex-cement ratio; it underestimated the degree of dispersion when this ratio was high, especially > 0.2.

ACKNOWLEDGEMENT

This work was supported in part by National Science Foundation.

REFERENCES

1.   P. CHEN, X. FU and D.D.L. CHUNG, *ACI Materials J.* **94** (1997) 147.
2.   X. FU and D.D.L. CHUNG, *Cem. Concr. Res.* **26** (1996) 69.
3.   P. SOROUSHIAN, F. AOUADI and M. NAGI, *ACI Materials J.* **88** (1991) 11.
4.   X. FU and D.D.L. CHUNG, *Cem. Concr. Res.* **26** (1996) 189.
5.   X. FU, W. LU and D.D.L. CHUNG, *Cem. Concr. Res.* **26** (1996) 1007.

ELECTROCHEMICAL CORROSION OF ELECTRODES IN A SIMULATED NUCLEAR WASTE GLASS MELT

S.K. SUNDARAM
Pacific Northwest National Laboratory
Richland, WA 99352, sk.sundaram@pnl.gov

ABSTRACT

Corrosion of potential candidate electrode materials, molybdenum and tantalum, in a simulated nuclear waste glass melt was investigated using electrochemical (dc-powered) and non-powered tests. Electrochemical corrosion data showed that tantalum was more corrosion-resistant than molybdenum in this melt. Tantalum also showed passivation. The non-powered test data showed that tantalum corroded more than molybdenum. This was attributed to penetration of protective passivation layer at the tantalum-glass interface by the glass melt. Microstructural features and chemistry across selected electrode-glass interfacial regions supported these results.

INTRODUCTION

An integrated waste approach was evaluated under the Department of Energy's Landfill Focus Area (formerly the Minimum Additive Waste Stabilization or MAWS program). The approach consisted of blending waste streams of different compositions to reduce the need for additives to produce a stable, high quality waste form. The program considered vitrification technology for stabilization of these wastes.

Vitrification is a well-established technology for processing temperatures below 1150°C due to service temperature limitation of the commonly used electrode, Inconel 690. As some of the blended wastes processed at temperatures higher than 1200°C, an investigation was undertaken in the FY 95 to extend the waste vitrification technology to high temperatures. Additionally, greater electrode life at processing temperature in the range of 1100°C was also considered.

Corrosion of materials of construction of melter for nuclear waste vitrification had been investigated since the 70's[1-14]. Electrochemical corrosion of electrodes for high temperature vitrification was first reported in 1995[15]. The authors demonstrated the prospect of electrochemical protection of electrode materials in contact with highly corrosive high-level waste simulated glass melts. In the present paper, electrochemical corrosion of candidate electrode materials in a simulated nuclear waste is reported and the electrochemical corrosion tests data are compared to non-powered tests data.

EXPERIMENTAL

Candidate Electrode Materials

Potential candidate electrode materials were molybdenum and tantalum. Molybdenum was chosen, as it was the most used melt-contact material for electrode/booster applications in commercial glass melters. Tantalum was chosen based on thermodynamic stability of its oxide and the prospect of electrochemical protection[15].

Glass Composition and Preparation

The selected glass composition represented a waste glass composition from Idaho site. The glass was prepared using oxides, carbonates, and hydroxides. The glass was melted in an electrically heated furnace and quenched by pouring on a stainless steel plate. Then, the glass was crushed into – 40 mesh and used for the corrosion tests. The targeted glass composition and corresponding chemical analytical results are presented in Table I.

Table I Glass Composition

Oxides	Target	Analysis (normalized)
Al_2O_3	9.91	12.22
B_2O_3	0.20	0.15
CaO	19.26	21.09
Fe_2O_3	3.71	3.80
K_2O	6.02	3.75
Li_2O	1.07	1.21
MgO	5.41	5.59
Na_2O	5.41	5.27
PbO	1.34	1.06
SiO_2	39.12	39.58
ZnO	4.53	4.75
Others	4.02	---
Total	100.00	---

Testing

Electrochemical (dc-powered): A three-electrode configuration was used. A molybdenum or tantalum rod was used as the working electrode. Two platinum rods were used as the counter and reference electrodes. The electrodes were passed through protective alumina tubes and one end of the tubes was sealed using a high temperature silicone sealant. At the other end, 1-cm long electrode ends were not protected for immersing them into a test glass melt. 150-g pre-melted glass powder was placed in an alumina crucible, which was, in turn, contained in a fused silica ceramic crucible. The double-crucible assembly was placed in a furnace. When the glass melt reached 1300°C, the electrodes assembly was lowered into the melt until the electrodes were about 0.5 in. from the bottom of the crucible. The assembly was then connected to a potentiostat (Solartron 1287) and a frequency response analyzer (Solartron 1260). First, open-circuit voltage (OCV) of the cell was measured for 60 minutes to ensure the stability of the cell. Then, impedance data were collected to determine the electrolytic resistance for correcting the potentiodynamic scans data. Finally, potentiodynamic scan data were collected to determine the electrochemical corrosion rate. The potentiodynamic data was plotted on linear scale and the polarization resistance was determined.

Non-powered: A molybdenum or tantalum rod was passed through a protective alumina tube and one of it was sealed using a high temperature sealant. Then, 1-cm long sample at the unsealed end of the protection tube was immersed into the test glass melt maintained at 1300°C and allowed to corrode for 24 and 72 hours of duration. The melt was contained in a Monofrax K3 crucible. After corrosion tests, the radius loss was measured. In the selected cases, the material-glass interface was characterized using scanning electron microscopy (SEM) and energy dispersive electron spectrometry (EDS). The radius loss data were converted into corrosion current density values using the equation, i_{ox} = (CR $_{weight}$ $_{loss}$ x d)/(3.27 x 10^{-3} x EW), where CR = corrosion rate in mm/year, d = density in gm/cm^3, and EW = equivalent weight. These values will be compared with current density values from electrochemical tests.

RESULTS

Potentiodynamic Scan Data

The potentiodynamic scan data with molybdenum and tantalum are presented in Figure 1. Molybdenum did not show any passivation. This was attributed to its thermodynamically less stable oxide, MoO$_3$. Th onset of passivation of tantalum at about –0.875 V (vs. reference) was observed. This data indicated that tantalum showed promise for electrochemical protection by passivation in this simulated waste glass melt.

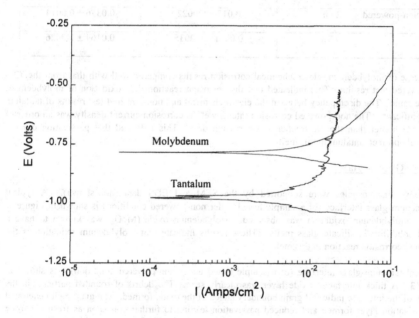

Figure 1. Polarization Curves of Candidate Electrode Materials

Electrochemical vs. Non-powered

The electrochemical test results were compared to the non-powered test results as shown in Table II. In cases of molybdenum and tantalum, the 72-hours non-powered test corrosion rate as well as corrosion current density was less than the values from 24-hours tests. This was attributed to rapid initial corrosion of electrodes followed by saturation of the melt with the corrosion product (oxide in this case) in the vicinity of electrode-glass interface and reduction of further corrosion.

Table II Electrochemical (dc-powered) Vs. Non-powered Corrosion Data

Corrosion	Molybdenum	Tantalum
Corrosion rate (mm/year)		
Electrochemical	61.88	22.17
Non-powered 24 h	131.40 ± 16.79	267.91± 9.86
72 h	54.75 ± 36.65	116.80 ± 18.98
Corrosion current density (A/cm^2)		
Electrochemical	0.008	0.003
Non-powered 24 h	0.017 ± 0.0022	0.0376 ± 0.0014
72 h	0.0071 ± 0.0035	0.0164 ± 0.0026

In the case of molybdenum, electrochemical corrosion results compared well with that from the 72-h non-powered test results. This indicated that the corrosion reaction, i.e., oxidation of molybdenum, was the same. The discrepancy between the electrochemical and non-powered test results of tantalum was significant. The non-powered corrosion rate as well as corrosion current density was an order of magnitude higher than the electrochemical corrosion data. This indicated that passivation did not effectively protect tantalum in this melt.

Electrode-Glass Interfaces

The above discrepancies were supported by the SEM and EDS data (not shown). A typical molybdenum-glass interface (of a sample tested under non-powered condition) is shown in Figure 2. Limited molybdenum oxidation was observed. Molybdenum oxide (MoO$_2$) was known to have a limited solubility in silicate glass melts. These results indicate that molybdenum oxidation is the governing corrosion reaction in this melt.

A typical tantalum-glass interface (of a sample tested under non-powered condition) was shown in Figure 3. A thick interfacial oxide layer was clearly seen. The debris of rounded particles in the vicinity of the interface indicated grain boundary attack of the oxide formed. The glass melt penetrated the passivation layer formed and reduced passivation leading to further corrosion as fresh electrode surface was exposed to the glass melt. This corroborates well with the observed higher corrosion of tantalum in the melt.

312

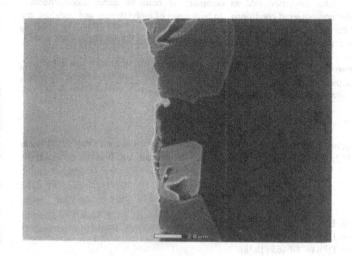

Figure 2 Molybdenum-Glass Interface
(Non-powered test, 1300°C, 72 h; Left: Molybdenum; Right: Glass)

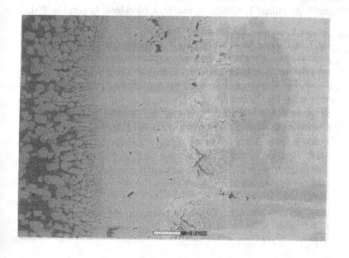

Figure 3 Tantalum-Glass Interface
(Non-powered test, 1300°C, 72 h; Left: Glass; Right: Tantalum)

CONCLUSIONS

1. Molybdenum showed higher corrosion rate as compared to tantalum under electrochemical conditions, but not under non-powered conditions. In both cases of molybdenum and tantalum, the reduction in observed corrosion rate with time in non-powered tests was due to saturation of the melt with the corrosion reaction product in the vicinity of the electrode-glass interface.

2. Though tantalum showed passivation in this melt, the tantalum-glass interface indicated grain boundary attack of the passivation layer formed at the interface. Penetration of the layer by the glass melt reduced passivation resulting in higher corrosion.

ACKNOWLEDGEMENTS

Financial support by Associated Western Universities, Inc., Northwest Division (AWU NW) under Grant DE-FG06-89ER-75522 or DF-F51 with the U.S. Department of Energy is gratefully acknowledged.

REFERENCES

1. G. G. Wicks, Report No. DPST-77-374 (1977).
2. G. G. Wicks, Report No. DP-MS-78-90 (1978).
3. G. G. Wicks, Report No. DPST-78-0465 (1978).
4. G. G. Wicks, Report No. DPST-79-526 (1979).
5. R. A. Palmer, Report No. RHO-SA-173 (1980).
6. G. G. Wicks, Report No. DPST-79-0580 (1980).
7. W. N. Rankin, Report No. DPST-81-0344 (1981).
8. M. J. Plodnic and K. R. Routt, Report No. DPST-80-494 (1980).
9. M. J. Plodnic and K. R. Routt, Report No. DPST-80-654 (1980).
10. R. D. Dierks, G. B. Mellinger, F. A. Miller, T. A. Nelson, and W. J. Bjorklund, Report No. PNL-3406 (1980).
11. C. L. Timmerman, Report No. D7H35-84-13 (1984).
12. P. J. Hayward, I. M. George, M. P. Woods, and T. S. Busby, Glass Tech., 28, 43 (1987).
13. A. S. Polyakov, G. B. Borisov, Z. S. Khasanov, and N. I. Moiseenko in Waste Management'89, (Vol.1, Editors: R. G. Post and M. E. Wacks, 1989), p.37.
14. G. B. Borisov, A. S. Polyakov, and Z. S. Khasanov in Waste Management'90, (Vol.1, Editors: R. G. Post and M. E. Wacks, 1990), p.39.
15. S. K. Sundaram, C. J. Freeman, and D. A. Lamar in Corrosion of Materials by Molten Glass, (Editor: G. A. Pecoraro, The American Ceramic Society, Ohio, 1995), pp. 311-412.

EXTENDED D.C. ELECTRICAL TRANSPORT MEASUREMENTS ON THE MIXED CONDUCTOR CU₃CS₂

P. K. LEMAIRE[*], J. BENOIT[*][#], AND R. SPEEL[**]
[*] Central Connecticut State University, New Britain CT 06050
[#] United Technlogies Research Center, East Hartford CT 06108
[**] Greenwich Air Services, East Granby CT 06026

ABSTRACT

D.C. electrical transport measurements have been done over the temperature range 200 K to 450 K on the mixed conductor $Cu_{3.0}CS_2$. This work extends the original work done on Cu_xCS_2 over the temperature range 260 K to 350 K. Above 220 K, the voltage versus time curves follow the Yokota model for mixed conductors. Below 220K, the voltage versus time curves were practically constant, suggesting very little ionic transport below this temperature, and an electronic conductivity of the order of 10^{-5} $(\Omega \, cm)^{-1}$ at 200 K. At ambient temperatures, the ionic conductivity and electronic conductivity were both of the order of 10^{-3} $(\Omega \, cm)^{-1}$, and the chemical diffusion coefficient found to be of the order of 10^{-6} $cm^2 s^{-1}$, in agreement with earlier work on Cu_3CS_2. Above 220 K, the ionic conductivity versus temperature plots were of the Arrhenius form with an activation energy of about 0.36 eV. The jump time and residence time were estimated to be of the order of 10^{-12}s and 10^{-6}s respectively, confirming hopping as the mode of ionic transport. The electronic conductivity versus temperature plot confirmed thermal activation as the mode of electronic transport. The results suggest Cu_xCS_2 to be very stable and the Yokota model, with very little modification, to be very reliable for the analysis of these mixed conductors.

INTRODUCTION

The study of solid electrolytes and mixed conductors continue to produce interesting science and useful new technologies. Applications include high specific energy batteries and fuel cells, gas sensors, ion-selective membranes, phosphors and laser materials [1,2,3,4]. One of the best known application of solid electrolytes is the sodium-sulfur battery which uses sodium β-alumina as the electrolyte and operates at about 300°C and above[1]. There is the obvious need for a room temperature high specific energy battery based on ambient temperature solid electrolytes and mixed-conducting materials. The mixed conductor, Cu_3CS_2, presented in this work is one of good ambient temperature mixed conductors, M_xCS_2 [5,6], being studied in our laboratory. Kuo, Cappelletti and Pechan [6] measured the D.C. transport properties of Cu_xCS_2 over the temperature range 260 K to 350 K and analyzed data following Yokota's model for mixed conductors as modified by Dudley and Steele [7]. They found Cu_xCS_2 (x = 2.8 to 3.6) to have room temperature ionic conductivities of about $3X10^{-3}$ $(\Omega \, cm)^{-1}$ which was somewhat dependent on copper concentration. The ionic conduction and electronic conduction activation energies for Cu_xCS_2 were found to be about 0.40 eV and 0.30 eV respectively. The ionic conduction activation energy for Cu_xCS_2 was also independent of x. This work extends the measurements of Cu_3CS_2 over a wider temperature range, 200 K to 450 K, clarifies the transport mechanisms, and also addresses the stability of these materials. It is also shown that the data can be analyzed simply following Yokota [8].

EXPERIMENTAL

The experimental set-up consisted of a CTI-Cryogenics closed cycle refrigerator, a Keithley 220 Programmable Current Source, a Keithley 196 Digital Multimeter, a Keithley 705 Programmable Scanner, a Lakeshore Cryotronics Model 330 Autotuning Temperature controller, all connected to a computer controller through a GPIB 488 bus. The routines for data acquisition and device control were home written in the National Instruments LabVIEW environment. Trial measurements were conducted under computer control to verify the overall software and instrument operation as well as the accuracy of the timed intervals. Confirmation of the timing intervals was necessary since a run of the sample from 250 K to 450 K required 28 days of mostly unattended operation.

The sample holder comprised of a mica - copper base, and spring loaded platinum current and voltage probes. For good thermal contact as well as electrical insulation between the sample and the cold finger of the closed cycle refrigerator, the sample was mounted on mica which in turn was mounted on a copper base. The sample was 4.57 mm in length and 1.92×0.54 mm in cross section. The current density was kept below approximately 1×10^{-6} A/cm^2 to prevent the plating out of the conducting ions on the electrodes. The distance between the voltage probes was 2.16 mm and they were approximately centered between the current probes to allow for a more accurate four point voltage measuring technique and to avoid the areas close to the current electrodes which may exhibit any unusual effects. Voltage versus time measurements were carried out for an applied constant current. After the voltage had reached steady state, the applied current was shut off and the decay voltage characteristics versus time were measured. Measurements were taken from 200 K to 450 K in 10 degree increments on the sample of Cu_3CS_2. The sample was allowed to equilibrate at each temperature set point for one hour prior to any measurements being taken. The sample used was one of the same samples produced by Kuo's group in 1983.

RESULTS AND DISCUSSION

There are a number of well established methods for studying transport mechanisms in ionic conductors. The one used in this project is the polarization technique based on a theory originally developed by Yokota [8]. In this method one can measure the chemical diffusion coefficient, the total conductivity and the partial conductivities of the mobile species. The technique also lends itself readily to automatic control which is especially indispensable for the study of rapidly diffusing ions. Electronic probes (platinum) were used instead of ionic probes since ionic probes are susceptible to decomposition and often lead to large errors in the measured quantities [9]. Furthermore, measurement of the ionic conductivity component using electronic probes have been shown to give the same results as the ionic probes [7, 9, 10].

The Yokota model suggests that for a small constant current density in a mixed conductor with electronic probes, the voltage across the material is given by:

$$V(t) = V_m - A \exp(-t/\tau) \qquad \text{eq. (1)}$$

316

where V(t) is the time dependent voltage, V_m is the maximum voltage, t is the time in seconds and τ is the time constant. When the current is turned off after reaching steady state, the voltage decay across the material is given by:

$$V(t) = V' \exp(-t/\tau). \hspace{4cm} \text{eq. (2)}$$

The total conductivity (σ_o) is given by:

$$\sigma_o = jl/V_o = \sigma_i + \sigma_e \hspace{4cm} \text{eq. (3)}$$

where, j is the current density, V_o is the voltage at t = 0 when a non-zero current is applied to the sample, l is the distance between the voltage probes, σ_i, and σ_e are the ionic, and electronic conductivities respectively. The electronic component of the total conductivity can be calculated using:

$$\sigma_e = jl/V_m \hspace{4cm} \text{eq. (4)}$$

Finally the chemical diffusion coefficient, D_e , can be obtained from the relation:

$$D_e = L^2/(\tau\pi^2) \hspace{4cm} \text{eq. (5)}$$

where L is the length of the sample. Thus, the Yokota model provides a simple and effective method of determining the total, ionic, electronic conductivities as well as the diffusion coefficient of some mixed conductors.

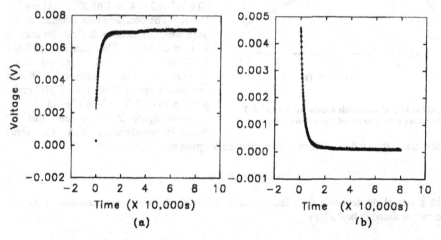

Figure 1: (a) Voltage vs. time for the build-up case. (b) Voltage vs. time for the decay case (circles are data points and lines are the fits)

As seen from figure 1, the voltage versus time curves follow the Yokota model. Using Sigma Plot™ software, standard curve fitting routines were written and used to evaluate the time constants for the voltage-time relationships. The extracted time constants, τ, ranged from 1.42 X 10^4 s at 250K to 5.03 X 10^2 s at 450K. Taking V_o as the minimum voltage with current applied, the applied current, I, the distance between the voltage probes, l, and the cross sectional area, A, the total conductivities, σ_o, were calculated using equation 3. In the Dudley and Steele modification of the Yokota model it was suggested that V_o be taken as the voltage at t = $\tau/3$ since the exponential build up of the voltage begins around that time. In our work, the voltages were measured as soon as the current was turned on, and the onset of exponential build up of voltages were found to be about at the minimum voltage for the build up case. This experimental approach to measuring V_o we believe is accurate, easier to obtain and in line with the Yokota model. The total conductivities were found to be from 4.7 X 10^{-4} $(\Omega$ cm$)^{-1}$ at 250 K to 1.9 X 10^{-1} $(\Omega$ cm$)^{-1}$ at 450 K. Below 220 K the conductivities were electronic in nature since the voltage versus time plots were practically constant. The conductivity at 200 K was found to be of the order of 6 X 10^{-5} $(\Omega$ cm$)^{-1}$.

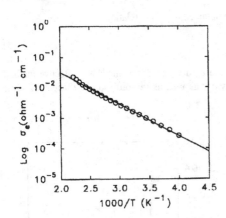

Figure 2: Log. of electronic conductivity vs 1000/T (circles are data points and line is curve fit)

Using the final voltage, V_m, the electronic conductivities, σ_e, were calculated using equation 4. The electronic conductivities ranged from 2.57 X 10^{-4} $(\Omega$ cm$)^{-1}$ at 250 K to 2.22 X 10^{-2} $(\Omega$ cm$)^{-1}$ at 450 K. The ionic conductivities were obtained by subtracting the electronic conductivities from the total conductivities. The ionic conductivities ranged 2.1 X 10^{-4} $(\Omega$ cm$)^{-1}$ at 250 K to 1.63 X 10^{-1} $(\Omega$ cm$)^{-1}$ at 450 K. The conductivities obtained in this work were consistent with those obtained by Kuo et.al. [6]. The chemical diffusion coefficients, D_c, were obtained using equation 5. The chemical diffusion coefficients ranged from 1.49 X 10^{-6} cm^{2}s^{-1} at 250 K to 4.22 X 10^{-5} cm^{2}s^{-1} at 450 K.

From figure 2, it is seen that the electronic conductivity over the entire temperature range follows a thermal activation model given by:

$$\sigma_e = \sigma_\infty \exp(-E/kT)   \qquad \text{eq. (6)}$$

with a very slight deviation from linearity around 400 K. The electronic activation activation energy was found to be 0.21 eV.

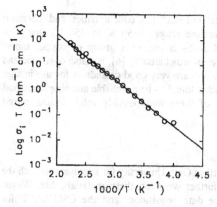

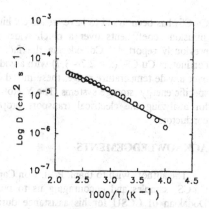

Figure 3: Log. of ionic conductivity x temperature vs. 1000/T (circles are data points and line is the curve fit)

Figure 4: Log. of chemical diffusion coefficient vs. 1000/T (circles are data points and line is curve fit)

Figure 3 shows the ionic conductivity versus temperature plot to be of the Arrhenius form and is given by:

$$\sigma_i T = C \exp (-E/kT)$$ eq. (7)

suggesting hopping as the mode of ionic transport, with an activation energy $E = 0.36$ eV which compares well with 0.4 eV obtained by Kuo et al. A slight change in the variation of ionic conductivity with temperature is observed around 400 K. Using the activation energy, the pre-exponent of the ionic conductivity C obtained from the ionic conductivity versus temperature plot, and basic statistical physics and assuming a simple harmonic approximation, a jump time and residence time of the order of 10^{-12} s and 10^{-6} s respectively were calculated, which also support hopping as the mode of ionic transport. From figure 4, the chemical diffusion coefficient versus temperature fits an Arrhenius equation of the form:

$$D_c = D_{co} \exp(-E/kT)$$ eq. (8)

where D_{co} is the chemical diffusion coefficient prefactor, E is the activation energy, k is Boltzmann's constant. D_{co} and E were found to be 4.2×10^2 cm^{2}s^{-1} and 0.15 eV respectively. The diffusion coefficient versus temperature plot also shows a slight change in its variation at about 400 K. At the low temperature end, the onset of a drop in ionic diffusion is seen.

The change in variation in the ionic conductivity, electronic conductivity, and chemical diffusion coefficient with temperature at 400 K suggests a possible solid to solid transition at this temperature. A DSC measurement is being done to verify this observation.

CONCLUSIONS

Cu_3CS_2 has been found to have relatively high ionic and electronic conductivities, and chemical diffusion coefficients over a much wider temperature range (250 K to 450 K) than was previously reported. Considering the fact that Cu_3CS_2 is one of a group of copper mixed conductors Cu_xCS_2 (x = 2.7 - 3.7) with a wide range of stoichiometry [6], is stable over time and over a wide temperature range, these mixed conductors are very good candidates for use in high specific energy storage systems. The Yokota model is found to be a reliable and simple method for analyzing the electrical transport properties of these and possibly other similar mixed conductors.

ACKNOWLEDGEMENTS

The authors wish to thank Prof. Ron Cappelletti of Ohio University for supplying us with the Cu_xCS_2 samples and encouraging us to pursue further work on these materials, Mr. Wayne Dodakian of CCSU for his assistance during the data acquisition, and the CSU/AAUP for providing a grant in support of this work.

REFERENCES

1. D.F. Shriver and G.C. Farrington, *C&EN*, May 20, (1985) 42.

2. J.T. Brown, *IEEE Trans. Energy Conv.*, **3**, 2, (1988) 193.

3. W. Drenckhahn, *Power Eng. J.*, April, (1996) 67.

4. A. Jones, M. Donoghue, "United States and Russia To Co-Develop Clean, Efficient Fuel Cell Technologies" http://www.etde.org/html/doe/whatsnew/pressrel/pr96137.html September, (1996).

5. P.K. LeMaire, E.R. Hunt and R.L. Cappelletti, *Solid State Ionics*, **34**, (1989) 69.

6. H. Kuo and R.L. Cappelletti, *Solid State Ionics*, **24**, (1987) 315.

7. G. J. Dudley and B.C. H. Steele, *J. Sol. St. Chem.*, **31**, (1980) 233.

8. I. Yokota, *J. Phys. Soc. Japan*, **16**, 11, (1961) 2213.

9. G.J.Dudley and B.C.H.Steele, *J. Sol. St. Chem.*, **21**, (1977) 1

10. Isaaki Yokota, *J. Phys. Soc. Japan*, **8**, (1953) 595

STABILITY AND CONDUCTIVITY OF $Gd_2((Mo_{1/3}Mn_{2/3})_xTi_{1-x})_2O_7$

J.J. Sprague, O. Porat, H. L. Tuller
Crystal Physics and Electroceramics Laboratory, Department of Materials Science and
Engineering, Massachusetts Institute of Technology, Cambridge, MA, 02139, USA

ABSTRACT

A composite solid state electrochemical device, with $(Gd_{1-x}Ca_x)_2Ti_2O_7$ serving as the
electrolyte and $Gd_2(Ti_{1-x}Mo_x)_2O_7$ (GT-Mo) as the anode has recently been proposed. The latter
exhibits high levels of mixed conduction under reducing atmospheres, but decomposes at high
P_{O2}. We have recently succeeded in extending the stability limits of the GT-Mo to higher P_{O2}
with the addition of Mn. In this study, we report on the conductivity and stability of
$Gd_2((Mo_{1/3}Mn_{2/3})_xTi_{1-x})_2O_7$ (GMMT) as a function of P_{O2}, T, and composition utilizing
impedance spectroscopy and x-ray diffraction. The addition of Mn extends the stability region
of the material to $P_{O2} = 1$ atm with little change in the magnitude of the conductivity. Defect
models explaining the dependence of the conductivity on oxygen partial pressure are presented.
Preliminary results from the use of an electronic blocking sandwich cell used to isolate the
ionic conductivity of GMMT are also presented.

INTRODUCTION

In recent years, the development of a monolithic solid oxide fuel cell has received
considerable attention [1-3]. In the monolithic design, a specific crystalline structure and
phase serves as a template for the cell. The composition is then spatially modulated to achieve
the desired functionality for each component of the cell. This kind of cell offers enhanced
thermal, mechanical, and chemical stability with respect to traditional multi-phase cells.

Past work [4-6] has shown that the pyrochlore system, $Gd_2Ti_2O_7$ (GT), exhibits the
versatility needed to make this monolithic cell. It has been shown that acceptor doped GT
(with Ca for example) shows high ionic conductivity ($\sigma_i \geq 10^{-2}$ at 1000 °C) with negligible
electronic conductivity over a very wide range of oxygen partial pressure (P_{O_2}) and temperature
[4], making it suitable as the electrolyte. On the other hand, recent work on Mo doped GT
(GT:Mo) indicates that this material exhibits high electronic conductivity ($\sigma_e \geq 10^{1.5}$ at 1000
°C) in addition to high ionic conductivity ($\sigma_i \geq 10^{-1}$ at 1000 °C) [6]. Mixed ionic and
electronic conduction (MIEC) has been found to be desirable for achieving good electrode
performance. GT:Mo was found to be stable under reducing atmospheres ($P_{O_2} \leq 10^{-15}$ atm),
making it a suitable choice as anode. However, it decomposed under the oxidizing conditions
necessary for the cathode. This decomposition was attributed to the variable valent Mo
oxidizing from +4 to +6 accompanied by a sharp increase in oxygen interstitial concentration.

The goal of this work is to fabricate and investigate stable pyrochlore materials with
adequate levels of MIEC that remain stable in oxidizing environments and are suitable as
cathode materials. With GT:Mo as the starting material, we co-dope with variably valent Mn
to stabilize the Mo oxidation at high P_{O_2}. The goal is to compensate the transition of Mo^{4+} to
Mo^{6+} with the transition of valence in Mn^{4+} to Mn^{3+}, which gives $Gd_2(Mo_{1/3}Mn_{2/3})_xTi_{1-x})_2O_7$
(GMMT). This compensation suppresses the oxygen interstitial level and allows the material
to remain stable to higher P_{O_2}. To maintain charge neutrality, twice the concentration of Mn is

added as compared to Mo (hence the Mo is 1/3 of the total dopant concentration, Mn is 2/3). At the stoichiometric point of the material, the two defects fully compensate, e.g.:

$$2\left[Mo_n^{\bullet\bullet}\right] = \left[Mn_n^{'}\right] \tag{1}$$

In order to characterize and understand the conduction mechanisms in GMMT, 2 probe AC impedance scans were made as a function of P_{O_2}, T, and dopant concentration, x. These are then compared against a theoretically generated defect model for the system to ascertain the microscopic conduction mechanisms.

EXPERIMENT

Powders of $Gd_2((Mo_{1/3}Mn_{2/3})_xTi_{1-x})_2O_7$ were made via a citric acid based liquid organic preparation route [7] for the compositions x = 0.05, 0.1, and 0.2. Disk shaped samples (average area of 0.3 cm^2 and length of 0.2 cm) were first uni-axially pressed at 5000 psi for 15 minutes, isostatically pressed at 40,000 psi for 10 minutes, and then sintered at 1600 °C for 16 hours in air. Electrodes were constructed by applying layers of Pt paste and firing at 1000 °C for 20-30 minutes.

AC impedance scans were performed with a Solartron 1260 Frequency Response Analyzer over the frequency regime, 0.1 Hz to 1 MHz and under the conditions 600 °C $\leq$ T $\leq$ 1000 °C and 10^{-25} atm $\leq P_{O_2} \leq$ 1 atm. Fitting of the AC impedance data was performed with the commercially available "Zview" software program from Scribner Associates, Inc.

Stability investigations were performed by x-ray powder diffraction. Patterns were recorded on all samples after being sintered in air. For the x = 0.1 and x = 0.2 samples, patterns were taken on samples quenched after an extended anneal at 800 °C at 10^{-20} atm.

RESULTS

The total conductivity of GMMT is shown as a function of T and P_{O_2} for x=0.05 (figure 1), x=0.1 (figure 2), and x=0.2 (figure 3). Under oxidizing conditions, the conductivities of the x = 0.1 and 0.2 samples are roughly P_{O_2} independent, while the conductivity of the x = 0.05 conductivity drops with a slope of approximately -1/2. At intermediate P_{O_2}'s (10^{-5} atm < P_{O_2} < 10^{-15}), the x = 0.05 and 0.1 compositions exhibit a P_{O_2} independent conductivity, while the x = 0.2 composition shows a decline in conductivity. At sufficiently low P_{O_2}, the x = 0.05 and 0.1 both show increasing conductivity with decreasing P_{O_2} with the x = 0.1 sample showing an approximate slope of -1/6.

The temperature dependent conductivities of all three compositions measured at 10^{-5} atm (in the P_{O_2} independent regime) are shown in figure 4. The x = 0.05 and x = 0.1 samples show an activation energy of approximately 0.55 eV. The x = 0.2 sample lies about an order of magnitude higher in conductivity than either the x = 0.1 or the x = 0.05 compositions and has an activation energy of 0.44 eV.

X-ray diffraction patterns of samples with x = 0.05, 0.1, and 0.2 show them to be pyrochlore after sintering in air. Measurements on a x = 0.3 sample, however, found it to be significantly multi-phase. The x = 0.1 and x = 0.2 samples were also quenched from 800 °C and 10^{-20} atm to determine phase stability under reducing conditions. The x = 0.1 remained pyrochlore, while the x = 0.2 sample showed significant second phase (tentatively assigned to $GdTiO_3$). In addition, after the x = 0.2 sample was reoxidized in air at 800 °C, and the original conductivity under these conditions was not recovered.

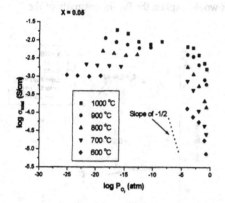 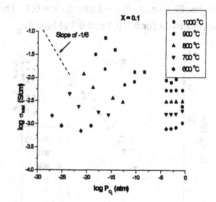

Figure 1. Plot of total conductivity of GMMT x=0.05 as a function of T and P_{O_2}. The slope at high P_{O_2} corresponds roughly to -1/2.

Figure 2. Plot of total conductivity of GMMT with x=0.1 as a function of T and P_{O_2} ; P_{O_2} independent at high P_{O_2}, and with a slope of -1/6 at low P_{O_2}.

DISCUSSION

The addition of Mn to GT:Mo (forming GMMT) extends the stability of the material system to much higher P_{O_2}. The materials studied here (x = 0.05, x = 0.1, x = 0.2) all have a stable pyrochlore phase in air as confirmed by x-ray diffraction, whereas the GT:Mo decomposed at $P_{O_2} > 10^{-15}$ atm [6] for Mo concentrations above 10%. The highest Mo concentration in GT that was still stable in air was 10%. It's conductivity at 900° C was about $10^{-2.5}$ in air. As seen in figure 3, the highest attainable conductivity in air has been extended by over an order of magnitude (with the x = 0.2 GMMT).

Defect chemical models were developed to aid in understanding the P_{O_2} and T-dependent conductivites observed in figures 1-4. We take into account the major defect reactions (Frenkel, Redox, Mn and Mo ionization, and intrinsic electron-hole generation) and charge neutrality to solve for the various defect concentrations as a function of P_{O_2}, T, and composition. Previous data on doped GT were used to make approximations for the various equilibrium constants [8]. The analytic approach used by Porat and Tuller [8] was used to solve for the various defect concentrations. The results are shown in figure 5 for x = 0.05 at 1000 °C.

We now examine the data for x = 0.05 (figure 1) in light of the defect diagram (figure 5). The key features of the conductivity data are a P_{O_2}-independent plateau bounded at high and low P_{O_2} by P_{O_2}-dependent regime with negative slopes. The charge carriers with these

323

characteristics (see figure 5) include oxygen vacancies ($V_O^{\bullet\bullet}$) and electrons (n). The plateau region at 1000 °C for vacancies is predicted to fall at $10^{-28} \leq P_{O_2} \leq 10^{-22}$ atm while that for electrons at $10^{-20} \leq P_{O_2} \leq 10^{-7}$ atm. The data of figure 1 is in much better agreement with predictions for electrons than vacancies.

Extending the model to higher values of x predicts that the I-II boundary should shift to higher P_{O_2}, i.e., to $P_{O_2} \approx 1$ atm for x = 0.1. This would explain the P_{O_2}-insensitivity of the conductivity for x = 0.1 and 0.2 at high P_{O_2}.

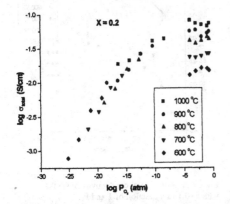 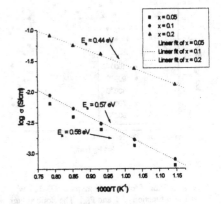

Figure 3. Plot of the total conductivity of GMMT with x = 0.2 as a function of P_{O_2} and T. The conductivity is P_{O_2} independent at high pressures. The decline under reducing conditions is due to phase decomposition of the material.

Figure 4. Arrhenius plot of the conductivity of GMMT for x = 0.05, 0.1, 0.2 taken at 10^{-5} atm (Regime II in figure 5). The linear fits provide the following activation energies shown.

We note from figure 4 that $\sigma(x = 0.1) \approx 1.34 \, \sigma(x = 0.05)$. If Mn is viewed as a deep donor in the gap and weakly ionized, then n should follow be given by [9]:

$$n = \left(\frac{N_C N_D}{2}\right)^{1/2} \exp\left(-E_D/2kT\right) \tag{2}$$

where N_C is the effective density of conduction band states, N_D is the dopant density, and E_D is the donor ionization energy level. According to this equation, n should be proportional to $[Mn_{Ti}]_{Total}^{1/2}$. In the present case, the carrier ratio should be $\sqrt{2} = 1.41$ close to the value of 1.34 above. Also, we can extract the position of the Mn level relative to the conduction band edge, i.e. $E_{a, meas.} = E_D/2$, or $E_D = 1.1$ eV.

However, in going from x = 0.1 to 0.2, the conductivity jumps by a factor of approximately 10. We suspect that at this juncture, a shift in mechanism occurs. At x = 0.2, the Mn impurity band apparently becomes sufficiently wide to support conduction within the band. Under these circumstances, the measured activation energy of 0.44 eV would represent a hopping energy rather than an ionization energy. This hypothesis requires further investigation.

Although the above defect model is generally in agreement with the experimental data, there are some problems. The first is that the slope for electrons in figure 1 at high P_{O_2} (-1/2) is steeper than that predicted in the model (-1/4). A steeper slope would be consistent with the oxygen interstials not being fully ionized as is commonly assumed. In addition, the reason why the x = 0.05 sample does not exhibit a stronger P_{O_2} dependence at lower P_{O_2} while the x = 0.1 sample does needs further explaination.

Conduction in GMMT has been shown to be primarily electronic in nature. Preliminary measurements using an electron blocking cell to isolate the ionic conductivity indicate that the ionic conductivity represents about 10% of the total conductivity. Further, we find the ionic conductivity to increase with increasing P_{O_2}, consistent with an interstitial mechanism.

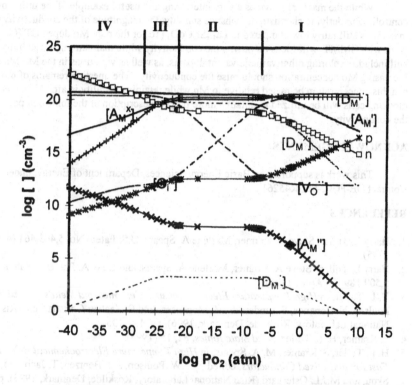

Figure 5. Defect model of GMMT for x = 0.05 at 1000 °C. The concentration of various defects is plotted vs. P_{O_2}. The various charge dominant charge neutrality regimes are shown: I. $\left[Mo_{Ti}^{\bullet\bullet}\right] \approx \left[O_i^{\prime\prime}\right]$, II. $2\left[Mo_{Ti}^{\bullet\bullet}\right] \approx \left[Mn_{Ti}^{\prime}\right]$, III. $2\left[V_O^{\bullet\bullet}\right] \approx \left[Mn_{Ti}^{\prime}\right]$, and IV. $n \approx 2\left[V_O^{\bullet\bullet}\right]$. It is explained in the text that the conductivity is most likely electronic and mirrors the dependency of n. The slope of n in the various regimes is: I. -1 /4, II. 0, III. -1/4. IV. -1/6.

CONCLUSION

A conductive pyrochlore material stable in air, $Gd_2((Mo_{1/3}Mn_{2/3})_xTi_{1-x})_2O_7$, has been fabricated and examined. Conductivity values as high as 10^{-1} S/cm were achieved for $x = 0.2$ at 900 °C in air, substantially higher than achieved with a Mo substituted pyrochlore material in air. The Mn is proposed to provide a buffering action for the Mo oxidation, thereby suppressing the oxygen interstial concentration at high P_{O_2} and effectively extending the stability regime of the material to higher P_{O_2}. A defect model, largely consistent with electrical conductivity data, was developed which points to Mn as a donor state 1.1 eV below the conduction band. A sharp jump in the conductivity between $x = 0.1$ and 0.2 suggests the formation of an impurity band at higher x values.

While the material provides a very interesting and useful example of the utility of controlling underlying chemistry to enhance stability, the magnitude of the conductivity of the material is still fairly low compared to $La_{1-x}Sr_xCoO_3$ [10] or the pure Mo doped GT [6]. Future work with GMMT will focus on raising the conductivity while maintaining it's stability. This will include exploring other variable valent dopants, as well as variations in the Mo/Mn ratio. Increasing Mo concentration should raise the conductivity. The question remains of how high can this concentration be raised relative to Mn while retaining stability in air. Also, electron blocking cells will be utilized to obtain a more precise description of the ionic component of the conductivity.

ACKNOWLEDGEMENTS

This work is supported by Basic Energy Sciences, Department of Energy, under Contract DE-FG02-96ER45261.

REFERENCES

1. Harry L. Tuller, Steve A. Kramer, Marlene A. Spears, U.S. Patent No. 5,403,461 (4 April, 1995).
2. Harry L. Tuller, Steve A. Kramer, Marlene A. Spears, and Uday A. Pal, U.S. Patent No. 5,509,189 (23 April, 1996).
3. H. L. Tuller, in *High Temperature Electrochemistry: Ceramics and Metals* , edited by F.W. Poulsen, N. Bonanos, S. Linderoth, M. Mogensen, and B. Zachau-Christiansen (Risø National Laboratory, Roskilde, Denmark, 1996), pp. 139-135.
4. S. Kramer, H. L. Tuller, *Solid State Ionics*, **82**, 15 (1995).
5. H. L. Tuller, S. Kramer, M. A. Spears, in *High Temperature Electrochemical Behavior of Fast Ion and Mixed Conductors*, edited by F.W. Poulson, J. J. Bentzen, T. Jacobsen, E. Skou, and M.J.L. Ostergard (Risø National Laboratory, Roskilde, Denmark, 1993), pp. 151-173.
6. O. Porat, C. Heremans, H.L. Tuller, *Solid State Ionics*, **94**, 75 (1997).
7. M. P. Pechini, U.S. Patent No. 3,330,697 (11 July, 1967).
8. O. Porat and H. L. Tuller, *Journal of Electroceramics*, **1**, 41 (1997).
9. Shyh Wang, Fundamentals of Semiconductor Theory and Device Physics, 1989, Prentice Hall, New Jersey, 1989, p.207.
10. V. V. Kharton, E.N. Naumovich, A.A. Vecher, and A.V. Nikolaev, *Journal of Solid State Chemistry*, **120**, 128 (1995).

COMPOSITION-DEPENDENT ELECTRICAL CONDUCTIVITY OF IONIC-ELECTRONIC COMPOSITE

M. PARK, G. M. CHOI
ɔhang University of Science and Technology, Dept. of Materials Science and Engineering,
ɔhang, 790-784, Korea

BSTRACT

Composition, dependence of electrical conductivity of ionic-electronic composite was
ɟamined using yttria(8mol%) stabilized zirconia-NiO composites. The contributions of
ectronic and ionic charge carriers to the electrical conductivity were determined by Hebb-
/agner polarization technique and electromotive force measurement of galvanic cell. Up to 6
ol% NiO addition, the conductivity decreased since the electronic NiO acted as an insulator in
ɔnic matrix. However the ionic transport was dominant until NiO content reaches 26 vol%.
ʌixed conduction was observed between 26 and 68 vol% of NiO. The effects of composition on
ιe electrical properties were explained by the microstructure and thus by the distribution of two
hases.

ʌTRODUCTION

Mixed conducting oxides which conduct both ions and electrons have a wide range of
ɩpplications such as electrodes in solid oxide fuel cells, oxygen permeable membranes and
ɔxygen sensors [1]. Two different types of mixed conducting oxides may be classified. One is
he single-phase material that shows both ionic and electronic conductivity. The other is the two-
ɔhase material composed of an electronic and an ionic conductor.

The composite employing oxygen ion conducting yttria-stabilized zirconia (YSZ) and p-
ype semiconducting NiO is an example of ionic-electronic mixture. There are several
ɩdvantages of choosing NiO as an electronic conductor and YSZ as an ionic conductor. NiO and
YSZ are stable at high temperature, their solubility in counterpart phase is limited, and doping of
ɩcceptors such as Li_2O easily controls the electrical conductivity of NiO [2,3].

In this work, mixed conducting YSZ-NiO composites were prepared in full composition
range and the electrical conductivity of the composites was measured by a.c. impedance and 4-
probe d.c. conductivity to determine the compositional dependence of conductivity and grain
boundary contribution to total conductivity. Hebb-Wagner polarization method and EMF
measurement of the galvanic cell have also been used to determine the contribution of the ionic
and electronic charge carriers on the conductivity. The distribution of two phases is studied as a
function of composition.

EXPERIMENT

NiO (99.9%, High Purity Chemicals) and YSZ (8 mol% Y_2O_3 doped ZrO_2, TZ8Y,
TOSHO) powders were used to prepare YSZ-NiO composites with NiO content varying 0~100
mol%. Pellets were formed by die pressing and then isostatically cold pressed into a disc at
200MPa. The compacted discs were then sintered at 1600°C for 4 h in air.

The total electrical conductivity of the YSZ-NiO composites was measured using 4-probe
d.c. technique. Specimens were cut into rectangular bars of 2.8x2.5x10mm³. Platinum paste
(Engelhard 6082) was applied as electrodes. The electrical conductivity measurement was
performed at temperatures between 400 and 1000°C in air. Impedance was measured from 200

to 600°C using impedance analyzers (LF Impedance Analyzer, Hewlett-Packard model 4192A and Gain-Phase Analyzer, Schlumberger model SI 1260).

The EMF cells were used to determine the oxygen ion transference numbers. They were calculated from the open-circuit voltages of EMF cells using YSZ-NiO composite pellets as electrolytes. The specimens (about 2cm in diameter and 2mm in thickness) were glass sealed to alumina tube. Air (Po_2=0.21) flowed outside the alumina tube and oxygen (Po_2=1.0) inside. Hebb-Wagner polarization method was used to determine the electronic transference number in ionic conduction dominant composition. The blocking electrode was prepared by sealing a platinum foil onto the sample surface with a pyrex glass.

RESULTS

No other phase except NiO and YSZ was detected in all composition range from x-ray diffraction spectra of materials with nominal compositions xNiO-(1-x) YSZ(hereafter 100xNYSZ). The relative sintered density of mixtures was greater than 97% of theoretical density. However the density of pure NiO was about 91%.

Fig.1 shows the 4-probe d.c. electrical conductivity with varying NiO content, measured between 400 and 1000°C. Since NiO is electrically more conductive than YSZ, the electrical conductivity of NiO-rich composition is higher than that of YSZ-rich composition. However, with increasing temperature, the difference in the electrical conductivity was reduced and the conductivity was nearly the same in most composition range above 600 °C.

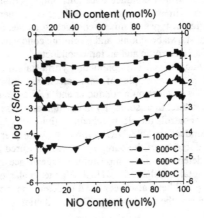

Fig.1 Composition dependence of electrical conductivity measured at between 400 and 1000 °C in air.

Up to 1.1 vol% of NiO addition, the electrical conductivity decreased due to Ni dissolution into YSZ matrix. In this region, additional oxygen vacancies generated by NiO addition in YSZ form vacancy-vacancy associates, which lead to the reduction of the electrical conductivity. The solubility limit of NiO in YSZ(8 mol% Y_2O_3 doped Zirconia) was determined to be about 2 mol%($\cong$1.1 vol%) from the measurement of lattice parameter and also confirmed by other study [3].

When NiO was added above solubility limit, the electrical conductivity further decreases since the electronic NiO particles, randomly distributed, block the movement of ionic charge carriers in the composites. When electronic NiO particles are discretely embedded in ionic YSZ matrix, NiO particles act as an insulator. In NiO, the ionic transference number is nearly zero, thus NiO blocks ionic charge carriers (oxygen ions).

When NiO content is above 26 vol% (40 mol%), the electrical conductivity of composites is shown to increase clearly with NiO content at 400°C. It is believed due to the three dimensional connection of NiO phase which is more conductive than YSZ phase. With NiO content below 26 vol%, NiO exists as discrete particles and samples are electrically conducted by oxygen ions in YSZ. With NiO content above 26 vol%, electron holes in NiO start to percolate and participate in the electrical conduction. With increasing amount of percolated NiO, the electrical conduction mechanism is expected to change from the ionic to the electronic.

Fig.2 shows the temperature dependence of the electrical conductivity. Up to 40 mol% of NiO, the curve shows a nearly straight line. However with increasing NiO content, a curvature was shown, indicating conduction mechanism change. When temperature is between 400 and 600 °C, the activation energy of electrical conduction decreases with increasing NiO content. In this temperature range, the activation energy of YSZ is about 1.0eV and that of NiO is about 0.3eV. Thus three-dimensional connection of NiO reduces the activation energy.

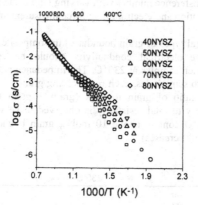

Fig.2 Temperature dependence of the electrical conductivity, measured by 4-probe d.c. method in air. yNYSZ represents y mol% NiO-(100- y) mol% YSZ.

Fig.3 shows the ionic and electronic transference numbers with varying NiO content and thus the change of conduction mechanism from ionic to electronic. Up to 26 vol% of NiO, t_i (ionic transference number) was nearly one, however further increase of NiO content decreased the t_i value. When NiO content was 68 vol%, t_{el} (electronic transference number) was nearly one. From this data, three types of conduction region were classified, i.e., electronic, ionic and mixed conduction region. Up to 26 vol% of NiO, the ionic conduction dominates. When NiO content is between 26 and 68, the composites are mixed conducting by oxygen ions in YSZ and holes in NiO. With further NiO content increase, the electron holes of NiO are main charge carriers.

In ionic conduction region, t_{el} increased with NiO content, however, the magnitude of t_{el} was less than 0.01, thus total electrical conductivity was determined by oxygen ion in YSZ. Ionic transference number was obtained by Hebb-Wagner polarization method in ionic conduction region. In mixed conduction region where the ionic and electronic conductivity were comparable, electromotive force measurement of galvanic cell was used to calculate t_i.

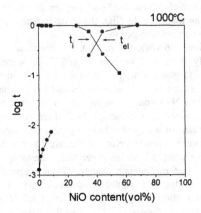

Fig.3 Ionic and electronic transference number with varying NiO content at 1000 °C, obtained by Hebb-Wagner polarization and electromotive force measurement of galvanic cell.

There are possibly several types of grain boundaries in composite material. Thus the effect of grain boundaries on the electrical conductivity should be considered. Fig.4 shows impedance patterns of YSZ-rich samples at 230 °C. The impedance patterns were analyzed and shown to be composed of two semicircles, each representing grain and grain boundary process. Fig.5 shows the resistance ratio of grain to bulk (grain + grain boundary). In YSZ, the contribution of grain boundary to total resistivity is observed to be very small above 230 °C. However, with increasing NiO content above 10 mol%, grain boundary contribution clearly increase (or grain contribution decreases).

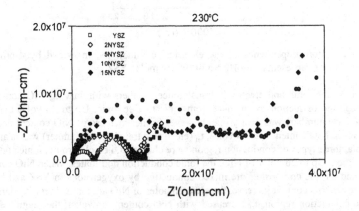

Fig.4 Impedance plots measured at 230 °C in air

With increasing temperature, the grain boundary contribution decreases and thus grain resistance becomes dominant in all composition range as shown in Fig.5. Thus, high temperature (above 600°C) conductivity data reflect the geometrical arrangement of NiO and YSZ grains.

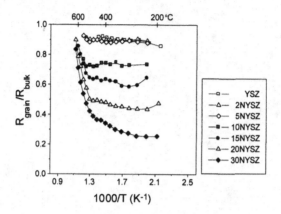

Fig.5 The ratio of grain to total resistivity obtained by 2-probe impedance measurement.

CONCLUSIONS

Below 26 vol%(40 mol%) NiO content, the ionic conduction by YSZ prevails and above 68 vol%(80 mol%) NiO content, the electronic conduction by NiO prevails. The percolation by both NiO and YSZ and thus mixed conduction was observed for the samples with NiO content between 26 and 68 vol%. Above 600 °C the grain boundary effect on electrical conductivity was small.

ACKNOWLEDGEMENTS

This work was supported by the Korea Science and Engineering Foundation(KOSEF) through the Center for Interface Science and Engineering of Materials at Korea Advanced Institute of Science and Technology (KAIST).

REFERENCES

1. W. Göpel and H. D. Wiemhöfer, Ber. Bunsenges. Phys. Chem., **94**, p. 981 (1990).

2. S. V. Houten, "Semiconduction in $Li_xNi_{(1-x)}O$," J. Phys. Chem. Solids, **17**, p. 7 (1960).

3. A. Kuzjukevics and S. Linderoth, "Interaction of NiO with Yttria-Stabilized Zirconia," Sol. St. Ionics, **93**, p. 255 (1997).

Part VII

Composites and Percolation Systems

Comparison of Techniques for Microwave Characterization of Percolating Dielectric –Metallic Media and Resolution of Discrepancies in Measured Data

Rick Moore, Lisa Lust, Edward Hopkins and Paul Friederich

Signature Technology Laboratory, Georgia Tech Research Institute, Georgia Institute of Technology, Atlanta, GA 30332 (ricky.moore@gtri.gatech.edu)

ABSTRACT

The measured microwave effective dielectric properties of metal-dielectric composites show discrepancies when data from free space, resonant cavity or transmission line measurements are compared. Discrepancies are especially evident for materials where the metallic concentration is near the percolation threshold. This paper presents theory and measured data which highlight and resolve these discrepancies. Electrical correlation length is the relevant parameter which must be considered in choice of measurement technique. Agreement between effective medium models and measurement are best when focussed beam measurement techniques are utilized so as to produce planar wave-fronts whose extent is 3-4 freespace wavelengths when incident on the sample.

INTRODUCTION AND PROBLEM STATEMENT

Electromagnetic transmission line measurement techniques have been the preferred methods to perform constitutive parameter measurements $(\varepsilon, \mu, \sigma)$ of iso or anisotropic scale invariant homogeneous materials [1,2] for frequencies from 50 MHz to 18 GHz. However, constitutive parameter measurements are often found to depend on the dimensionality of the test fixture, sample (sample thickness and/or sample-test fixture cross sectional dimensions) and electromagnetic wavelength when the transmission line techniques are applied in the measurement of artificial materials. Composites are composed of dielectric matrices, $\varepsilon_d(f) = \varepsilon_{dr} + j\sigma_d / 2\pi f \varepsilon_0$ and metallic or semi-conducting inclusions of conductivity σ_1. The measurement of their effective electromagnetic properties applies only when the electromagnetic wavelength within the material, λ_M, is much greater than characteristic dimensions of sample inhomogeneity.

The sources of observed measurement variation derive from electrical percolation threshold and relative electrical conductor correlation, $\xi(f) = \sqrt{\varepsilon_s(f)\mu_s(f)} a_0 |p - p_c|^{-\upsilon} / \lambda_0$, within the material. Here a_0 is the conducting inclusion size; p is the inclusion volumetric concentration; p_c is the electrical percolation threshold; $\varepsilon_s(f)\mu_s(f)$ are permittivity and permeability of the media surrounding the inclusion at the frequency f with free space wavelength λ_0 and υ is the characteristic critical exponent [3] (4/3 in 2D and 0.88 in 3D). Measurement errors which result from the correlation length and wavelength dependence are discussed below.

Measurement variations deriving from sample finite thickness, 2 vs 3 dimensions

Unlike homogeneous electrically lossy dielectric materials, the effective complex conductivity, $\Sigma(f)$ or permittivity $\varepsilon(f) = j\omega\varepsilon_0\Sigma(f)$ can vary with thickness or sample cross section. For p near but less that p_c,

$$\Sigma(f,p) \approx \left(\sigma_1\{j\frac{f}{f_1}(1 - j\frac{f_0}{f})\}^{\frac{-t}{s+t}}\right) Y\left\{|p - p_c|\left(j\frac{f}{f_1}(1 - j\frac{f_0}{f})\right)^{\frac{-1}{s+t}}\right\} \qquad 1.$$

The characteristic frequencies are $f_1 = \sigma_1 / 2\pi\varepsilon_{dr}\varepsilon_0$, $f_0 = \sigma_d / 2\pi\varepsilon_{dr}\varepsilon_0$. Y is a conductivity scaling functional [3], $j\frac{f}{f_1}(1 - j\frac{f_0}{f})$ is $\frac{\omega}{\omega_c}$ of reference 3 with s and t critical transport exponents [1 or 2]

The value of p_c is known to be a function of sample scaled thickness a_0/T [4,5]. Neimark[4] derived and approximate relationship between the percolation threshold of a material of finite thickness, T,

$$p_c = p_{c3} + (p_{c2} - p_{c3})(\frac{a_0}{T})^{\frac{\psi}{i+i}}$$ to the 2 and 3D values p_{c2}, p_{c3} and the relation was qualitatively confirmed in

numerical simulation by Zhang and Stroud [5]. The change in threshold reflects when the correlation length exceeds the material thickness T. The application of this relation and Equation 1 for materials of thickness T ($p_c \Rightarrow p_e$) shows that the effect finite thickness is to introduce a third characteristic frequency, $f_1^T = (a_0/T)^\psi f_1$ where ψ is near 4.8. This frequency should have an effect in the microwave-millimeter wave measurements for conducting particulates ($\sigma_1 \approx 10^6 - 10^7 mho/m$) and a_0/T between 5 and 10. Material samples of identical dielectric-metallic volume fractions but different thickness can show different values of permittivity at the same frequency. Thus, the experimenter must be conscious of thickness as a cause of measurement variability for identical compositions but of differing thickness. These variations will be apparent in the following measured data.

Measurement variations deriving from sample and sample-test fixture cross section and Measurement Configurations.

The percolation threshold also scales for finite cross section materials, e.g. $p_c - p_e \propto L^{-1/v}$ where p_e is the percolation threshold which might be observed in the measurement of a finite sample [6]. The relation reflects the physical requirement that the correlation length $\xi = a_0|p - p_c|^{-v} \ll L$. In AC measurements this relation must be replaced by a three fold condition on correlation length, wavelength and cross sectional dimension, e.g. $L >> \lambda >> \xi(f)$. The first inequality is required to allow a reflection and transmission coefficient of the material to be defined and the second to meet criteria for measurement of effective electromagnetic constitutive parameters.

Transmission line measurements have sought to meet the first criteria by placing small samples (typically less than $\lambda_0/2$ in size) within a coaxial line, waveguide or other transmission line which is constructed of conducting boundaries. A coaxial line measurement configuration is shown below. S_{11} and S_{12} represent reflection and transmission coefficients measurement paths. The fixture geometry meets electrical boundary conditions for simulation of plane wave propagation through and reflection from the material sample via electrical imaging of the sample within the transmission line walls. However, this is only true *for materials which perfectly fit the transmission line and have perfect electrical contact at the sample transmission line interface.* Analytical procedures are available to correct for imperfect fit of nonconducting materials [2] and conducting fillers have been used to successfully fill small gaps transmission line walls and nonconductive materials sides. However neither of these approaches can correctly reproduce the electrical continuity which is within a percolating artificial material.

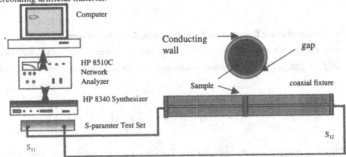

The transmission line measurements require the sample to be cut. Cutting the conducting-dielectric composite to a finite size (possibly smaller than $\xi(f)$) perturb the electrical path lengths within the sample and thus intrinsic polarizability and electrical dissipation are disturbed. Even a perfect fit percolating sample may not image boundary conditions exactly in the waveguide walls. If small gaps exist between the percolating material and transmission line wall (indicated above for the coaxial fixture) charge will accumulate at the gap and thus introduce additional polarizability while inhibiting current continuity. The expected result is that a measurement may yield a large polarizability (large real part of $\varepsilon(f)$) but lower electrical loss (small imaginary part of $\varepsilon(f)$). These expectations will be apparent in discussions of measured data.

MEASUREMENT CONFIGURATION FOR PROBLEM RESOLUTION

In order to better meet the inequality on a sample size, wavelength and correlation length, a free space measurement procedure has been adopted and is recommended for measurement of conductor-dielectric composites. To insure plane wave propagation a symetric lens configuration (shown below) is used to produce a focussed gaussian beam which is incident on the sample under test. The dimension of the gaussian beam is 3-4 free space wavelengths with measured phase of the incident electromagnetic field within 5 degrees of planarity. This configuration allows large sample sizes and can be used to scan the area of a sample to check for variations of a sample surface.

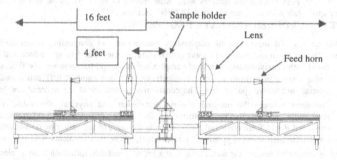

Sample descriptions

Samples were made from volumetric distributions of microspheres (nominally 30 μm in diameter) which have a conductor totally or partially coating their exterior surface. Composites were made from hollow silica spheres (wall thickness 5 μm) with three distinct coatings; a 100% surface coat of 0.1 μm thick Ag, a 63% surface coating of Ag and a 60% surface coat of Ag. Ag was applied via a electroless plating process. The figure below indicates a SEM photo of a sphere's surface. This sphere type has a 60% surface coating made up of Ag particulates approximately 60 nm in size. As the figure shows surface particulates are interconnected.

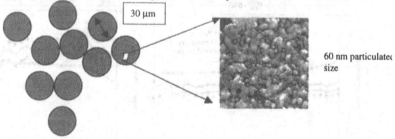

Figure 1 Spherical inclusions with a partially coated conducting surface were used in formulating the test composites.

The variation in surface conducting particulate fraction is reflected in a variation in sphere to sphere surface contact conductivity when a composite is made and thus the volumetric percolation threshold of the composite can be varied. By controlling the surface concentration, the volumetric value of p_c could be varied between 3D like, $p_{c3D} \approx 0.31$ (measured for the fully coated spheres) and 2D like values, $p_{c2D} \approx 0.5$ for the partially coated (60% coat) spheres. Analysis of measured data for the three surface coatings demonstrated that a scaling of the form $p_{cm} = p_{c3D} / A_c$ (where A_c was the coated surface fraction) could predict the measured volumetric percolation threshold for each of the different sphere surface types.

One hundred and eight composite samples were formed in 24" x 24" sheets. Three microsphere coatings were used in two dielectric binders (urethane and epoxy); six volume fractions (28, 31, 35, 43, 47 and 51%) and three thicknesses (0.020", 0.050" and 0.15"). Several additional thicknesses were made at a few concentrations to evaluate scaling of electrical parameters. Volumetric concentrations (as above 28 to 51%) were made and measured for each of the Ag surface coatings. These data allowed the measurement of p_{c3D} as a function of the Ag surface fractions on the sphere surfaces. Three thickness of the composite were chosen to measure the change in critical threshold with thickness, i.e. ($a_0 / T \approx$ 3, 7 and 20). Three transmission line sample were cut from each of the 108 panels and reflection/transmission coefficients were measured from each material for 0.10 to 18 GHz. Free space measurements were made from 2 to 100 GHz on the full size panels. Transmission and free space data were compared in the 4-18 GHz spectrum.

MEASUREMENT PROCEDURES AND DATA PROCESSING

These discussions on material measurements emphasizes measurement of conducting sphere-dielectric panel transmission coefficients at cm and millimeter wave (mmW) frequencies. The free space focussed beam measured data is compared with coaxial measurements of complex reflection and transmission coefficients and their inversion to determine dispersive permittivity, $\varepsilon(f)$, conductivity, $\sigma(f)$, permeability, $\mu(f)$, and impedance, Z_S (f) as composite material constitutive parameters. The constitutive parameters of the material are derived from the same basic relations between the intrinsic material parameters and physical observables of the measurement. The observables, or measureable quantities are the voltage reflection, R, and transmission coefficients, T. The normalized R and T are commonly referred to as elements of the scattering or S-matrix; the "front" side reflection and transmission are S_{11} and S_{21} respectively. The measurements with the material sample are referenced, or calibrated against, the measurement of a known standards, reflection from an electrical short or matched load and transmission through free space. Data processing is described in references 1,2.

MEASURED DATA

The difficulties in performing transmission line measurements of the composites is apparent in all measured data where volume fractions are within about 5% of the critical fraction. Transmission line and freespace permittivities agree within concentrations are at least 5% below percolation. Figure 2-4 shows some example data. The differences between free space and transmission line measurements are consistent with the arguments presented in the second section.

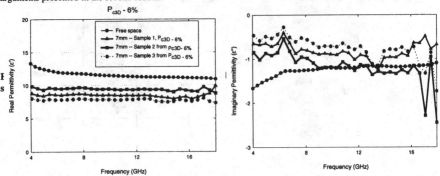

Figure 2 Measured permittivity at 6% below percolation (51%) for the 60% Ag surface coating. 7mm samples were cut from the free space sample after free space data were taken.

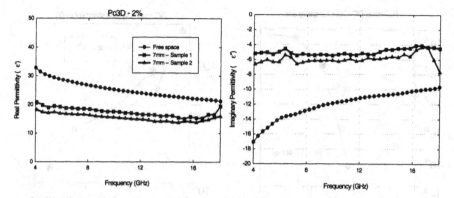

Figure 3 Measured permittivity at 2% below percolation for the surface coating of Figure 2. 7mm samples were cut from the free space sample after free space data were taken.

As the critical point is approached the impact of the imperfect surface contact in the transmission line dominates errors in measured data. Figure 4 shows transmission coefficient amplitude for the 100% sphere coating and a concentration at and 5% above the critical volume fraction. Transmission line data indicate a large capacitance with low loss (data at percolation is almost identical to that 5% above) while free space data demonstrate that the compositions are strongly attenuating.

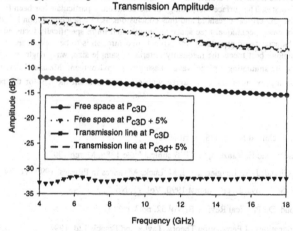

Figure 4 Comparison of transmission loss for free space vs. transmission line.

Thickness scaling of the critical threshold and therefore electromagnetic parameters is demonstrated in Figures 5. Relative conductivity for a fixed thickness is shown on the left for a thick film. The measured microwave impedance is shown on the right as calculated from the imaginary part of the permittivity near 10 GHz. Data are shown for each of the sphere types as a function of the scaled composite thickness.

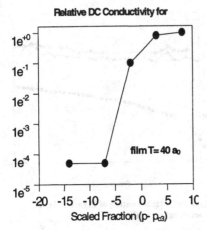

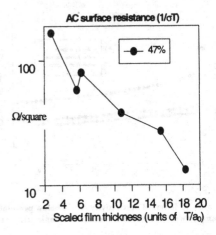

Figure 5 These figures show variation of relative conductivity for a thick film and impedance variations for films of various scaled thicknesses.

CONCLUSIONS

Transmission line measurement techniques should not be used for RF characterization of conductor-dielectric composites when the composites are known to have conductor concentrations which are within about 5% of the critical volume fraction. Further, the experimentalist must be conscious of the surface morphology of the conducting particulates. The surface morphology of the conducting particulate has been found to impact the expected value of the 3D critical threshold and thus conclusions as to when correlation length may impact the validity of the measurement technique. Free space techniques and more specifically focussed beam approaches should be used to characterize composites if the critical concentration is to be approached during measurment. This free space technique best meets the inequality criteria on sample size, wavelength and correlation length, $L \gg \lambda \gg \xi(f)$ while simulating a plane wave measurement environment. Measured data from the focussed beam system has been previously shown to agree with effective medium models for the permittivity of the composite media [7].

References

1. R. Geyer, NIST Technical Note 1338, April 1990.

2. C. Weil, M. Janezic and E. Vanzura, NIST Technical Note 1386, March 1997.

3. J. P. Clerc, G. Giraud, J. M. Laugier, and J. M. Luck, Advances in Physics, 1990, Vol 39., No. 3, 191-309.

4. Neimark, A. V.,Sov. Phys. JETP, August 1990, Vol. 71, No. 2, 341-348.

5. Zhang, X.; Stroud, D., Physical Review B, Vol.52, no.3, p.2131-7

6. M. Sahimi, Applications of Percolation Theory, Taylor and Francis Ltd, 1994.

7. P. Kemper , R. Moore, Materials Research Soc. 1994 Symposium Proceedings

DIELECTRIC SPECTROSCOPY OF INSULATOR/CONDUCTOR COMPOSITES

Julie Runyan Kokan and Rosario Gerhardt
School of Materials Science and Engineering, Georgia Institute of Technology, Atlanta, GA 30332-0245
Robert Ruh
Universal Technology Corporation, Beavercreek, Ohio 45431-1600
David S. McLachlan
Physics Department, University of the Witwatersrand, PO Wits 2050, Johannesburg, South Africa

ABSTRACT

The dielectric properties of composites are affected in different ways by a number of parameters. These include the electrical properties of both the filler and matrix, the wetting properties of the matrix on the filler, the size and shape of the filler, and the amount of the filler. Most composite systems have been studied via dc resistivity measurements which clearly show the effect of the addition of a second phase to an insulating matrix. However, frequency dependent measurements can provide additional insight into the mechanisms controlling the electrical response. The frequency dependence of the dielectric properties of composites will be shown. The permittivity and admittance plots of BN/B$_4$C composites will be given and the relevance of the trends seen in them will be discussed.

INTRODUCTION

Modeling of the electrical properties of composites is desirable, not only in being able to predict the electrical properties, but may also be used to predict the mechanical properties. By using the electrical properties to determine the microstructure, the mechanical properties can therefore be predicted through the microstructural models.

Most modeling of electrical properties of composites has been done on the dc conductivity/resistivity. Many mixing models exist which allow for property prediction. Most of these models work better for dilute composites. These models include the following
Parallel Model:

$$\sigma_m = v_i \sigma_i + v_c \sigma_c$$

Series Model:

$$\frac{1}{\sigma_m} = \frac{v_i}{\sigma_i} + \frac{v_c}{\sigma_c}$$

and Lichteneckers Rule:

$$\log \sigma_m = v_i \log \sigma_i + v_c \log \sigma_c$$

where σ_c and σ_i are the conductivities of the conducting and insulating phase respectively, v_c and v_i are the volume fractions of the conducting and insulating phases, respectively, and σ_m is the conductivity of the composite [1].

Another popular method of predicting the properties of composites is percolation theory. Percolation theory is based on the idea that a large change of properties will occur when the second phase is totally connected from one side of the composite to the other. The volume fraction at which this occurs is called the percolation threshold. It depends on many factors including the connectivity of the phases, the size of each phase, the shape of each phase, and the wetting behavior of the phases. Percolation models allow for a large (orders of magnitude) change of properties over a very small concentration range [2].

Recently[3,4], one of the authors has proposed an equation which incorporates the mixing rules[1], most aspects of the Bruggeman Effective Media theory[1], and percolation theory[3,4]. Formerly known as the General Effective Media equation (GEM), the McLachlan equation, which models the complex conductivity (dielectric constant) over the entire volume fraction and a limited frequency range, is:

$$(v_i)\frac{\sigma_i^{\frac{1}{t}}-\sigma_m^{\frac{1}{t}}}{\sigma_i^{\frac{1}{t}}+\left(\frac{1}{f_c}-1\right)\sigma_m^{\frac{1}{t}}}+(v_c)\frac{\sigma_c^{\frac{1}{t}}-\sigma_m^{\frac{1}{t}}}{\sigma_c^{\frac{1}{t}}+\left(\frac{1}{f_c}-1\right)\sigma_m^{\frac{1}{t}}}=0$$

where all t and s are modeling parameters, f_c is the critical volume fraction, and all other variables are defined as above [1].

Impedance spectroscopy is an established method for the study of composites. It is sensitive to interfaces as space charges build up at the interfaces between the conductor and insulator. These space charges affect the ac properties of the composite. Using the different dielectric functions-- impedance, admittance, permittivity, and modulus-- different effects can be seen.

FREQUENCY DEPENDENCE OF DIELECTRIC PROPERTIES

An important method for studying the electrical behavior of composites is to use all of the dielectric functions [5]. The frequency dependence has been found to be different in the different functions[6,7]. The functions are as follows:

$$\varepsilon^* = \varepsilon' - j\varepsilon''$$
$$M^* = M' + jM''$$
$$Z^* = Z' - jZ''$$
$$Y^* = Y' + jY''$$
$$\tan\delta = \frac{\varepsilon''}{\varepsilon'} = \frac{M''}{M'} = \frac{Z'}{Z''} = \frac{Y'}{Y''}$$
$$\varepsilon^* = \frac{1}{M^*} = \frac{1}{j\omega C_0 Z^*} = j\omega C_0 Y^*$$

where ε^* is the complex permittivity, M^* is the complex modulus, Z^* is the complex impedance, Y^* is the complex admittance, and ω is the angular frequency. Figure 1 shows the ideal frequency dependence of the real part of these functions as a composite goes from an insulator to a conductor. The original conductivity curves were generated using the McLachlan equation with s=1, t=2, and the electrical properties of graphite and boron nitride as the conductor and insulator, respectively[8]. The rest of the curves were generated by converting these calculated data into the other dielectric functions.

The curves in figure 1 were generated using an ideal dielectric (no dispersion, no loss component, which is equivalent to an ideal capacitor). In the conducting phase, the dielectric term is neglected, which is valid over the frequency range used. As the material gets more conducting, the permittivity increases as does the frequency dispersion. The permittivity increases due to space charge build up at the interfaces between the conducting phase and the insulating due to the difference in conductivity of the two phases[9]. The frequency dispersion increases due to an increase in the loss at lower frequencies. The lowering of the permittivity above the percolation threshold is due to a decreasing number of conductor/insulator interfaces as a result of conducting particle merging.

The admittance shows the opposite effect in regards to frequency dispersion. However, for a conductor containing an insulator at high enough frequencies, the displacement current becomes significant and the media becomes dispersed. Obviously, the admittance decreases with the addition of an insulating phase, as it is just the conductivity without the geometric factor. The modulus shows the same effect as the permittivity does, but in reverse. This is also true for the impedance as compared to the admittance. This is to be expected as the modulus is the inverse of the permittivity, and the impedance is the inverse of the admittance. The tan δ is interesting as it predicts a decreasing value for the conducting samples and an increasing value for the insulating samples while the percolation threshold is flat with increasing frequency.

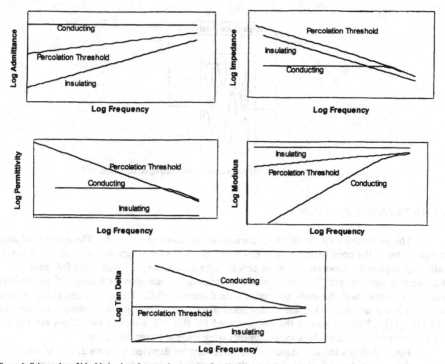

Figure 1: Schematics of ideal behavior of composite samples for all of the real dielectric functions. These were calculated from model conductivity calculations of ideal BN/Graphite composite samples with s=1 and t=2 that were very close to the percolation threshold [8].

EXPERIMENTAL

The data presented in this paper are from BN/B₄C composites fabricated as described in reference 8. Samples from 0% to 100% BN were tested. Table 1 is a list of all of the samples tested. BN exists in platelets, as is shown in figure 2[10].

The frequency range used was 10Hz to 10MHz with an input voltage of 0.5V. This was done with an HP4192A Impedance Analyzer. The direction of the electric field makes a difference, however, due to space constraints, only the data taken parallel to the hot pressing direction (or perpendicular to the face of the BN platelets) as illustrated in figure 2 is reported.

Table I: BN/B₄C Samples that were tested.

Volume Fraction BN	Number of Samples
.00	1
.10	2
.20	2
.40	2
.60	6
.80	2
1.00	1

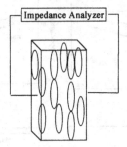

Figure 2: Schematic of BN/B₄C samples and the testing arrangement.

RESULTS AND DISCUSSION

The permittivity of the BN/B$_4$C composites is shown in figure 3. The more insulating samples, that is the ones with the most BN have little to no frequency dependence. This is just what was expected. However, the more conducting samples, the ones with more B$_4$C have larger frequency dispersion, just like predicted. One thing these data show is a range of permittivity values for samples with the same nominal volume fraction of B$_4$C. This must mean that there is a variation in the samples. This will be discussed more later. The reported dielectric constant of BN is7.1[11]. This is close to the one obtained for this BN, although the values are slightly higher presumably because these measurements were made at lower frequencies.

Figure 4 shows the admittance of the same samples depicted in figure 3. These data also follow the expected trends of frequency dispersion and magnitude. The higher conductivity samples have little or no frequency dispersion. The more insulating samples have lower admittance (therefore lower conductivity) just as predicted.

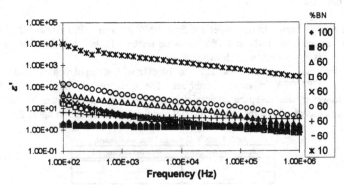

Figure 3: The dielectric constant of BN/B₄C composites as a function of frequency for different volume fractions of BN

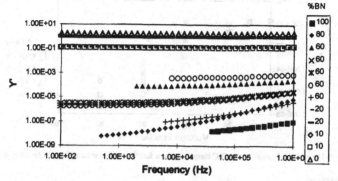

Figure 4: The real admittance of BN/B₄C composites as a function of frequency for different volume fractions of BN.

Figures 5 and 6 show the permittivity and the admittance as a function of volume fraction of BN at different frequencies. In all cases one can see a tremendous change in the properties as the composition changes from conducting B_4C to insulating BN. The shapes of the Y' curves are as described for DC resistivity of composites [1] especially at the lower frequencies. However, the different markers show the different frequencies indicating that the properties vary with frequency. Multiples of the same markers at the same volume fraction represent different samples with the same volume fraction at the same frequency. This shows some sample to sample variability. This is especially evident at 60% BN, and also seen in the frequency explicit plots of figures 3 and 4. This range is near the percolation threshold where minute differences in the microstructure or connectivity lead to large changes in the transport properties. Therefore, some of the 60%BN samples may be above ϕ_c, that is a path of B_4C goes clear through the specimen, while others are not.

Another interesting feature on the volume fraction plots is the frequency dispersion, and lack there of, that is visible. In the permittivity plots, above 60% BN, there is no frequency dispersion. That is to say, all frequencies give the same number for the dielectric constant. In the admittance plot, frequency dispersion can only be seen at 60% and above. It is also interesting to note that the more conducting samples have less frequency dispersion than the less conducting ones. All of this is on agreement with the theory and remarks made at the end of the previous paragraph.

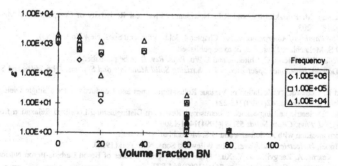

Figure 5: The permittivity of BN/B₄C composites as a function of BN volume fraction for three frequencies.

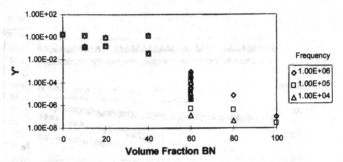

Figure 6: The real admittance for BN/B₄C composites as a function of BN volume fraction for three frequencies.

CONCLUSIONS

The presented data show clearly the importance of looking at the frequency dependence when dealing with composites. First, depending on the frequency at which the composite will be used, the electrical properties can be drastically different. This is especially important when the volume fraction of the conducting phase is near the percolation threshold. Also the frequency dependence varies depending on the conductivity of the composite. For instance, in the permittivity plane, insulating samples are almost frequency independent, but conducting samples have a large frequency dependence. In the admittance plane, the opposite is true—the conducting samples are frequency independent and the insulating samples have a large frequency dispersion.

The other important point that can be surmised from this data is that there are other factors than volume fraction which determine the properties of insulator/conductor composites. This is shown by the differences in the samples with the same volume fractions, especially when their composition lies near the percolation threshold. It is expected that the connectivity of the phase is the controlling factor.

The electrical properties of composites are frequency dependent . This fact needs to be taken into account when designing and using these materials.

ACKNOWLEDGEMENTS

Part of this work was funded by NASA NGT 52830.

REFERENCES

1. D.S. McLachlan, M. Blaszkiewicz, and R.E. Newnham, "Electrical Resistivity of Composites," *J. Am. Ceram. Soc.*, 73[8](1990)2187-2203.
2. R. Zallen, *The Physics of Amorphous Solids*, Chapter 4, John Wiley and Sons, New York (1983).
3. J.Wu and D.S. McLachlan, *Phys. Rev. B*, to be published.
4. D.S. McLachlan, W.D. Heiss, C. Chiteme, and J. Wu, *Phys. Rev. B*, to be published.
5. J. Ross McDonald, *Impedance Spectroscopy Emphasizing Solid Materials and Systems*, John Wiley and Sons, New York (1987).
6. W. Cao and R. Gerhardt, "Calculation of Various Relaxation Times and Conductivity for a Single Dielectric Relaxation Process," *Solid State Ionics*, 42(1990)213-221.
7. R. Gerhardt, "Dielectric and Impedance Spectroscopy Revisited: Distinguishing Localized Relaxation from Long Range Conductivity," *J. Phys. Chem. Solids* 55[12](1994)1491-1506
8. Private communication with A. Sofianos and D.S. McLachlan
9. A.R. VonHippel, *Dielectrics and Waves*, John Wiley and Sons, New York(1953).
10. R. Ruh, M. Kearns, A. Zangvil, and Y. Xu, "Phase and Property Studies of Boron Carbide-Boron Nitride Composites," *Journal of the American Ceramics Society*, 75[4](1992)864-872.
11. *CRC Handbook of Chemistry and Physics*, ed. By D.R. Lide, CRC Press, Ann Arbor (1994)12-45.

COMPUTER SIMULATION OF IMPEDANCE FOR 2-D CONDUCTOR-INSULATOR COMPOSITE

Dae Gon HAN and Gyeong Man CHOI
Department of Materials Sci. & Eng., Pohang University of Science and Technology, Pohang 790-784, Korea

ABSTRACT

Using a computer simulation, impedance spectra of mixtures of conducting and insulating hard spheres which have random or regular arrangements of the components are studied. These simulations can be used to calculate the a.c. electrical properties of a multi-component composite using a personal computer. It is shown in this study that a.c. impedance spectra are sensitive functions of the filling fraction and the geometrical arrangement of the components, and especially, the impedance spectra of the composite show the abnormal arc originated from the isolated clusters in the composite.

INTRODUCTION

A composite is defined as a material made up of two or more different media which are arranged in a regular or irregular pattern. Composite materials are widely used in industrial applications as resistors, sensors, and transducers. It is well known that the conductivity of electrical materials is strongly affected by the atomistic mechanism (charge carrier mobility and concentration, hopping rate etc.) and the microstructure (grain size distribution, grain morphology, porosity etc.). The detailed geometrical arrangement of the constituents also plays a significant role in determining the bulk properties of the composites. Theoretical and experimental analyses of the conductivity of polycrystalline composites have been mostly focused on the identification of the underlying atomistic mechanism[1] and microstructure effects[2], whereas the geometrical arrangement effects have been scarcely studied.

Recently, with the use of the modern frequency analyzer that can do the fast and precise measurements, impedance spectroscopy (IS) is gaining popularity for studying the electrical properties of composite materials. Impedance plots are useful in determining the appropriate equivalent circuit and estimating the values of the circuit parameters[3]. Although the process is straightforward in some cases, it is not obvious what portion of the equivalent circuit corresponds to what component in materials. These problems are settled by various experimental methods for the given impedance plot. However, until now the effect of geometrical arrangement on impedance spectroscopy was hardly considered.

The main objective of this paper is to study geometrical arrangement effects on the impedance spectra of polycrystalline composite. In this paper, the composite composed of closed packed hard spheres of conductors and insulators was modeled on ideal polycrystalline composites made up of a regular hexagonal grain. We consider a situation where one phase, denoted by "1", is mixed with a second phase, denoted by "2".

METHODOLOGY

We have carried out numerical simulations of the systems of mixture of hard spheres, imagining the sample was placed between two electrodes. As shown for the simplified 3×4 network in Fig.1.a, the simulation system corresponds to the site percolation of triangular network. We used the transfer-matrix algorithm to calculate the d.c. and a.c. conductivity of each triangular network. This algorithm was originally developed by Derrida and coworkers[4] to calculate the d.c. conductivity of rectangular networks of bond percolation. The resistivity (ρ, Ω/sphere) and the permittivity (ε, F/sphere) are assigned as the resistance and capacitance of one sphere, respectively. For the simulation, the

347

values of ρ_1 and ρ_2 are given as 10^9 [Ω/sphere] and 10^3 [Ω/sphere], respectively. Both ε_1 and ε_2 are given as 8.854×10^{-11} [F/sphere]. Two kinds of spheres used in the simulations correspond to two RC parallel circuits as shown in Fig.1.b. The values of spheres were chosen to reflect the case where the conductive particles were introduced into the insulating matrix of the same permittivity. The diagonal admittance matrix is newly introduced for the triangular network, and all elements of matrix are complex numbers for a.c. impedance simulation. We used Marsaglia's algorithm[5] to generate random numbers. The transfer-matrix algorithm and Marsaglia's algorithm are very simple and need less memory so that they make the a.c. impedance simulation possible on the personal computer. The detailed algorithm was given elsewhere[6]. In this study, all simulations are made over a mesh of 30×100 (N×L) spheres by an IBM compatible personal computer.

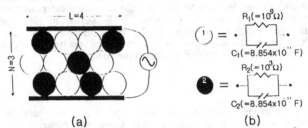

Fig.1 Closed packed hard spheres model (a) and the electrical components of spheres (b) used in simulations.

RESULTS AND DISCUSSION

The impedance spectra are presented in M-plots because M-plots have better resolution than Z-plots when the difference of the time constants is due to the difference of the resistance values. The frequency in simulation ranges from 10^{-2} Hz to 10^2 MHz and reflects the frequency range of many impedance analyzers. The simulated impedance spectra with "completely-random patterns" of two spheres are given in Fig.2.a. The shape of the impedance spectra is too complex to be analyzed clearly into two components. The third semicircles are clearly seen when the filling fraction p_2 is 0.2 and 0.4.

When the distribution of spheres has "no-contact random patterns" as shown in Fig.2.b, the impedance spectra can be clearly analyzed by a two-component series circuit model. Series equivalent circuit model is often used to analyze the electrical components of polycrystalline materials. When σ_1 and σ_2 are the conductivities of the basic constituent 1 and 2, respectively, and p_2 is the volume fraction of the constituent 2, the series circuit model gives the following equation:

$$\sigma_{tot}^{-1} = (1 - p_2)\sigma_1^{-1} + p_2\sigma_2^{-1} \qquad \cdots\cdots (1)$$

where σ_{tot} is the real or complex conductivity of the composite. In no-contact random patterns, black spheres are distributed randomly, but forbidden to contact each other, thus no black clusters form. The third semicircle is not observed although the filling fraction, p_2 reaches as high as 0.26. When the filling fraction increases further, the distribution loses randomness since the number of sites, which may be occupied by black spheres without contacting each other, is limited.

The analysis of above two patterns suggests that the appearance of the third semicircle is due to the clusters formed in the composites. When the horizontal-clusters are introduced as shown in Fig.2.c, however, the third semicircle is not observed. The ratio of two diameters of the semicircles is equal to the ratio of two volume fractions (p_1=0.818, p_2=0.182). So, this impedance spectrum is analyzed nicely by the two-phase series circuit model. However, the overlapping third semicircle appears when the oblique-clusters are introduced (Fig.2.d). The ratio of diameter

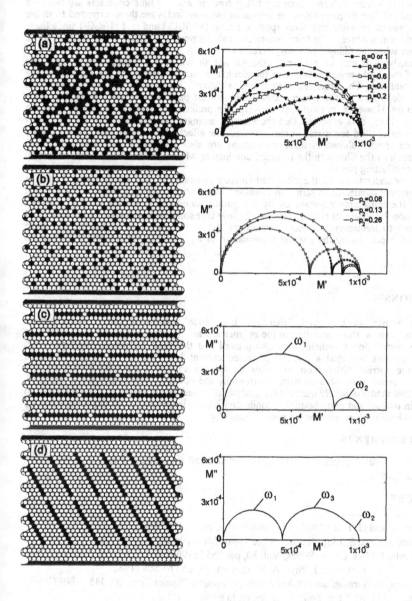

Fig.2 Distributions of spheres (left side) and the corresponding, simulated impedance spectra (right side).

of the semicircles is very different from the filling fraction and the time constants are resolved into ω_1, ω_2 and ω_3. The frequencies at the apexes of the semicircles are those expected from the semicircles representing white and black spheres, i.e. ω_1 $(=1/R_1C_1)$ and ω_2 $(=1/R_2C_2)$ since these are two component mixtures. The third semicircle with the peak frequency of ω_3, which is about $10^{-1} \times \omega_2$, grows when the filling fraction p_2 increases.

The analysis of various patterns, although not shown in this paper, suggests that the appearance of the third semicircle is due to the elongated clusters oriented in the direction of the preferred current line. The preferred current line is defined, here, as the shortest current path between two electrodes. In the present lattice, the shortest path between two electrodes is the oblique direction. However, in most real materials, the preferred current line is perpendicular to the electrodes. Thus, it is expected that the clusters arranged perpendicular to the electrodes generate the clearer third semicircle in real systems. The effects of the elongated clusters on the a.c. impedance spectra shown in these simulations are also expected to be observed in real composites, such as the film with the included conducting whisker, or the bulk composites with the included conducting plates.

We now understand that the elongated clusters generate the third semicircle as if there is another component connected in series, in addition to the known two components. It is possible to introduce the third time constant ω_3 by the parallel combination of several R_1C_1 parallel circuits with one R_2C_2 parallel circuit. Thus, we believe that some regions including the elongated cluster respond to frequency as a new component due to the distortion of current and these regions can be expressed by the parallel combination of R_1C_1 parallel circuit and R_2C_2 parallel circuit.

CONCLUSIONS

It is shown in this study that the impedance spectra of the two-component composites may show three arcs due to the geometrical arrangement effects in the vicinity of the effective percolation threshold. It is also proved that the third semicircle originates from the isolated clusters arranged along the preferred current line. The effect of the elongated clusters on the current distribution, and thus on the generation of the third semicircle was explained by a series of simulations using horizontally and obliquely oriented clusters. Although the series circuit model is widely used for the analysis of impedance of polycrystalline materials to separate grain interior and grain boundary conductivities, the interpretation of impedance spectra for the multi-phase mixture must be interpreted carefully.

ACKNOWLEDGMENTS

This work was supported by The Korea Science and Engineering Foundation (KOSEF), 1997 research fund.

REFERENCES

1. A. J. Bosman and H. J. van Daal, Advances in Physics 19, 1(1970).
2. M. P. Anderson and S. Ling, Journal of Electronic Materials 19, 1161(1990).
3. J. E. Bauerle, J. Phys. Chem. Solids, Vol. 30, pp. 2657 (1969)
4. B. Derrida, J. Vanniemenus, J. Phys. A: Math. Gen. 15, L557-L564 (1982).
5. G. Marsaglia, B. Narasimhan and Arif Zaman, Computer Physics Com. 60, 345 – 349 (1990)
6. D. G. Han and G. M. Choi, Solid State Ionics, in press.

SCALING BEHAVIOR OF THE COMPLEX CONDUCTIVITY OF GRAPHITE-BORON NITRIDE PERCOLATION SYSTEMS

JUNGIE WU AND D S MCLACHLAN
Physics Department, University of the Witwatersrand,
PO Wits 2050, Johannesburg, South Africa.

ABSTRACT

Measurements of both components of the complex AC conductivity σ^*_m of continuum percolation systems, based on Graphite and hexagonal Boron Nitride, over a large frequency range, for samples near the conductor insulator composition, are reported. The results of the real part of σ^*_m (σ_{mr}) above the critical volume fraction (ϕ_c) and the imaginary component ($\sigma_{mi} = i\omega\varepsilon_o\varepsilon_{mr}$) below ϕ_c are shown to have the correct power law dispersion behavior, but only if the non-universal exponents measured in previously described DC experiments are used. It is also shown that all the results can be scaled, as is predicted by percolation theory, so as to lie on two continuous curves, one below and one above ϕ_c. Unfortunately the actual ω_c values used to scale the experimental results are found not to be in good agreement with theoretical predictions.

INTRODUCTION.

Good percolation systems are characterized by a rapid change in a narrow range of conductor volume fractions. It has been well established, both experimentally and theoretically , that near the percolation threshold ϕ_c the DC conductivities are given by

$$\sigma_m(\phi,0)=\sigma_c((\phi-\phi_c)/(1-\phi_c))^t \text{ for } \phi>\phi_c \text{ and } \sigma_m(\phi,0)=\sigma_i((\phi_c-\phi)/\phi_c)^{-s} \text{ for } \phi<\phi_c. \quad (1)$$

Here ϕ is the conductor volume fraction, s and t the conductivity exponents for $\phi>\phi_c$ and $\phi<\phi_c$ respectively and σ_c and σ_i are the conductivities of the conducting and insulating components. The real part of the low frequency dielectric constant $\varepsilon_{mr}(\phi,\omega)$ of percolation systems has been shown to diverge as

$$\varepsilon_{mr}(\phi,\omega)=\varepsilon_i((\phi-\phi_c)/(1-\phi_c))^{-S^*}, \quad (2)$$

where ε_i is the dielectric constant of the insulating component. Neither of these equations predicts a dispersive media, other than the dipersion due to the components It was previously thought that the critical exponents s and t were universal , depending only on the dimension of the system, and that $s=s^*$, but this is now known not to be the case.

Major review articles, which include sections on the complex conductivity and scaling are [1-3]. Very close to ϕ_c more complex expressions (given below), involving both σ_c and σ_i or ε_r have to be used and the system becomes dispersive. Using these theoretical predictions, dispersion measurements enable one to tell how close to and on what side of ϕ_c a particular sample lies.

The AC conductivity $\sigma_m^*(\phi,\omega)$, or dielectric constant $\varepsilon_m^*(\phi,\omega)$ can be arrived at by a scaling ansatz [1-3] for the complex AC conductivity $(\sigma^*(\phi,\omega)=\sigma_{mr}(\phi,\omega)-i\omega\varepsilon_0\varepsilon_r(\phi,\omega))$. The form of this scaling ansatz is

$$\sigma_m^*(\phi,\omega) \; \alpha \; \sigma_c \; |\phi-\phi_c|^t \, F^*_{+ \, or \, -}(i\omega/\omega_c), \qquad (3)$$

where F^*_+ and F^*_- are the scaling functions above and below ϕ_c respectively, and ω_c is a critical or scaling frequency given by

$$\omega_c=(\sigma_c/(\varepsilon_0\varepsilon_r)) \; |\phi-\phi_c|^{\,t \, +S} \quad (\alpha \; \sigma(\phi,0)^{(t+S)/t}) \qquad (4)$$

Note that the imaginary parts of σ_c^* and ε_r^* of the conducting and insulating components respectively, have been assumed to be zero (ideal materials) in eqns. (3) and (4), therefore $\sigma_c^* = \sigma_c$ and $\varepsilon_i^*=\varepsilon_{ri}$. At high frequencies near ϕ_c eqn. (3) reduces to [1-3],

$$\sigma_{mr}(\phi,\omega) \; \alpha \; (\omega/\omega_c)^u \text{ and } \varepsilon_{mr}(\phi,\omega) \; \alpha \; (\omega/\omega_c)^{-v}, \qquad (5)$$

where $u=t/(t+s)$, $v=s/(t+s)$ and $u+v=1$. The following expression is used to obtain curves

for F_+ and F_-, upon which the experimental results could be scaled,

$$F_+(x\equiv\omega/\omega_{c+}) = \sigma_{GM}/\sigma_c((\phi-\phi_c)/(1-\phi_c))^t \quad \& \quad F_-(x\equiv\omega/\omega_{c-}) = \sigma_{GM}/\sigma_c((\phi_c-\phi)/\phi_c), \qquad (6)$$

where σ_{GM} is given by [4],

$$(1-\phi)(\sigma_i^{1/S} - \sigma_{GM}^{1/S})/(\sigma_i^{1/S} + A\sigma_{GM}^{1/S}) + \phi(\sigma_c^{1/t} -\sigma_{GM}^{1/t})/(\sigma_c^{1/t} + A\sigma_{GM}^{1/t}) = 0, \qquad (7)$$

and $A=(1-\phi_c)/\phi_c$. A review of an earlier version of this equation is given in [5]. Although this is an expression involving only real quantities, and scales DC conductivity results, if $\omega\varepsilon_0\varepsilon_r$ is substituted for σ_i in $x_+ = (\sigma_i/\sigma_c)((\phi-\phi_c)/(1-\phi_c))^{S+t} \equiv \omega/\omega_c$ and $x_- = (\sigma_i/\sigma_c)((\phi_c-\phi)/\phi_c)^{S+t} \equiv \omega/\omega_c$ in the eqn. (7) for F_+ and F_-. The resulting expressions fit the scaled data very well and have the limiting slopes for $F_+(x_+)$ and $F_-(x_-)$, predicted by theory [1-3], for x_+ or ω/ω_{c+} and x_- or ω/ω_c both greater and less than one. No other expression exists which will fit the data over the full range of x_+ (ω/ω_c) and x_- (ω/ω_c), including the region where x_+ and x_- are close to one. The equations given in [1-3] also all involve arbitrary constants.

RESULTS AND DISCUSSION

The systems measured in this study were a series of compressed Graphite-Boron Nitride discs and loosely packed powders consisting of 50%G50%BN and 55%G45%BN, undergoing compression. The experimental details are given in [4,6].

All three systems showed extremely sharp conductor-insulator transitions [4.6] which gave very consistent and accurate (DC) values of ϕ_c, s and t when fitted to eqns. (1) or (7). Low frequency dielectric measurements of ε_r $(\phi,\omega\approx0)$ gave very similar values for ϕ_c but not s*, which violates the concept of a universal value for s, which requires that s=s*. For the remainder of this paper unless otherwise specified the value of t used is the DC one and the value of s the AC one s* and all the Figures given are for the 55%G powder. Figures 1a and 1b show the ε_r and σ_r dispersion results obtained for the 55%G powder near ϕ_c. From these figures it can be seen that for $\phi > \phi_c$ the low frequency conductivity results are dispersion free, as are the $\phi < \phi_c$ results for ε_r, as required in eqns. (1), but that for $|\phi-\phi_c|$ very small there is the expected dispersion. The slopes (u=0.87 & v=0.10) for the samples very close to ϕ_c are in agreement, within the experimental error, with the u=0.87 and v=0.13 values calculated from the DC t=4.8 and the AC s^*= 0.72 values. The only other work which measured the s and t and u and v exponents independently is that of Chen and Johnson [7] who studied filimentary Ni-F and modular Ni-M nickel particles embedded in polypropylene, they obtained u=0.88 & 0.81 and v =0.14 & 0.15 respectively. The values calculated from their independently measured s and t values are u=0.88 & 0.78 and v=0.15 & 0.22 respectively, which also agree within the experimental error. As the accepted universal values are t=2.0 and s =0.87 which give u=0.58 and v=0.42, both the results, given here an those of Chen and Johnson, are only consistent if the independantly measured non universal values of s & t are compared with the u & v measured in dipersion experiments.

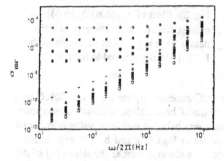

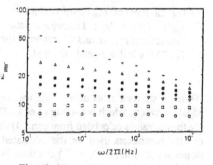

Fig. 1a. Plots of the conductivity $\sigma_{mr}(\phi,\omega)$ against frequency, for a 55%G-45%BN powder, on a log log scale, for various values of ϕ.

Fig. 1b. Plots of the real part of the dielectric constant ε_{mr} (ϕ,ω) against frequency, for a 55%G-45%BN powder, on a log log scale for various values of ϕ.

In Figs. 2a and b the scaled experimental curves and the theoretical ones, obtained form eqns. (6) and (7), are given. The theoretical curves are obtained using the parameters measured from DC and low frequency AC experiments of ϕ_m and ε_r as functions of ϕ, using eqns. (1) and (2), in eqns. (6) and (7). The actual values used to obtain the curves are σ_c =3126 $(\Omega$-m$)^{-1}$, ε_I =3.3, ε_0=8.85x10^{-12} C^2/N.m^2, t=4.8, s= 0.72, ϕ_c=0.123 and the actual value of ϕ, with ω covering the experimental range from 30 Hz to 100 Mhz. Naturally as these are scaling curves, the various sections of the curves generated for actual ϕ and ω values overlap each other. The experimental results for $\phi>\phi_c$ are first divided by the actual measured $\sigma_m(\phi,0)$ results and the $\phi<\phi_c$ results are normalized by the value for σ_m (ϕ,0) calculated form eqn. (1) using the relevent ϕ value and the parameters given above. Both results were then scaled along the ω/ω_{c+} or ω/ω_{c-} axis so as to form a continuous curve which fitted onto or close to the curves as shown. The distance the results had to be scaled along these axis enabled ω_{c+} and ω_{c-} to be calculated.

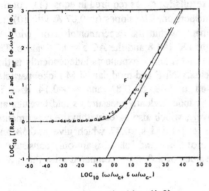

Fig. 2a. Plots of σ_{mr} (ϕ,ω)/σ_m (ϕ,0) against ω/ω_{c+} (ω/ω_{c-}). The experimental data is from Fig. 1a and the origin of the F$_+$ (F$_-$) line is discussed in the text.

Fig. 2b. Plots of σ_{mi} (ϕ,ω)/ϕ_m (ϕ,0) against ω/ω_{c-} (ω/ω_{c+}). The experimental data is from Fig. 1b and the origin of the F$_-$ (F$_+$) line is discussed in the text.

Figure 3 shows a plots of ω_{c+} the measured DC conductivity for the discs, 50%G and 55%G powders, which according to eqn. (4) should all have a slope of (t+s)/t. Also shown are the ω_{c+} values calculated from eqn. (4) using using both the measured value of σ_m (ϕ,0) and the parameters used for the calculated curves in Figs. 2 a and b. The experimental slopes are (q=1.03, 0.84, 0.82) and there is clearly a disagreement as the slopes (q) of the experimental ω_{c+} curves are less than the ones calculated from the measured DC conductivity and 1/u from AC experiments. In the case of the powders an even more alarming difference is the fact that the ω_{c+}s differ by more than an order of magnitude. Other workers who successfully scaled their AC conductivity results [8-10], but did not independently measure s and t, also obtained a slope from their ω_{c+} against $\sigma_m(\phi,0)$ plots, less than their measured values of 1/u=(s+t)/t, but did not comment on the fact that their measured values of ω_+ are very different in magnitude from the calculated ones. Huntley and Zettl [8] obtained a q of 0.82 for, presumably 2d gold films, while Benguigui [9] and Chakrabarty et al [10] both obtained q=1.1 for mixtures of glass and iron balls and carbon wax mixtures respectively.

Plots of ω_{c-} against a $\sigma_m(\phi,0)$ calculated from eqn. 1 and the parameters previously used also gave slopes for q (1.16, 1.07 &1.14), also lower than the calculated ones (1.20,1.15, & 1.06). These slopes differ from each other by less than in the previous case. Unfortunately there was still a large difference in the magnitudes of the measured and calculated ω_c values. The only other person to scale his σ_{mi} or ϵ_{mr} results was Benguigui [9]. Unfortunately his system of glass and iron balls was not large enough (did not contain enough particles) to obtain reliable results near ϕ_c, where there is a diverging coherence length [1-3]. Therefore we believe that these results are the most complete and reliable ones yet measured.

Fig. 3. Plots of the log of the experimentally measured and calculated values of $\omega_{c+}/2\pi$ against the measured value of $\sigma_m (\phi,0)$. [x disc -calculated =1.2, □ disc -measured = 1.03, ☐ 55%G- calculated = 1.05, △55%G -measured =0.82, * 50%G -calculated =1.06 & ∇ 50%G -measured =0.84]

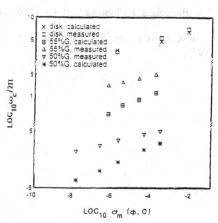

$$LOG_{10} \; \sigma_m \; (\phi, 0)$$

CONCLUSIONS

The positive side of this paper is that it has been shown that good experimental results on an excellent percolation system can be scaled onto single curves near the percolation threshold, but only when the previously measured, as a function of ϕ, values of s, t and ϕ_c are used. This correspondence between the parameters obtained from measurements as a function of ϕ and as a function of ω is also very gratifying. On the negative side is the fact that we do not yet understand the values obtained for s, s*, t, u and v, nor the behaviour and magnitudes of the experimental values of the critical frequencies ω_+ and ω_-.

REFERENCES

1. J.P. Clerc, G. Girand, J.M. Langier and J.M. Luck, Advances in Physics **39**, 191 (1990).

2. D.J. Bergman and D. Stroud, Solid State Physics **46**, edited by H. Ehrenreich and D. Turnbull (Academic Press, San Diego,1992), p147.

3. Ce-Wen Nan, Prog. in Materials Science **37**, 1 (1993).

4. Jungie Wu and D.S. McLachlan, Phys Rev. **B56**, 1236 (1997).

5. D.S. McLachlan, M. Blaszkiewicz and R.E. Newnham, J. Am. Ceram.Soc. **73**, 2187 (1990).

6. Jungie Wu, Ph D Thesis, University of the Witwatersrand (1997).

7. I.G. Chen and W.B. Johnson, J. Mat. Science **26**, 1565 (1991).

8. M.F. Hundley and A Zettl, Phys. Rev. **B38**, 10290 (1988).

9. L. Benguigui, J. Physique Lett. 46, **L-1015** (1985).

10. R.K. Chakrabarty, K.K. Bardham and A. Basu, J. Phys.: Condensed Matter **5**, 2377 (1993).

CONDUCTIVITY AND NOISE MEASUREMENTS IN 3D PERCOLATIVE CELLULAR STRUCTURES

C. CHITEME and D.S. MCLACHLAN

Physics Department, University of the Witwatersrand, Johannesburg, SA.

Abstract

Conductivity results and 1/f noise (S_{av}) measurements from some systems with a cellular structure (composites in which small conductor particles embed on the surface of larger and regular insulator particles) are given. The usual DC percolation parameters (ϕ_c, t & s) were obtained from fitting the results to the Percolation equations. ϕ_c values for the systems have been found to lie in the range 0.01 - 0.07, while both non-universal and close to universal values have been measured for the exponents s and t. In addition, 1/f or flicker noise results on the systems give an additional exponent ω from the relationship $S_{av}/V_{dc}^2 = KR^\omega$. For the systems measured so far, the exponent ω is observed to take different values ω_1 close to and ω_2 further away from the conductor-insulator transition, but on the conducting side ($\phi > \phi_c$). The very different values (s, t & ω), obtained for the various conducting powders, in the same macroscopic structure, indicates that the way the powders distribute themselves on the insulating particles is a major factor in determining the exponents.

Introduction

Since the 1970s extensive conductivity and dielectric studies of 2D (thin films) and 3D percolation systems have been made [1,2,3,4,5,6]. In some 3D systems the conducting particles tend to form three dimensional cellular structures, but no systematic study of this structure has previously been done. In this paper the first systematic study of the percolation properties of a series of 3D composites with a cellular structure (similar to the conductor being the 'soap film' of bubbles) will be presented, with the goal of learning more about what determines the percolation exponents. The composite consists of small particles of conductor coating a regular insulator matrix, made from larger particles.

Theory

The critical volume fraction or percolation threshold (ϕ_c) of a conducting component is where a conductor-insulator transition occurs. The percolation conductivity equations that are used to characterise this behaviour are:

$$\sigma_m = \sigma_c((\phi - \phi_c)/(1-\phi_c))^t \quad ; \text{ conducting region } (\phi > \phi_c) \qquad (1a)$$

$$\sigma_m = \sigma_i((\phi_c - \phi)/\phi_c)^{-s} \quad ; \text{ insulating region } (\phi < \phi_c) \qquad (1b)$$

$$\sigma_m \sim |\phi - \phi_c|^{-t/(t+s)} \quad ; \text{ defines the crossover region } (\phi \sim \phi_c);$$

where ϕ is the volume fraction of the conductor component, σ_c and σ_i are the conductivities of the conducting and insulating components respectively; and σ_m is the conductivity of the composite. The exponents t and s are for the conducting and insulating regions respectively. The modified GEM equation [7,8,9]:

$$\{(1-\phi)(\sigma_i^{1/s} - \sigma_m^{1/s})/(\sigma_i^{1/s} + A\sigma_m^{1/s})\} + \{\phi(\sigma_c^{1/t} - \sigma_m 1/t)/(\sigma_c^{1/t} + A\sigma_m^{1/t})\} = 0 \qquad (2)$$

where $A = (1 - \phi_c)/\phi_c$, can equally well be used to fit the data and incorporates and interpolates between eqs. (1a) and (1b), where σ_i/σ_c is not extremely small($< 10^{-9}$).

Several models have appeared in the literature explaining the behaviour of percolation systems near the critical or percolation threshold [10]. The Kusy model (Figure 1(a)) shows that ϕ_c decreases considerably as the ratio of the insulator radius to conductor radius (R_i/R_c) increases. Whereas the mechanisms giving rise to particular values of ϕ_c are now fairly understood [8,9,10,11], the relationship between the microstructure and various s and t is not. Experiments designed to give a particular macroscopic microstructure, where ϕ_c is in principle calculable, are therefore necessary to provide an insight into how the nature of the conducting powder affects the various percolation exponents, this being the main objective for studying this series of Cellular Structures (Figure 1(b)).

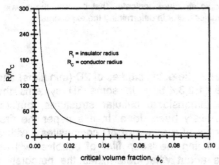

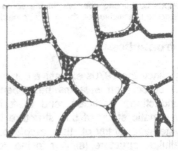

Fig. 1(a) Showing the Kusy model($\phi_c= 1/(1+ K*B)$) simulation curve with $B = R_i/R_c$ and $K = 1.27$

Fig. 1(b) The Cellular microstructure

conductor ● insulator ◯

Furthermore, the use of 1/f noise (S_{av}) in the investigation of percolation systems [12 and references therein] has stimulated a lot of interest because of the ability of the noise power to probe the inhomogeneities present in the microstructure. Hooge's empirical formula (1969) [13] for homogeneous samples gives:

$$S_{av}(f) = \alpha V_{dc}^2 \Delta f /(Nf^\gamma) \qquad (3)$$

Here, V_{dc} is the voltage across the sample, α is a dimensionless constant and N is the total number of charge carriers. For 1/f or flicker noise, $\gamma \sim 1$. In 1985, Rammal etal [14] introduced a new scaling exponent κ to describe the divergence of the normalised noise spectrum S_{av}/V_{dc}^2 near ϕ_c but only on the conducting side of the percolation threshold:

$$S_{av}/V_{dc}^2 \; \alpha \; (\phi - \phi_c)^{-\kappa} \tag{4}$$

Recalling that R varies as $(\phi - \phi_c)^{-t}$ (from percolation theory, eq.1(a)) and combining this with (4) gives [15]:

$$S_{av}/V_{dc}^2 \; \alpha \; R^\omega \tag{5};$$

where $\kappa = \omega t$. Equation (5) is a more reliable fitting relation than (4), given the uncertainity involved in the determination of ϕ and ϕ_c values.

Experimental procedure

Talc powder coated with 4% wax (which was used as a common insulating matrix) was mixed in varying proportions with each of the following conducting powders; ground Carbon Black (gCB) ~8μm, NbC (superconductor) ~3μm, Nickel (magnetic) ~3μm, ground Graphite (gG) ~10μm, ground Graphite/Boron Nitride mixture (gG/BN with the combined ϕ fixed at 0.15) ~10μm). The talc-wax particles are spherical and large (~300μm) compared to the conductor particles which allows application of the Kusy Model (curve) in Figure 1(a). The powder mixtures were carefully mixed to allow the small conductor particles to be embedded on the surface of the large and regular insulator particles. Two mixing methods(planetary and tumbling)were employed. Mixtures of the soft particles (gCB, gG & gG/BN) and talc-wax were mixed in the planetary mill for 20 minutes, but without the usual agate balls to minimise penetration of the conductor particles into the talc-wax. The hard particles (NbC & Ni) were mixed by tumbling in a container (rotating bottle with glass rods running parallel to its axis) in a slowly rotating drum for 20 minutes. About 4 grams of each mixture was used to make the pellet (26 mm dia. & ~3mm thick) by forming in a press at 1800 psi for one and half hours.

After forming the pellet, electrodes were painted on using silver paste and DC conductivity measurements were done using either a Keithley model 617electrometer (insulating samples >10^5-10^{16} Ω) or an LR 400 Resistance bridge (conducting samples 10^{-4}-10^5 Ω). Noise measurements were done using an HP3562A Spectrum Analyser and a low noise Stanford (SR 560) preamplifier.

Results

The DC conductivity results for gCB and NbC composites are shown in Figures 2(a) & 2(b). Values of ϕ_c, t and s for all the systems are given in Table I. The ϕ_c values are small (all under 10%) as expected from theory [4,5] which predicts a low value of the threshold (compared to 0.16 for the random continuum system when $R_c = R_i$) for conductor particles much smaller than the insulator. A lower threshold ϕ_c for graphite compared to that of NbC (which has a smaller size of particles) is probably partially due to the elongated (platelet) shape of the graphite particles. Such a geometry has been observed to reduce ϕ_c considerably [11]. In addition, it is suggested that the hard NbC particles are knocked into the surface of the soft talc-wax particles, reducing the effective contact probability between the conductor particles (note also the gradual increase in σ_m with increased NbC concentration after ϕ_c, indicating a high t exponent).

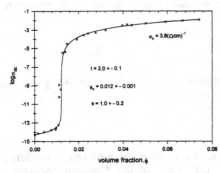 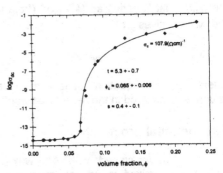

Fig. 2(a) Room temperature DC conductivity results for Ground Carbon Black/Talc-wax. Parameters obtained from fitting to the combined Percolation equations.

Fig. 2(b) Room temperature DC conductivity results for the Niobium Carbide/Talc-wax Parameters obtained by fitting to the combined Percolation equations.

TABLE I: DC Percolation Parameters

COMPOSITE SYSTEM	ϕ_c	t	s
Ground-Carbon Black/Talc-wax	0.012±0.001	2.06±0.10	1.0±0.2
Ground-Graphite/Talc-wax	0.035±0.001	1.93±0.06	0.6±0.1
Ground-Graphite/BN(15%)/Talc-wax	0.033±0.001	2.50±0.10	1.3±0.2
Nickel(magnetic)/Talc-wax	0.025±0.003	1.50±0.10	1.1±0.6
NbC(superconductor)/Talc-wax	0.065±0.006	5.30±0.70	0.4±0.1

Only two systems have been identified (Graphite/talc-wax & Carbon Black/talc-wax) whose exponents (s, t) lie in the universal range. Anomalous values of the exponents have been observed in the other systems. A particularly high t value and a small s have been observed in the NbC/talc-wax system. The high t value appears to be consistent with the touching particles model proposed for the loosely packed Graphite/BN powders [16]. The smallest t value of 1.5, observed in the magnetic (Nickel/talc-wax) system, is not unusual in continuum systems, where values of 1.5-2.5 are common place. The three-component system (Graphite/BN/talc-wax, where the volume fraction of Graphite+BN is kept constant at 0.15) highlights how s and t depend on the distribution of the conducting particles even in systems, with the same ϕ_c. The BN powder (which is a mechanical but not electrical isomorph of graphite) was deliberately introduced to see the effect of changing the distribution of the conductor particles on the insulator. The value of ϕ_c remained unchanged (within experimental error) but s and t changed with respect to the values of the Graphite/talc-wax system.

The noise measurements of the systems done to date seem reasonable when compared with previous work [12,13,15,17]. Values of γ ~1 were found as expected for

1/f noise. The exponent ω values are shown in Table II and Figure 3 shows the corresponding plots. In our systems, the exponent ω takes different values, ω_1 close to and ω_2 further away from the percolation threshold, which can be attributed to the difference in the nature and number of the interparticle contacts [17] as one moves from the transition to the conducting region. The ω ~ 3 is associated with the Maxwell conduction regime while ω ~ 1.5 observed in some of the systems is consistent with the Sharvin-type metallic junctions. Wu & Mclachlan [16] observed ω = 1.47 in compacted Graphite/BN. Values close to unity normally indicate noisy contacts which limit the intrinsic conduction. Since we deliberately avoided oxide-coated conductors, tunneling through an oxide barrier can be ruled out. It is only reasonable therefore, to suppose that the intermediate values of ω (which are not predicted but observed) arise from a superposition of two or more effects of these types of contacts(Maxwell and Sharvin-type)which are randomly distributed, especially close to the transition region. The results also highlight the role played by the microgeometry in determining the exponents.

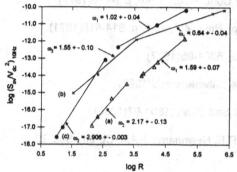

Fig. 3 Showing the Normalised Noise dependence on resistance(R) for:
(a) ground Carbon Black/Talc-wax   Δ
(b) Niobium Carbide/Talc-wax   x
(c) ground Graphite/BNTalc-wax   ●
The solid lines are linear fits to the data to give the $\omega_{1,2}$.

TABLE II: Noise exponents

COMPOSITE SYSTEM	ω_1	ω_2
Ground-Carbon Black/Talc-wax	1.59±0.07	2.17±0.13
Ground-Graphite/BN(15%)/Talc-wax	1.02±0.04	2.906±0.003
NbC(superconductor)/Talc-wax	0.64±0.04	1.55±0.10

Conclusion

An attempt is being made to characterise percolation cellular structures through the determination of the critical concentration ϕ_c and the exponents t and s. While the ϕ_c values agree reasonably well with the Kusy model, it is still too early however, to predict a definite trend in the s and t exponents as only the DC results have been fully investigated so far. Further measurements (more 1/f noise and new AC measurements on all the systems) need to be done to obtain the corresponding exponents which may give further insight into the structure of the conducting component networks and hence the microstructure. Low temperature measurements will also be done to establish the conduction mechanisms (probably variable range hopping), when $\phi < \phi_c$.

Acknowledgements

The authors would like to gratefully acknowledge support from Professor M Hoch (for the use of the high pressure press), Professor Ian Macleod and Ravid Goldstein (for the use of the HP3562A Spectrum Analyser), Professor Moys and David Whitefield (for doing the Particle Size Analysis), Dr. M. Witcomb and the staff of the Electron Microscopy Unit who were always ready to assist with the operation of the JSM 840 SEM.

References

1. C. W. Nan, Prog. in Mat. Sci. 37,1-116(1993).

2. B. Abeles, App. Sol. Stat. Sci., vol.6, 1-109, Acad. Press(1976).

3. J.C. Garland and D. B. Tanner, AIP Conf. Proc. no. 40, p 2-416(1977)

4. A. Malliaris and D.T. Turner ,J. Appl. Phys. 42 ,no. 2, p. 614-618(1971).

5. R.P. Kusy , J. Appl. Phys. 48, no. 12, 5301-05(1977) .

6. J. Wu , PhD thesis, University of the Witwatersrand(1997).

7. D. S. Mclachlan , J. Phys. C18, Sol. Stat. Phys.,1891-97(1985).

8. D.S. Mclachlan , Blaszkiewicz and R.E. Newnham , J. Am. Ceram. Soc. 73, 2187-2203(1990).

9. D.S. Maclachlan , MRS proceedings, v 411, p309-320 (1996).

10. F. Flux , J. Mater. Sci. 28, 285-301(1993).

11. I. Balberg , Phil. Mag. B, vol. 56, no. 6, 991-1003(1987).

12. Y. Yagil, G. Deutscher and D. J. Bergman, Int. J. Mod. Phys. B, vol.7, no. 19, 3353-74(1993).

13. P. Dutta and P. M. Horn , Rev. Mod. Phys., vol. 53, no. 3, 497-516(1981).

14. R. Rammal, C.Tannous, P. Breton, and A.M.S. Tremblay , Phys. Rev. Lett. 54, no.15 , 1718-21(1985).

15. C. Chen and Y. C. Chou , Phys. Rev. Lett. 54, no. 23, 2529-32(1985).

16. J. Wu and D.S. Mclachlan , Phys. Rev. B 56,1236(1997).

17. R. Rammal etal, Phys. Rev. B, vol.42, no. 6, 3386-94(1990).

AUTHOR INDEX

SUBJECT INDEX

365

tantalum, 309
Ta_2O_5, 101
TEOS silicon dioxide, 97
textured 0.58% Y_2O_3 doped CeO_2 films, 279
TFT(-), 119
    LCD, 113
thermal stresses, 253
3-D composite materials, 235
TiAl multilayer, 125
time domain, 203
topology, 291

transformer pressboard, 29
transmission line measurement, 335

universal, 167, 357

varistor(s), 213, 221, 235, 253
very lossy, 167
vitrification, 309

ZnO, 213, 221, 253

Printed in the United States
By Bookmasters